Infectious Disease

Series Editor

Vassil St. Georgiev
National Institute of Health Dept. Health & Human Services, Bethesda, MD, USA

The Infectious Disease series provides the best resources in cutting-edge research and technology.

More information about this series at http://www.springer.com/series/7646

Manjunath P. Pai · Jennifer J. Kiser
Paul O. Gubbins · Keith A. Rodvold
Editors

Drug Interactions in Infectious Diseases: Mechanisms and Models of Drug Interactions

Fourth Edition

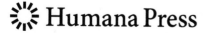
Humana Press

Editors
Manjunath P. Pai
College of Pharmacy
University of Michigan
Ann Arbor, MI, USA

Paul O. Gubbins
Division of Pharmacy Practice and
Administration
UMKC School of Pharmacy at MSU
Springfield, MO, USA

Jennifer J. Kiser
Department of Pharmaceutical Sciences
Skaggs School of Pharmacy and
Pharmaceutical Sciences
University of Colorado
Aurora, CO, USA

Keith A. Rodvold
College of Pharmacy and Medicine
University of Illinois at Chicago
Chicago, IL, USA

Infectious Disease
ISBN 978-3-030-10198-5   ISBN 978-3-319-72422-5   (eBook)
https://doi.org/10.1007/978-3-319-72422-5

This Humana Press imprint is published by Springer Nature
The registered company is Springer International Publishing AG
The registered company address is: Gewerbestrasse 11, 6330 Cham, Switzerland

Foreword

In the 1950s and 1960s, there was euphoria that antibacterial drugs had been discovered, which seemed to have the potential to eliminate the major role infectious diseases had in reducing the quality and duration of human life. Penicillins, cephalosporins, macrolides, tetracyclines, and aminoglycosides were a small but manageable armamentarium, which seemed destined to solve many human challenges.

Since the 1960s and 1970s, we have recognized how readily most infectious agents learn to become resistant to the anti-infective agents to which they are exposed. Methicillin-resistant *Staphylococcus aureus* (MRSA), vancomycin-resistant *Enterococcus faecium* (VRE), carbapenemase-producing *Klebsiella* (KPC), azole-resistant *Candida*, and acyclovir-resistant herpes simplex have been examples of how much urgency there is to create new drugs which will have activity against organisms that have learned to evade currently available anti-infective agents.

We have also developed new classes of drugs for more recently recognized pathogens such as human immunodeficiency virus (HIV) and hepatitis C. These older and newer drugs are given to patients who are receiving a rapidly expanding armamentarium of molecules to treat their chronic and acute underlying conditions.

Healthcare providers are well aware that drugs are only effective and safe if administered with tactical and strategic planning. The right dose, given at the right time, to the right patient is a foundation for effective and safe care. However, as patients are administered more and more agents for a wide range of health challenges, interactions among drugs become more and more likely.

Every experienced clinician has anecdotes of unanticipated drug interactions that affected clinical outcome. Drug interactions can have a major negative impact on drug efficacy and can greatly enhance toxicity for the antimicrobial agent being focused on or for concurrent drugs that may be life-sustaining.

This fourth edition of *Drug Interactions in Infectious Diseases* provides healthcare providers with a unique resource for both understanding basic principles and finding important information. Volume 1 on Mechanisms and Models of Drug

Interactions and Volume 2 on Antimicrobial Drug Interactions are well organized for providers to quickly find practical information. This resource maximizes the likelihood that the healthcare team can optimize efficacy and safety in this era when patients are so often receiving multiple drugs.

Henry Masur, MD
Chief, Critical Care Medicine Department
NIH-Clinical Center
Bethesda, MD, USA

Editors' Preface

The benefits of new medical therapies in infectious diseases cannot be appreciated without understanding and mitigating risk. Drug interactions in infectious diseases are a major source of medical harm that can be prevented. Over the past two decades, we have witnessed a major expansion in our anti-infective armamentarium. This expansion has been coupled with an improved understanding of drug interaction mechanisms and scientific approaches to measure them. Our transformation of the fourth edition of this text to a two-volume series is a direct reflection of the growing knowledge in this domain. Volume 1 provides a mechanistic profile of drug interactions as well as in vitro, in vivo, in silico, and clinical methods to evaluate these interactions. Volume 2 is structured by anti-infective class to provide clinicians, researchers, and academicians a useful resource to meet their practical needs.

Given the scale of this field of study, no comprehensive reviews on antimicrobial drug-drug interactions can be easily published through journals. Software programs and deep learning algorithms that can integrate the effects of all known covariates of drug-drug interaction are in development but have as yet not entered clinical practice. Hence, clinical intuition and vigilance remain key defenses against untoward drug-drug interactions. Since the last publication in 2011, several new antimicrobials have received regulatory approval. The chapters have been updated to reflect these new additions. Three distinct chapters related to the pharmacologic management of human immunodeficiency virus- and hepatitis C virus-related infections have been added in response to recent drug approvals.

The strength of the textbook lies not only in the fact that it is a comprehensive reference book on drug interactions but it also has chapters that provide insights that are difficult to find in the medical literature. We are confident that the information provided in the detailed tables and text will increase the acumen of the practicing clinician, the academic instructor, and the infectious disease researcher.

As the editors of the fourth edition of *Drug Interactions in Infectious Diseases*, we are thrilled to deliver a text that will enhance your clinical knowledge of the complex mechanisms, risks, and consequences of drug interactions associated with antimicrobials, infection, and inflammation. The quality and depth of the information provided would not be possible without the contributions of an excellent number of authors. We are indebted to our authors for their time and diligence to ensure that this textbook remains a primary reference for those engaged in the field of infectious diseases. Finally, we thank our families for their support and encouragement throughout this endeavor.

Ann Arbor, MI, USA

Aurora, CO, USA

Springfield, MO, USA

Chicago, IL, USA

Manjunath P. Pai

Jennifer J. Kiser

Paul O. Gubbins

Keith A. Rodvold

Author's Preface

It is well known that drug interactions pose a major risk to patients. Even a cursory look at approved drug product labels for anti-infective drugs, such as HIV drugs, direct-acting antivirals for HCV, azole antifungal drugs, and anti-mycobacterial agents, reveals that drug interactions present a huge challenge for patients and their healthcare providers. However, before a drug reaches patients, drug development scientists have the opportunity to define the potential for drug interactions. The work of these scientists and the regulatory scientists responsible for drug approval results in information available to healthcare providers and patients.

Concerns related to drug interactions grow as the knowledge of pharmacology advances. The interactions may be due to CYP enzymes, non-CYP enzymes, the ever-growing list of drug transporters, changes in gastric pH, and more. It is easy to be overwhelmed by the scope of the issue. How do you develop an informative and efficient drug interaction program? What drugs are likely perpetrators or victims of interactions? Do you have to study all potential interactions? This textbook helps answer those questions. The chapters address general drug interaction concepts, specific classes of anti-infective drugs, and application of the concepts to drug development. Together, the information helps one focus on the overarching goals of a drug interaction program, determine the potential for clinically significant drug interactions, and develop management strategies for the interactions. The first goal can be divided into four questions. Does the investigational drug alter the pharmacokinetics of other drugs? Do other drugs alter the pharmacokinetics of the investigational drug? What is the magnitude of the change? Is the change clinically significant?

As indicated in the initial chapters of this book, there are many potential mechanisms for drug interactions. Also, concerns go beyond interactions between small molecules. Other considerations include interactions due to biologic products, food components, and herbal medications. However, the bulk of drug interaction evaluations involve investigation of CYP enzyme- or transporter-based interactions. Drug development programs include multiple steps to evaluate the potential for these interactions. For both CYP enzyme and transporter interactions, programs often begin with in vitro evaluations that screen for interactions. If the in vitro evaluations reveal potential interactions, additional evaluations, usually clinical studies with pharmacokinetic endpoints, follow. In some situations, model-based simulations can replace clinical studies or help refine their design [1]. Scientific quality and rigor is essential for all studies. The methods and interpretation of in vitro metabolism and transporter studies must follow best practices because the results may screen out the need for clinical evaluations [2]. Each clinical study should be designed to address the goal of the study. Some clinical studies, referred to as index studies, use perpetrators (inhibitors or inducers) or substrates (victims) with well-known pharmacokinetic and drug interaction properties [1]. Results of the index studies can be extrapolated to other drug combinations and inform the need for additional studies. The design of index studies should maximize the potential to detect an interaction. In contrast to index studies, concomitant use studies investigate drug interactions between the investigational drug and other drugs used in the target population [2]. Results of concomitant use studies provide useful information for the healthcare provider and patient.

The progression from in vitro to index and then concomitant use studies is a common drug development path. However, there are other options. In silico studies that use physiologically based pharmacokinetic (PBPK) methods may substitute for some clinical studies [1]. Instead of dedicated drug interaction studies, prospectively planned evaluations nested within a larger clinical trial may provide useful drug interaction information in the intended patient population. The nested studies often use population pharmacokinetic methods. The in silico and population PK evaluations should be carefully designed to address their specific goals.

Two draft guidance documents from the US Food and Drug Administration provide more details about in vitro and in vivo drug interaction studies: In Vitro Metabolism- and Transporter-Mediated Drug-Drug Interaction Studies Guidance for Industry [3] and Clinical Drug Interaction Studies – Study Design, Data Analysis, and Clinical Implications Guidance for Industry [4].

The progression of drug interaction evaluations that determine the presence and magnitude of pharmacokinetic changes forms the foundation for the next questions: Is the interaction clinically significant? How are clinically significant interactions managed? Thus, solid knowledge regarding general drug interaction concepts, issues related to specific classes of anti-infective drugs, and application of the concepts to drug development are essential to the development of anti-infective drugs.

Kellie Schoolar Reynolds, PharmD
Deputy Director, Division of Clinical Pharmacology IV
Office of Clinical Pharmacology, Office of Translational Sciences
Center for Drug Evaluation and Research, US Food and Drug Administration

References

1. Rekic D, Reynolds KS, Zhao P, et al (2017) Clinical drug-drug interaction evaluations to inform drug use and enable drug access. J Pharm Sci 106:2214–2218
2. Yoshida K, Zhao P, Zhang L, et al (2017) In Vitro-in vivo extrapolation of metabolism and transporter-mediated drug-drug interactions- overview of basic prediction methods. J Pharm Sci 106:2209–2213
3. US Department of Health and Human Services, Food and Drug Administration, Center for Drug Evaluation and Research (CDER) (2017) Draft guidance for industry. In Vitro Metabolism- and Transporter Mediated Drug-Drug Interactions. Available at https://www.fda.gov/downloads/Drugs/GuidanceCompliance RegulatoryInformation/Guidances/UCM581965.pdf. Accessed 30 Oct 2017
4. US Department of Health and Human Services, Food and Drug Administration, Center for Drug Evaluation and Research (CDER) (2017) Draft guidance for industry. Clinical Drug Interaction Studies – Study Design, Data Analysis, and Clinical Implications. Available at https://www.fda.gov/downloads/Drugs/ GuidanceComplianceRegulatoryInformation/Guidances/UCM292362.pdf. Accessed 30 Oct 2017

Contents

Contributors

Rob E. Aarnoutse, PharmD, PhD Department of Pharmacy, Radboud University Medical Center, Nijmegen, The Netherlands

Joseph S. Bertino Jr, PharmD College of Physicians & Surgeons, Columbia University, New York, NY, USA

Bertino Consulting, Schenectady, NY, USA

David M. Burger, PharmD, PhD Department of Pharmacy, Radboud University Medical Center, Nijmegen, The Netherlands

Keith Gallicano, PhD Novum Pharmaceutical Research Services, Murrieta, CA, USA

Grant T. Generaux, MS Pharm New Hope, PA, USA

Kerry B. Goralski, PhD Department of Pharmacology, Dalhousie University, Halifax, NS, Canada

College of Pharmacy, Faculty of Health Professions, Dalhousie University, Halifax, NS, Canada

Paul O. Gubbins, PharmD Division of Pharmacy Practice and Administration, UMKC School of Pharmacy at MSU, Springfield, MO, USA

Jennifer J. Kiser, PharmD Department of Pharmaceutical Sciences, Skaggs School of Pharmacy and Pharmaceutical Sciences, University of Colorado, Aurora, CO, USA

Matthew A. Ladda, BSc Pharm College of Pharmacy, Faculty of Health Professions, Dalhousie University, Halifax, NS, Canada

Surulivelrajan Mallayasamy, MPharm, PhD Department of Pharmacy Practice, MCOPS, Manipal University, Manipal, India

Department of Pharmacotherapy, University of North Texas System College of Pharmacy, Fort Worth, TX, USA

Catia Marzolini, PhD School of Pharmacy, Keele University, Keele, UK

Division of Infectious Diseases and Hospital Epidemiology, University Hospital Basel and University of Basel, Basel, Switzerland

Jenna O. McNeil, MD Department of Family Medicine, Dalhousie University, Halifax, NS, Canada

Darren Michael Moss, PhD Department of Molecular and Clinical Pharmacology, University of Liverpool, Liverpool, UK

School of Pharmacy, Keele University, Keele, UK

Anne N. Nafziger, MD, PhD, MHS Bertino Consulting, Schenectady, NY, USA

School of Pharmacy & Pharmaceutical Sciences, Department of Pharmacy Practice, University at Buffalo, State University of New York, Buffalo, NY, USA

David E. Nix, PharmD The University of Arizona College of Pharmacy, Tucson, AZ, USA

Manjunath P. Pai, PharmD College of Pharmacy, University of Michigan, Ann Arbor, MI, USA

Scott R. Penzak, PharmD, FCP Department of Pharmacotherapy, University of North Texas System College of Pharmacy, Fort Worth, TX, USA

Keith A. Rodvold, PharmD, FCCP, FIDSA Colleges of Pharmacy and Medicine, University of Illinois at Chicago, Chicago, IL, USA

Marco Siccardi, PhD Department of Molecular and Clinical Pharmacology, University of Liverpool, Liverpool, UK

Kelly Sprandel-Harris, PharmD, BCPS Achaogen, Inc, San Fransisco, CA, USA

Lindsey H. M. te Brake, PhD Department of Pharmacy, Radboud University Medical Center, Nijmegen, The Netherlands

Eric Wenzler, PharmD, BCPS University of Illinois at Chicago, College of Pharmacy, Chicago, IL, USA

Chapter 1
Introduction to Drug-Drug Interactions

Manjunath P. Pai, Jennifer J. Kiser, Paul O. Gubbins, and Keith A. Rodvold

1.1 Introduction

A recent investigation by the *Chicago Tribune* revealed that 52% of the 255 tested pharmacies in Illinois failed to stop dispensations of drugs with known serious drug-drug interactions [1]. This failure occurred when given two mock prescriptions of agents with well-documented instances of adverse drug interactions. Three of five prescriptions included antimicrobials, with pairs such as clarithromycin-ergotamine, clarithromycin-statin, and ciprofloxacin-tizanidine that have the potential to result in serious harm. Stricter counseling standards have now been added to Illinois's pharmacy practice laws, and this bad publicity has led to updates in the software systems used to avert drug-drug interactions [2]. This case in Illinois is likely not unique. National surveillance studies estimate close to 900,000 adverse drug events on the use of key anticoagulants, diabetes, and opiate-related medications alone [3]. Recent population estimations have demonstrated that the current use of prescription medications and dietary supplements has significantly increased in older adults between 62 and 85 years old [4]. Among these older adults in the United States, the potential for major drug-drug interactions had increased from

M.P. Pai (✉)
College of Pharmacy, University of Michigan, Ann Arbor, MI, USA
e-mail: amitpai@med.umich.edu

J.J. Kiser
Department of Pharmaceutical Sciences, Skaggs School of Pharmacy and Pharmaceutical Sciences, University of Colorado, Aurora, CO, USA

P.O. Gubbins
Division of Pharmacy Practice and Administration, UMKC School of Pharmacy at MSU
Springfield, MO, USA

K.A. Rodvold
Colleges of Pharmacy and Medicine, University of Illinois at Chicago, Chicago, IL, USA

© Springer International Publishing AG 2018
M.P. Pai et al. (eds.), *Drug Interactions in Infectious Diseases: Mechanisms and Models of Drug Interactions*, Infectious Disease,
https://doi.org/10.1007/978-3-319-72422-5_1

1

approximately 8.4% in 2005–2006 to 15.1% in 2010–2011 [4]. This recent exposition highlights that serious drug-drug interactions remain a major public health threat that can only be mitigated through improvements in our healthcare delivery networks [5].

Given that our pharmacopeia expands on an annual basis, sophisticated computer algorithms that can rapidly integrate new information are likely to be a key solution to this problem. However, the design of such algorithms face a major challenge, the net probability of an adverse event extends beyond simple pairwise comparisons of drugs [6]. The individual pharmacologic effect of drugs is influenced by intrinsic factors [body composition and pharmacogenomics] and extrinsic factors such as food, beverages, pollutants, disease states, and concomitantly administered drugs. These interactions between drugs and intrinsic/extrinsic factors are most often not serious but in rare cases can be life threatening. Given that estimated costs of adverse events that occur in part due to drug interactions exceed $136 billion in the United States alone, a clear incentive exists to solve this challenge [7]. Improved understanding of the underlying mechanisms that impact drug disposition and ascribing appropriate actions to drug interactions are critical to the optimal delivery of healthcare.

Our current approach to understanding these interactions is defined through the lens of perpetrator and victim [8]. This approach includes the controlled evaluation of the pharmacokinetic or pharmacodynamic profile of a drug in the presence and absence of that environmental condition. The pharmacokinetic effect includes evaluation of systemic exposure that is altered by changes in the absorption, distribution, metabolism, and excretion of the victim drug. The pharmacodynamic effect includes measurement of the additivity, synergy, or antagonism between a drug and the environmental condition that is driven by their influence on the same or complementary receptor sites. This methodological approach provides confidence in the level of expected interaction but may not adequately reflect the true effect in the complex clinical milieu [9]. Variables such as age, sex, body composition, and pharmacogenomics layer on additional intrinsic dimensions that influence the degree and consequences of drug-drug interactions. For example, the net effect of a drug in a patient that is characterized to be a poor metabolizer of it is difficult to discern when they are acutely infected, have a history of chronic alcoholism and smoking, are septic, and are developing liver failure, while receiving another agent known to induce the metabolism of the drug. Sophisticated computer algorithms at the cutting edge of artificial intelligence will likely be employed in the coming decades to meet these challenging clinical scenarios [10]. However, it is likely that some form of bioanalytic measurement will still be necessary to tailor dosage recommendations in patients under these dynamic physiologic, pharmacological, and environmental conditions.

In the interim, designing recommendations that avert serious medical harm due to drug interactions is paramount. The general approach to qualifying a response to drug interactions includes:

1. Determining whether the interaction is large enough to necessitate a dosage adjustment either due to a large change in exposure or achievement of exposures predictive of toxicity

2. Whether the risk of a rare but adverse consequence cannot be easily mitigated by dosage modification
3. Whether therapeutic drug monitoring is necessary to overcome poor predictions of existing quantitative models

Current compendia that compile drug-drug interactions rely on individual product labels or small datasets of case reports or case series to ascribe a severity of interaction. These compendia require regular updates and are expected to lag behind new drug approvals and clinical experience. Over 1500 new molecular entities (NMEs) are currently approved, and an additional 20–60 NMEs are approved each year [11]. As of February 2017, the largest knowledge base was Micromedex with 13,133 unique drugs with listings of 4.5 million drug pairs [12]. Furthermore, drug-drug interactions in the field of infectious diseases continues to expand as new drugs are approved, metabolic enzymes and transporters are identified, and recommendations for dosage adjustment due to altered susceptibility profiles are revised. Most of the new drug development in infectious diseases has centered on the management of hepatitis C and human immunodeficiency virus (HIV) [13]. More recently, global efforts have centered on reigniting drug development against multidrug-resistant pathogens. However, the introduction of antimicrobials that are NMEs remains sparse due to limited programs and economic incentives to commit resources to fully synthetic antimicrobial development [14, 15]. Ensuring that available antimicrobial agents are used optimally requires a concerted understanding of their pharmacology and drug interaction potential. This book is divided into a volume dedicated to mechanisms and models of drug-drug interactions and a volume with granularity on individual therapeutic class effects. This introductory chapter provides a broad overview of absorption, distribution, metabolism, and excretion [ADME] to support your understanding of clinical pharmacology and the basis for drug-drug interactions.

1.2 Absorption

Most antimicrobials are administered intravenously for acute infections with a transition to an oral formulation once an adequate response is observed [16]. Several antimicrobials do not have an oral formulation due to poor or unpredictable bioavailability [16]. The use of alternate routes of administration such as intramuscular, subcutaneous, sublingual, transdermal, etc. are often not feasible due to the relatively large dose (nonmammalian target site) that is necessary to derive therapeutic effect [17]. Thus alterations to the drug absorption processes in the gastrointestinal tract can impact the systemic concentration-time profile of an antimicrobial and its pharmacodynamic effects. Solid oral dosage forms must first go through a dissolution process that releases the active pharmaceutical ingredient [API] into the gastrointestinal (GI) tract [18]. Once solubilized, the API must traverse the enterocyte cell membrane barrier to enter the bloodstream.

As expected, the rate of dissolution is dependent on the surface area of dissolving solid which can be manipulated during the formulation process [19]. Several antimicrobials such as griseofulvin, halofantrine, ketoconazole, posaconazole, and nitrofurantoin serve as examples of agents with dissolution problems that limit bioavailability. Nitrofurantoin is available as a macrocrystalline formulation as well as a monohydrate formulation that nicely illustrates the impact of dissolution on drug absorption. Nitrofurantoin in its macrocrystal (Macrodantin®) form is more rapidly solubilized while the monohydrate form (Macrobid®) forms a gel-like matrix in the GI tract that slows its release [16]. This permits twice daily administration of Macrobid® but requires four times a day administration of Macrodantin ®. A reduction in solubility of fluoroquinolones and tetracyclines through heavy metal chelation interactions serve as key examples of how altered dissolution impacts the rate and extent of absorption [20, 21].

Once solubilized, drugs can enter systemic circulation through the transcellular or paracellular route [22]. Compounds that rely on the paracellular route have a limited absorption window of 4–6 h because pore sizes are largest in the jejunum and smallest in the colon [23]. In contrast, compounds that can traverse the GI barrier transcellularly technically can be absorbed throughout the gastrointestinal tract. Hydrophilic compounds such as the aminoglycosides and certain beta-lactams are poorly absorbed because they rely on paracellular pathways [23]. There is no simple relationship that can universally ascribe molecular weight, and likelihood of drug absorption though Lipinski's "rule of five" is often used as a general guide where a MW >500 is associated with poor permeability [24]. The gastrointestinal tract also contains numerous enzymes and transporters that can attenuate systemic availability. The cytochrome P450 (CYP) isoenzyme 3A4 (CYP3A4) system plays a major role in drug metabolism within the enterocyte that is often coupled with the efflux transporter, p-glycoprotein [25]. Glucuronyltransferases and sulfonyltransferases play a major role in limiting the passage of intact drug across the gastrointestinal tract [26]. Inhibition of these pathways supports bioavailability of several compounds and has been used to improve the bioavailability of HIV protease inhibitors [27]. Alternate transporter such as the intestinal peptide transporter (PEPT1) serve to support the fivefold enhancement in bioavailability of the acyclovir prodrug, valacyclovir [28]. Similarly, most of the well-absorbed beta-lactam antibiotics have been shown to be substrates of PEPT1 [29]. Alterations in the solubility, enterocyte metabolism, and transport serve as key variables that influence oral bioavailability that can be influenced by environmental conditions and drug-drug interactions.

1.3 Distribution

Once a compound enters systemic circulation, distribution across membranes follows passive diffusion with the rate of target organ entry driven by blood flow rates and capillary junctional dimensions (5 nm diameter) that are large enough for most drugs [30]. Even large molecules such as vancomycin with a molecular weight of

1449 daltons has a dimension of 3.3×2.2 nm that permits paracellular transport [31]. The exception to these capillary dimensions are the blood-brain barrier, retinal blood barrier, and testicular blood barrier [32]. For intracellular targets, only the free or unbound drug can traverse the interstitial fluid compartments between plasma and tissue [33]. Extravascular albumin plays a key role in drug-protein binding, while proteins such as ligandin, myosin, actin, and melanin influence intracellular binding [33]. Compounds like fluconazole serve as exemplars of highly permeable drugs that are mainly unbound in plasma [34]. Concentrations of fluconazole in vaginal secretions, breast milk, saliva, sputum, prostatic fluid, seminal vesicle fluid, and cerebrospinal fluid are similar to plasma concentrations [35]. Altered distributions of drugs due to protein binding displacement are considered to be temporal blips that rarely influence the safety or efficacy of antimicrobials [36].

The degree of drug distribution is often expressed using the pharmacokinetic term volume of distribution that is a value not truly reflective of a physiologic space [37]. The affinity for albumin relative to phospholipid membranes is influenced by the charge of antimicrobials in aqueous environments [38]. Antimicrobials that are bases have a higher affinity for cell membranes and alpha-1 acid glycoproteins relative to albumin. Bases also tend to accumulate within lysosomes through ion trapping due to the lower pH in this intracellular environment [39]. Macrolides, lincosamides, and aminoglycosides are key classes of antimicrobials that are basic relative to the vast majority, which are acidic [40]. Acids tend to have lower affinities for membranes and higher affinities for albumin [40]. The clearest distinction in these profiles can be seen when comparing azithromycin to erythromycin. Azithromycin has a second basic center in the macrolide aglycone ring which increases the free [unbound] volume of distribution from 4.8 to 62 L/kg [41]. This relatively small molecular modification dramatically increases tissue retention and the half-life of azithromycin. Transporters can alter tissue distribution of antimicrobials, but the scale of this site of drug-drug interaction is considered limited to date and may be a consequence of the difficultly associated with measuring tissue concentrations beyond the standard matrices of blood and plasma [42]. A key example includes p-glycoprotein inhibition by clarithromycin leading to enhancement of oxycarbazepine biodistribution in to the brain with neurotoxicity as a result [43].

1.4   Metabolism

The structure of the glomerulus permits all xenobiotics [unbound state] including relatively large nanoparticles to essentially be filtered with the rate of reabsorption into systemic circulation dependent on their lipophilicity [44]. This degree of lipophilicity also governs the propensity of these compounds to undergo metabolic transformation [45]. Although the kidney plays a role in phase 2 metabolism [conjugation], the principle site of phase 1 [oxidation, reduction, hydrolysis] and phase 2 metabolism is the liver [45]. Drug clearance occurs through the liver and kidneys, and these phases of metabolism occur in parallel and do not have to be sequential.

The CYP system is a heme containing superfamily of enzymes that drives a variety of oxidative interactions [46]. The CYP isoenzymes are localized primarily in the endoplasmic reticulum, and several transporters regulate the influx and efflux of drugs into hepatocytes. The isoenzymes CYP3A4 (neutral, acidic, and basic drugs), CYP2D6 (basic drugs), and CYP2C9 (neutral and acidic drugs) are responsible for metabolism of three-quarters of all drugs [47]. The isoenzyme CYP1A2 affects neutral to lipophilic planar molecules that are basic such as caffeine, theophylline, and tizanidine and is inhibited by agents like ciprofloxacin [48]. The isoenzyme CYP2E1 targets small (<200 daltons) lipophilic linear molecules such as halothane and acetaminophen and is inhibited by isoniazid [48]. Substrates of CYP2D6 include tricyclic antidepressants, beta-blockers, and class 1 antiarrhythmics and can be inhibited by ritonavir [48]. Metabolism through CYP2D6 is easily saturable and is absent in 7% of Caucasians due to genetic polymorphisms that can lead to a high risk for overexposure and toxicity with certain drug-drug interactions [48]. Substrates of CYP2C9 include several nonsteroidal anti-inflammatory agents, phenytoin, [S]-warfarin, and sulfonylureas [48]. This isoenzyme is also highly polymorphic leading to significant variability in drug exposure as seen with the triazole, voriconazole. The dominant isoenzyme is CYP3A4 that is responsible for metabolism of 60% of available drugs [48]. The isoenzyme has a relatively large probe-accessible pocket which permits simultaneous metabolism of multiple substrate molecules at a time [49]. Inhibition of the CYP3A4 pathway can have a profound effect on several classes of drugs as seen with macrolides, triazoles, and HIV protease inhibitors.

Metabolism through the phase 2 pathway includes conjugation through multiple enzyme systems that can lead to addition of a glucuronide, glycine, N-methylation, O-methylation, acetylation, and sulfation or addition of mercapturic acid [50]. Some of these enzyme systems (sulfation and glutathione conjugation) are capacity limited so it can manifest Michaelis-Menten kinetic profiles at high doses [51]. Genetic polymorphisms have also been clearly demonstrated with the N-acetyl transferases (NAT), whereby populations can be divided into slow and fast acetylators [52]. The acetylation pathway has been implicated with key toxicities [52]. Compounds such as sulfanilamide (first antibacterial agent) and sulfamethoxazole undergo acetylation to produce less soluble metabolites that can precipitate in the renal tubules causing crystalluria and kidney injury [53]. Peripheral neuropathy secondary to isoniazid has also been attributed to NAT genotype [54].

1.5 Excretion

Excretion or elimination refers to the final transit of unchanged drug or drug in metabolite form out of the body. The principal routes of excretion include urine, feces, bile, saliva, perspiration, respiration, tears, and milk. The predominant route of excretion where drug-drug interactions tend to occur includes the renal and biliary elimination pathways [55]. Renal elimination is governed by the glomerular filtration rate [GFR], tubular secretion, and tubular reabsorption. Approximately 1 liter of blood flows through the kidneys each minute, so the total blood volume passes through the renal circulation every 5 min [56]. Approximately, a twelfth of this volume is filtered by the glomerulus yielding an expected average GFR estimate of 120 mL/min [56]. The molecular weight cutoff is considered to be 30–50 kDa and so for all practical purposes, all non-plasma protein-bound antimicrobials are freely filtered [44]. Tubular reabsorption of drugs principally occurs through passive diffusion and is dependent on the lipophilicity of agent in the renal tubule [44]. Various transport proteins are also present that are involved in the basolateral and apical movement of compounds. These include organic anion transporters (OAT), organic cation transporters (OCT), p-glycoprotein also referred to as MDR1 after the multidrug resistance gene, and the multidrug and toxin extrusion (MATE) transporter [57].

Changes to cardiac output or renal blood flow most immediately alters the GFR and can impact drug elimination [58]. Drugs like amphotericin B can reduce renal blood flow, for example, and in theory reduce renal clearance of other drugs [59]. Inhibition of tubular secretion by the agent probenecid serves as a classic example of inhibition of tubular secretion that has been used as a beneficial drug-drug interaction to boost the systemic exposures of penicillins [60]. The proximal renal tubule also contains the major efflux transporter, p-glycoprotein, which impacts several drugs [61]. Inhibition of this elimination pathway by antimicrobials such as the macrolides, triazoles, and certain HIV protease inhibitors has been well documented to lead to several drug interactions [61]. Tubular reabsorption mechanisms are reliant on passive diffusion processes that are influenced by alterations in urinary pH. In contrast to renal excretion, biliary excretion tends to primarily occur with excretion of conjugated metabolites into the gut lumen. As noted above, a similar diversity of transporters exists for the canalicular transport of drugs and metabolites from the hepatocyte into bile. Certain microorganisms in the gut can hydrolyze these conjugated substrates leading to reformation of native drug that can be reabsorbed in this more lipophilic state [62]. This reentry of parent compound into the hepatoportal system is referred to as enterohepatic recycling and can prolong systemic exposure of certain drugs [62]. Estrogen and progestin derivatives in oral contraceptives undergo conjugation and enterohepatic conjugation. Alteration in microbial flora by certain antimicrobials can theoretically reduce the effectiveness of oral contraceptives by reducing enterohepatic recycling, though clinical data supporting this mechanism is sparse [63].

1.6 Evaluation of Clinical Drug Interactions

Specific guidance is provided by regulatory bodies on the design and data analysis of drug interaction studies that can have implications for dosing and labeling [64–67]. The process of design includes gauging the potential for interaction by first characterizing the routes of elimination and the contribution of enzymes and transporters on drug disposition. Given that the potential for an interaction is theoretically possible with any drug, several decision trees have been created to define the evaluation pathway [68]. In vitro studies serve as the first screening system to identify whether the drug is a substrate, inducer, or inhibitor of a metabolizing enzyme, most commonly through evaluation of effects on CYP. Evaluation of inhibition through liver microsomal studies is simpler and less cumbersome than the evaluation of induction that requires cultured hepatocytes. Metabolism is considered significant if $\geq 25\%$ of drug elimination is attributed to this pathway [68]. The predictive value of in vitro studies on the degree of interaction remains compound specific [69]. However, the information gained from these studies helps to support the design of more focused clinical trials that are expensive and time-consuming and may carry more than minimal risk to healthy volunteer participants. The probability of a drug-drug interaction can be gauged by the in vitro inhibition constant (Ki). When the ratio of the in vivo concentration of the inhibitor to Ki is <0.1, the need for a clinical trial assessing that interaction potential is expected to be low [70]. In contrast, when that ratio is >10.0, a clinical drug-drug interaction study is often necessary [71].

The clinical drug-drug interaction study includes comparison of the area under the concentration-time curve [AUC] of a probe substrate or drug under evaluation with and without the "perpetrator" drug [if compared to probe] or a known inhibitor or inducer through a cross-over design. The implication of the difference in AUC using subjects as their own controls is dependent on the therapeutic-toxicity window for each drug. When evaluating the effect of a drug on a probe substrate, the geometric mean ratio (GMR) is used to classify the strength of inhibition. A weak inhibitor changes the GMR by 1.25–2.0-fold; a 2.0–5.0-fold change is considered moderate inhibition; and a >5.0-fold change is considered strong inhibition [71]. Similarly inducers are classified as strong, moderate, and weak based on $\geq 80\%$, 50–<80%, and 20–<50% reductions in the AUC of the substrate [64]. Alternate approaches through the use of multiple probe substrates, or drug cocktail studies, have been proposed and used to evaluate multiple CYP metabolic pathways simultaneously [71]. Similarly, physiologic-based pharmacokinetic models have been applied to predict the potential for interaction by incorporating drug physiochemical properties, in vitro derived pharmacologic constants, and clinical population pharmacokinetic models [9].

1.7   Sources of Information for Drug-Drug Interactions

An objective review of the drug-drug interaction potential for an individual case scenario often requires use of screening software. As expected, a review of the primary literature is also essential because a lag time is expected between entry of new information into the public domain and incorporation into a secondary or tertiary reference source. An important source of information for new drugs includes a review of the "Drug Approval Package" submitted to regulatory bodies such as the US Food and Drug Administration that is accessible through Drugs@FDA [72]. Specifically, the "Clinical Pharmacology and Biopharmaceutics Reviews" by regulatory agents often contains links to study reports or study designs employed to qualify the interaction potential of the new drug under review.

For healthcare providers who require information for speedier clinical decisions, clinical pharmacology software platforms are essential. Seven key resources are currently available that include Lexicomp® Interactions module, Micromedex® Drug Interactions, Clinical Pharmacology Drug Interactions Report, Facts and Comparisons® eAnswers, Stockley's Drug Interactions, Drug Interactions Analysis and Management, and Drug Interaction Facts™ [73]. These resources were recently evaluated for scope (i.e., does the resource contain the entry?) and completeness in describing the mechanism, severity, level of documentation, and course of action. This evaluation sampled 100 interactions that included a sample of 80–90 drug-drug interactions and 10–20 herb-drug interactions. Micromedex® Drug Interactions and Lexicomp® Interactions module ranked highest for completeness and were in the top four programs for scope. Newer algorithms in development such as convolutional neural networks that employ natural language processing extraction methods are likely to improve existing platforms in the near future [74, 75]. The clinical utility and impact of these newer tools remain to be defined.

This revised and up-to-date fourth edition of *Drug Interactions in Infectious Diseases* has progressed to a two volume textbook. Both volumes are dedicated to the delivery of clinical knowledge and relevant drug interactions associated with the use of anti-infective agents. It is our hope that these textbooks will continue to be another important source for information about drug interactions.

References

1. Roe S, Long R, King K (2016) Pharmacies miss half of dangerous drug combinations. Chicago Tribune, Chicago. Available from http://www.chicagotribune.com/news/watchdog/druginteractions/ct-drug-interactions-pharmacy-met-20161214-story.html. Accessed 15 Oct 2017
2. Long R, Roe S (2017) Illinois' new pharmacy rules to improve consumer safety may start today. Chicago Tribune, Chicago. Available from http://www.chicagotribune.com/news/watchdog/druginteractions/ct-drug-interactions-rauner-rules-met-20170818-story.html. Accessed 15 Oct 2017
3. Wang Y, Eldridge N, Metersky ML, Verzier NR, Meehan TP, Pandolfi MM et al (2014) National trends in patient safety for four common conditions, 2005-2011. N Engl J Med 370(4):341–351
4. Qato DM, Wilder J, Schumm LP, Gillet V, Alexander GC (2016) Changes in prescription and over-the-counter medication and dietary supplement use among older adults in the United States, 2005 vs 2011. JAMA Intern Med 176:473–482
5. U.S. Department of Health and Human Services, Office of Disease Prevention and Health Promotion (2014) National action plan for adverse drug event prevention. Author, Washington, DC

6. Plank-Kiegele B, Burkle T, Muller F, Patapovas A, Sonst A, Pfistermeister B et al (2017) Data requirements for the correct identification of medication errors and adverse drug events in patients presenting at an emergency department. Methods Inf Med 56(4):276–282
7. Johnson JA, Bootman JL (1995) Drug-related morbidity and mortality. A cost-of-illness model. Arch Intern Med 155(18):1949–1956
8. Chen F, Hu ZY, Jia WW, Lu JT, Zhao YS (2014) Quantitative evaluation of drug-drug interaction potentials by in vivo information- guided prediction approach. Curr Drug Metab 15(8):761–766
9. Varma MV, Pang KS, Isoherranen N, Zhao P (2015) Dealing with the complex drug-drug interactions: towards mechanistic models. Biopharm Drug Dispos 36(2):71–92
10. Rinner C, Grossmann W, Sauter SK, Wolzt M, Gall W (2015) Effects of shared electronic health record systems on drug-drug interaction and duplication warning detection. Biomed Res Int 2015:380497
11. Kinch MS, Haynesworth A, Kinch SL, Hoyer D (2014) An overview of FDA-approved new molecular entities: 1827-2013. Drug Discov Today 19(8):1033–1039
12. Fung KW, Kapusnik-Uner J, Cunningham J, Higby-Baker S, Bodenreider O (2017) Comparison of three commercial knowledge bases for detection of drug-drug interactions in clinical decision support. J Am Med Inform Assoc 24(4):806–812
13. De Clercq E, Li G (2016) Approved antiviral drugs over the past 50 years. Clin Microbiol Rev 29(3):695–747
14. Seiple IB, Zhang Z, Jakubec P, Langlois-Mercier A, Wright PM, Hog DT et al (2016) A platform for the discovery of new macrolide antibiotics. Nature 533(7603):338–345
15. Yan M, Baran PS (2016) Drug discovery: fighting evolution with chemical synthesis. Nature 533(7603):326–327
16. Pai MP, Bertino JS Jr (2014) Tables of antimicrobial agent pharmacology. In: Mandell GLBJ, Dolin RL, Blaser M (eds) Principles and practice of infectious diseases, 8th edn. Elsevier, Philadelphia, PA, USA
17. Silver LL (2011) Challenges of antibacterial discovery. Clin Microbiol Rev 24(1):71–109
18. Mudie DM, Amidon GL, Amidon GE (2010) Physiological parameters for oral delivery and in vitro testing. Mol Pharm 7(5):1388–1405
19. Al-Kassas R, Bansal M, Shaw J (2017) Nanosizing techniques for improving bioavailability of drugs. J Control Release 260:202–212
20. Tatro DS (1972) Tetracycline-antacid interactions. JAMA 220(4):586
21. Ogawa R, Echizen H (2011) Clinically significant drug interactions with antacids: an update. Drugs 71(14):1839–1864
22. Laksitorini M, Prasasty VD, Kiptoo PK, Siahaan TJ (2014) Pathways and progress in improving drug delivery through the intestinal mucosa and blood-brain barriers. Ther Deliv 5(10):1143–1163
23. Linnankoski J, Makela J, Palmgren J, Mauriala T, Vedin C, Ungell AL et al (2010) Paracellular porosity and pore size of the human intestinal epithelium in tissue and cell culture models. J Pharm Sci 99(4):2166–2175
24. Matsson P, Doak BC, Over B, Kihlberg J (2016) Cell permeability beyond the rule of 5. Adv Drug Deliv Rev 101:42–61
25. van Waterschoot RA, Schinkel AH (2011) A critical analysis of the interplay between cytochrome P450 3A and P-glycoprotein: recent insights from knockout and transgenic mice. Pharmacol Rev 63(2):390–410
26. Caldwell J, Gardner I, Swales N (1995) An introduction to drug disposition: the basic principles of absorption, distribution, metabolism, and excretion. Toxicol Pathol 23(2):102–114
27. Benet LZ, Izumi T, Zhang Y, Silverman JA, Wacher VJ (1999) Intestinal MDR transport proteins and P-450 enzymes as barriers to oral drug delivery. J Control Release 62(1–2):25–31
28. Yang B, Smith DE (2013) Significance of peptide transporter 1 in the intestinal permeability of valacyclovir in wild-type and PepT1 knockout mice. Drug Metab Dispos 41(3):608–614

29. Biegel A, Gebauer S, Hartrodt B, Brandsch M, Neubert K, Thondorf I (2005) Three-dimensional quantitative structure-activity relationship analyses of beta-lactam antibiotics and tripeptides as substrates of the mammalian H+/peptide cotransporter PEPT1. J Med Chem 48(13):4410–4419
30. Sarin H (2010) Physiologic upper limits of pore size of different blood capillary types and another perspective on the dual pore theory of microvascular permeability. J Angiogenes Res 2:14
31. Schafer M, Schneider TR, Sheldrick GM (1996) Crystal structure of vancomycin. Structure 4(12):1509–1515
32. Li MW, Mruk DD, Cheng CY (2012) Gap junctions and blood-tissue barriers. Adv Exp Med Biol 763:260–280
33. Smith DA, Di L, Kerns EH (2010) The effect of plasma protein binding on in vivo efficacy: misconceptions in drug discovery. Nat Rev Drug Discov 9(12):929–939
34. Fischman AJ, Alpert NM, Livni E, Ray S, Sinclair I, Callahan RJ et al (1993) Pharmacokinetics of 18F-labeled fluconazole in healthy human subjects by positron emission tomography. Antimicrob Agents Chemother 37(6):1270–1277
35. Debruyne D (1997) Clinical pharmacokinetics of fluconazole in superficial and systemic mycoses. Clin Pharmacokinet 33(1):52–77
36. McElnay JC, D'Arcy PF (1983) Protein binding displacement interactions and their clinical importance. Drugs 25(5):495–513
37. Toutain PL, Bousquet-Melou A (2004) Volumes of distribution. J Vet Pharmacol Ther 27(6):441–453
38. Hamilton JA (1989) Medium-chain fatty acid binding to albumin and transfer to phospholipid bilayers. Proc Natl Acad Sci U S A 86(8):2663–2667
39. Manallack DT, Prankerd RJ, Yuriev E, Oprea TI, Chalmers DK (2013) The significance of acid/base properties in drug discovery. Chem Soc Rev 42(2):485–496
40. Obach RS, Lombardo F, Waters NJ (2008) Trend analysis of a database of intravenous pharmacokinetic parameters in humans for 670 drug compounds. Drug Metab Dispos 36(7):1385–1405
41. Smith DA, Beaumont K, Maurer TS, Di L (2015) Volume of distribution in drug design. J Med Chem 58(15):5691–5698
42. Grover A, Benet LZ (2009) Effects of drug transporters on volume of distribution. AAPS J 11(2):250–261
43. Santucci R, Fothergill H, Laugel V, Perville A, De Saint Martin A, Gerout AC et al (2010) The onset of acute oxcarbazepine toxicity related to prescription of clarithromycin in a child with refractory epilepsy. Br J Clin Pharmacol 69(3):314–316
44. Ruggiero A, Villa CH, Bander E, Rey DA, Bergkvist M, Batt CA et al (2010) Paradoxical glomerular filtration of carbon nanotubes. Proc Natl Acad Sci U S A 107(27):12369–12374
45. Fagerholm U (2007) Prediction of human pharmacokinetics - renal metabolic and excretion clearance. J Pharm Pharmacol 59(11):1463–1471
46. Brodie BB, Axelrod J, Cooper JR, Gaudette L, La Du BN, Mitoma C et al (1955) Detoxication of drugs and other foreign compounds by liver microsomes. Science 121(3147):603–604
47. Charifson PS, Walters WP (2014) Acidic and basic drugs in medicinal chemistry: a perspective. J Med Chem 57(23):9701–9717
48. Zanger UM, Schwab M (2013) Cytochrome P450 enzymes in drug metabolism: regulation of gene expression, enzyme activities, and impact of genetic variation. Pharmacol Ther 138(1):103–141
49. Ohkura K, Kawaguchi Y, Watanabe Y, Masubuchi Y, Shinohara Y, Hori H (2009) Flexible structure of cytochrome P450: promiscuity of ligand binding in the CYP3A4 heme pocket. Anticancer Res 29(3):935–942
50. Coughtrie MW (2015) Ontogeny of human conjugating enzymes. Drug Metab Lett 9(2):99–108
51. Reith D, Medlicott NJ, Kumara De Silva R, Yang L, Hickling J, Zacharias M (2009) Simultaneous modelling of the Michaelis-Menten kinetics of paracetamol sulphation and glucuronidation. Clin Exp Pharmacol Physiol 36(1):35–42

52. Walker K, Ginsberg G, Hattis D, Johns DO, Guyton KZ, Sonawane B (2009) Genetic poly-morphism in N-Acetyltransferase [NAT]: population distribution of NAT1 and NAT2 activity. J Toxicol Environ Health B Crit Rev 12(5–6):440–472

53. Nelson E (1964) Kinetics of the acetylation and excretion of sulfonamides and a comparison of two models. Antibiot Chemother 12:29–40

54. Stettner M, Steinberger D, Hartmann CJ, Pabst T, Konta L, Hartung HP et al (2015) Isoniazid-induced polyneuropathy in a tuberculosis patient - implication for individual risk stratification with genotyping? Brain Behav 5(8):e00326

55. Masereeuw R, Russel FG (2001) Mechanisms and clinical implications of renal drug excretion. Drug Metab Rev 33(3–4):299–351

56. Barger AC, Herd JA (1971) The renal circulation. N Engl J Med 284(9):482–490

57. Ivanyuk A, Livio F, Biollaz J, Buclin T (2017) Renal drug transporters and drug interactions. Clin Pharmacokinet 56(8):825–892

58. Werko L, Ek J, Varnauskas E, Bucht H, Thomasson B, Eliasch H (1955) The relationship between renal blood flow, glomerular filtration rate and sodium excretion, cardiac output and pulmonary and systemic blood pressures in various heart disorders. Am Heart J 49(6): 823–837

59. Pai MP, Norenberg JP, Telepak RA, Sidney DS, Yang S (2005) Assessment of effective renal plasma flow, enzymuria, and cytokine release in healthy volunteers receiving a single dose of amphotericin B desoxycholate. Antimicrob Agents Chemother 49(9):3784–3788

60. Burnell JM, Kirby WM (1951) Effectiveness of a new compound, benemid, in elevating serum penicillin concentrations. J Clin Invest 30(7):697–700

61. Tanigawara Y (2000) Role of P-glycoprotein in drug disposition. Ther Drug Monit 22(1): 137–140

62. Klaassen CD, Cui JY (2015) Review: aechanisms of how the intestinal microbiota alters the effects of drugs and bile acids. Drug Metab Dispos 43(10):1505–1521

63. Zhanel GG, Siemens S, Slayter K, Mandell L (1999) Antibiotic and oral contraceptive drug interactions: is there a need for concern? Can J Infect Dis 10(6):429–433

64. U.S. Department of Health and Human Services, U.S. Food and Drug Administration (2015) Drug interactions & labeling. Washington, DC. Available from https://www.fda.gov/Drugs/DevelopmentApprovalProcess/DevelopmentResources/DrugInteractionsLabeling/default.htm. Accessed 15 Oct 2017

65. Jankovic SM (2014) Comparison of EMA and FDA guidelines for drug interactions: an overview. Clin Res Regul Aff 31:29–34

66. Prucksaritanont T, Chu X, Gibson C, Cui D, Yee KL, Ballard J, Cabalu T, Hochman J (2013) Drug-drug interaction studies: regulatory guidance and an industry perspective. AAPS Journal 15:629–645

67. Nagai N (2010) Drug interaction studies on new drug applications: current situations and regulatory views in Japan. Drug Metab Pharmacokinet 25(1):3–15

68. Zhang L, Zhang YD, Zhao P, Huang SM (2009) Predicting drug-drug interactions: an FDA perspective. AAPS J 11(2):300–306

69. Fowler S, Zhang H (2008) In vitro evaluation of reversible and irreversible cytochrome P450 inhibition: current status on methodologies and their utility for predicting drug-drug interactions. AAPS J 10(2):410–424

70. Huang SM, Strong JM, Zhang L, Reynolds KS, Nallani S, Temple R et al (2008) New era in drug interaction evaluation: US food and drug administration update on CYP enzymes, transporters, and the guidance process. J Clin Pharmacol 48(6):662–670

71. de Andres F, LL A (2016) Simultaneous determination of cytochrome P450 oxidation capacity in humans: a review on the phenotyping cocktail approach. Curr Pharm Biotechnol 17(13):1159–1180

72. U.S. Department of Health and Human Services, U.S. Food and Drug Administration (2015) Drug interactions & labeling. Washington, DC. Available from https://www.accessdata.fda.gov/scripts/cder/daf/. Accessed 15 Oct 2017

73. Patel RI, Beckett RD (2016) Evaluation of resources for analyzing drug interactions. J Med Libr Assoc 104(4):290–295
74. Liu S, Tang B, Chen Q, Wang X (2016) Drug-drug interaction extraction via convolutional neural networks. Comput Math Methods Med 2016:6918381
75. Ayvaz S, Horn J, Hassanzadeh O, Zhu Q, Stan J, Tatonetti NP et al (2015) Toward a complete dataset of drug-drug interaction information from publicly available sources. J Biomed Inform 55:206–217

Chapter 2
Mechanisms of Drug Interactions I: Absorption, Metabolism, and Excretion

David M. Burger, Lindsey H.M. te Brake, and Rob E. Aarnoutse

2.1 Introduction

It is difficult to assess the overall clinical importance of many drug interactions. Often, drug interaction reports are based on anecdotal or case reports, and the involved interaction mechanisms are not always clearly defined. In addition, determining clinical significance requires an assessment of the severity of potential harm. This makes an unequivocal determination of "clinically significant" difficult.

Drug interactions can be pharmacokinetic or pharmacodynamic in nature. Pharmacokinetic interactions result from alterations in a drug's absorption, distribution, metabolism, and/or excretion characteristics. Pharmacodynamic interactions are a result of the influence of combined treatment at a site of biological activity, and yield altered pharmacologic actions at standard plasma concentrations. Although drug interactions occur through a variety of mechanisms, the effects are the same: the potentiation or antagonism of the effects of drugs.

The mechanisms by which changes in absorption, distribution, metabolism, and excretion occur have been understood for decades. However, more recently developed technology has allowed for a more thorough understanding of drug-metabolizing isoforms and influences thereon. Much information has been published regarding drug interactions involving the cytochrome P450 (CYP_{450}) enzyme system [1–3]. This will be an important focus of this chapter, since the majority of currently available anti-infectives are metabolized by, or influence the activity of, the CYP_{450} system. This chapter provides a detailed review of the mechanisms by which clinically significant pharmacokinetic drug interactions occur. Drug transporter-based interactions will be mentioned where appropriate, but for a more detailed description, the reader is referred to Chap. 3.

D.M. Burger (✉) • L.H.M. te Brake • R.E. Aarnoutse
Department of Pharmacy, Radboud University Medical Center, Nijmegen, The Netherlands
e-mail: david.burger@radboudumc.nl

© Springer International Publishing AG 2018
M.P. Pai et al. (eds.), *Drug Interactions in Infectious Diseases: Mechanisms and Models of Drug Interactions*, Infectious Disease,
https://doi.org/10.1007/978-3-319-72422-5_2

2.2   Drug Interactions Affecting Absorption

Mechanisms of absorption include passive diffusion, convective transport, active transport, facilitated transport, ion-pair transport, and endocytosis. Certain drug combinations can affect the rate or extent of absorption of anti-infectives by interfering with one or more of these mechanisms. Generally, a change in the extent of a medication's absorption of greater than 20% may be considered clinically significant in case of drugs with a relatively narrow therapeutic index. The most common mechanisms of drug interactions affecting absorption are shown in Table 2.1.

2.2.1   Changes in pH

The rate of drug absorption by passive diffusion is limited by the solubility, or dissolution, of a compound in gastric fluid. Basic drugs are more soluble in acidic fluids and acidic drugs are more soluble in basic fluids. Therefore, compounds that create an environment with a specific pH may decrease (or increase) the solubility of compounds with pH-dependent absorption. However, drug solubility does not completely ensure absorption, since only un-ionized molecules are absorbed. Although acidic drugs are soluble in basic fluids, basic environments can also decrease the proportion of solubilized acidic molecules that are in an un-ionized state. Therefore, weak acids ($pK_a = 3$–8) may have limited absorption in an alkaline environment and weak bases ($pK_a = 5$–11) have limited absorption in an acidic environment.

Antacids, histamine receptor antagonists, and proton-pump inhibitors all raise gastric pH to varying degrees. Antacids transiently (0.5–2 h) raise gastric pH by 1–2 units [4], H_2-antagonists dose-dependently maintain gastric pH > 5 for many hours, and proton-pump inhibitors dose-dependently raise gastric pH > 5 for up to

Table 2.1 Potential mechanisms of drug interactions involving absorption and distribution

Absorption
Altered gastric pH
Chelation of compounds
Adsorption of compounds
Altered gastric emptying
Altered intestinal motility
Altered intestinal blood flow
Altered active and passive intestinal transport
Altered intestinal cytochrome P450 isozyme activity
Altered intestinal P-glycoprotein activity
Distribution
Altered protein binding

19 h [5]. The concomitant administration of these compounds leads to significant alterations in the extent of absorption of basic compounds [6].

These interactions can also be clinically significant. For example, when patients in the Hepatitis C Virus (HCV) Target study used a proton-pump inhibitor while starting HCV treatment with a ledipasvir-containing regimen, lower rates of sustained virological response were observed [7]. Ledipasvir is an NS5A-inhibitor of HCV replication that has poor solubility at pH >3.0. Similar effects have been seen for the HIV protease inhibitors indinavir and atazanavir [8] and the non-nucleoside reverse transcriptase inhibitor rilpivirine [9]. When combined, plasma concentrations of the antiretroviral agents may become subtherapeutic, and virological failure may occur [10]. Other examples of anti-infective agents known to require an acidic environment for dissolution are ketoconazole [11], itraconazole [12–15], posaconazole [16, 17], and dapsone [18, 19]. Because of large interindividual variability in the extent of altered gastric pH, significant interactions may not occur in all patients.

It must be noted here that pH-dependent effects may vary between different formulations of some of the abovementioned anti-infectives. For instance, posaconazole absorption is negatively influenced when the oral suspension is taken with acid-reducing agents, but this does not occur with posaconazole tablet formulation [20]. Likewise, itraconazole dissolution is affected by omeprazole when taken as capsules but not as oral solution which contains itraconazole already dissolved in cyclodextrins [21].

2.2.2 Chelation and Adsorption

Drugs may form insoluble complexes by chelation in the gastrointestinal tract. Chelation involves the formation of a ring structure between a metal ion (e.g., aluminum, magnesium, iron, and to a lesser degree calcium ions) and an organic molecule (e.g., anti-infective medication), which results in an insoluble compound that is unable to permeate the intestinal mucosa due to the lack of drug dissolution. High concentrations of cations are present in food supplements, including many multivitamin preparations, but also in some antacids. The latter can be confusing as both a pH effect and a chelation effect may occur after simultaneous intake with an organic molecule.

A number of examples of the influence on anti-infective exposure by this mechanism exist in the literature including the quinolone antibiotics in combination with magnesium and aluminum-containing antacids, sucralfate, ferrous sulfate, or certain buffers. These di- and trivalent cations complex with the 4-oxo and 3-carboxyl groups of the quinolones, resulting in clinically significant decreases in the quinolone area under the concentration–time curve (AUC) by 30–50% [22–24]. A second well-documented, clinically significant example of this type of interaction involves the complexation of tetracycline and iron. By this mechanism, tetracycline antibiotic

AUCs are decreased by up to 80% [25]. More recently, the absorption of members of the group of HIV-integrase inhibitors also appears to be harmed by concomitant intake of divalent cations, as has been demonstrated for raltegravir [26], elvitegravir [27], and dolutegravir [28].

Cations present in enteral feeding formulations do not appear to interfere significantly with the absorption of these compounds [29, 30].

Adsorption is the process of ion binding or hydrogen binding and may occur between anti-infectives such as penicillin G, cephalexin, sulfamethoxazole, or tetracycline and adsorbents such as cholestyramine. Since this process can significantly decrease antibiotic exposure, the concomitant administration of adsorbents and antibiotics should be avoided.

2.2.3   Changes in Gastric Emptying and Intestinal Motility

The presence or absence of food can affect the absorption of anti-infectives by a variety of mechanisms. High-fat meals can significantly increase the extent of absorption of fat soluble compounds such as griseofulvin, cefpodoxime, cefuroxime axetil, saquinavir, and rilpivirine. Prolonged stomach retention can cause excessive degradation of acid-labile compounds such as penicillin and erythromycin.

Since the primary location of drug absorption is the small intestine, changes in gastric emptying and gastrointestinal motility may have significant effects on drug exposure. Rapid gastrointestinal transit effected by prokinetic agents such as cisapride, metoclopramide, and domperidone may decrease the extent of absorption of poorly soluble drugs or drugs that are absorbed in a limited area of the intestine [31]. However, clinically significant effects on anti-infectives have not been documented.

2.2.4   Effects of Intestinal Blood Flow

Intestinal blood flow can be modulated by vasoactive agents and theoretically can affect the absorption of lipophilic compounds. However, there is no evidence to date that this results in clinically significant drug interactions.

2.2.5   Changes in Presystemic Clearance

The drug-metabolizing cytochromes P450 (CYP) 3A4 and 5 are expressed at high concentrations in the intestine and contribute to drug inactivation. P-glycoprotein is expressed at the lumenal surface of the intestinal epithelium and serves to extrude unchanged drug from the enterocyte into the lumen. Both CYP3A4/5 and

P-glycoprotein share a significant overlap in substrate specificity [32, 33], although there is no correlation between affinities [34]. Determining the relative contributions of intestinal CYP3A4/5 and P-glycoprotein activity to drug bioavailability and interactions is an active area of investigation. Potential drug interactions involving these mechanisms are discussed in detail below.

2.2.6 Cytochrome P450 Isozymes

Gastrointestinal cytochrome P450 isozymes, responsible for Phase I oxidative metabolism (for a more detailed discussion of CYP isoforms, see Sect. 2.4.1 Phase I Drug Metabolism), are most highly concentrated in the proximal two-thirds of the small intestine [35]. Two intestinal CYP isoforms, CYP3A4 and CYP3A5 (CYP3A4/5), account for approximately 70% of total intestinal P450 protein and are a major determinant of the systemic bioavailability of orally administered drugs [36–39].

For example, the benzodiazepine midazolam is a specific CYP3A4/5 substrate with no affinity for P-glycoprotein. An investigation of oral and intravenous midazolam plasma clearance in 20 healthy young volunteers [40] revealed an incomplete correlation between the two measures ($r = 0.70$). The large variability in midazolam oral clearance not accounted for by hepatic metabolism most likely represents the contribution of intestinal CYP3A4/5. Therefore, it appears that at least 30–40% of the clearance of many CYP3A metabolized compounds may be significantly influenced by CYP3A4/5 located in enterocytes. Since the activity of intestinal CYP3A4/5 can also be influenced by a variety of environmental factors, the potential for drug interactions to occur during drug absorption is great.

A good example of the significant effects of drug interactions occurring at the intestinal isozyme level involve the inhibition of CYP3A4/5 with grapefruit juice [41, 42]. Generally, this interaction results in a minimum threefold increase in the extent of absorption and toxicity of the concomitantly administered agent, but can also result in decreased efficacy of prodrugs needing CYP3A for conversion to active moieties. The concern of this interaction is strictly limited to orally administered agents, since the active components of grapefruit juice are either inactivated in the gut or are present in such minute quantities in the portal circulation that no effect on hepatic metabolism occurs. Clinical data available for anti-infective–grapefruit juice interactions include the protease inhibitor saquinavir [43], the antifungal agent itraconazole [44], and the macrolide clarithromycin [45], and there are also indications for effects on anthelmintics and antimalarials [42]. Whereas saquinavir AUC increases twofold with a single 400-mL dose of commercially available grapefruit juice, itraconazole and clarithromycin AUCs do not change significantly. The absence of an effect of grapefruit juice on the oral clearance of these latter two compounds suggests that their first-pass metabolism does not rely significantly on intestinal CYP3A4/5.

Anti-infectives can also inhibit intestinal CYP isozyme activity themselves. For example, the protease inhibitor ritonavir is a potent inhibitor of CYP3A4 activity [46, 47]. This characteristic can be clinically useful, as demonstrated by the increased bioavailability of several HIV protease inhibitors including saquinavir, lopinavir, atazanavir, and darunavir when given in combination with low-dose ritonavir [48]. This application is called "pharmaco-enhancement" or "boosting" and has now also been introduced in HCV therapy by the development of the HCV protease inhibitor paritaprevir that needs low-dose ritonavir to boost its plasma exposure and activity [49, 50].

Whereas the discovery of (low-dose) ritonavir as a pharmaco-enhancer can be seen as the direct consequence of the observed high drug interaction potential of this agent at its therapeutic dose, it is not a surprise that pharmaceutical companies have searched for non-therapeutic agents with similar pharmaco-enhancement profiles as ritonavir. Cobicistat is an agent chemically related to ritonavir but without its anti-HIV activity; its affinity for CYP3A is similar to ritonavir [51, 52].

Other CYP isozymes present in enterocytes may also influence drug absorption. Environmental factors may influence their activity as well, and drug–environment interactions may result in significantly altered absorption. However, further research is needed to better characterize these influences before specific interactions can be predicted.

2.2.7   Changes in Active and Passive Transport: P-Glycoprotein

A rapidly expanding field of research is that of intestinal transcellular transport. Over the past 20 years, multiple intestinal transporters located on the brush-border and basolateral membrane of the enterocyte have been identified [53–55]. The potential for competitive inhibition of these transporters with quinolone antibiotics, antiretroviral agents, and directly acting antivirals for HCV infection among others has been documented in many studies [56, 57]. This contributes an additional mechanism by which anti-infective drug interactions may occur.

The Caco-2 cell model is a human colonic cell line sharing similarities with enterocytes and is widely used as a model for oral absorption [58–60]. Investigations using this cell line have demonstrated that certain compounds can modulate the tight junctions of the intestinal epithelia and alter paracellular drug absorption. There is still incomplete understanding of the structure and function of tight junctions, which has limited the development of such modulating compounds to enhance paracellular absorption [61, 62].

Of the intestinal transporters, P-glycoprotein is probably the most relevant. This transporter is the product of the multidrug resistance 1 (MDR1) gene found in a variety of human tissues including the gastrointestinal epithelium [63, 64]. This efflux pump is expressed at the lumenal surface of the intestinal epithelium and opposes the absorption of unchanged drug by transporting lipophilic compounds out of enterocytes back into the gastrointestinal lumen. P-glycoprotein has demonstrated up to tenfold variability in activity between subjects [65] and has a significant role in oral drug absorption. Decreased bioavailability occurs because intact drug molecules are pumped back into the gastrointestinal tract lumen and exposed multiple times to enterocyte metabolism.

P-glycoprotein has broad substrate specificity, and inhibiting or inducing the activity of this protein can lead to significant alterations in drug exposure. P-glycoprotein genotype has also been associated with basal expression and induction of CYP3A4 [66]. However, because many drugs have affinities for both P-glycoprotein and CYP3A4/5, it is difficult to determine by what specific mechanism drug interactions occur. For some compounds, inhibition of both P-glycoprotein function and CYP3A4/5 activity may be required to produce clinically significant interactions.

Many anti-infectives have binding affinity for P-glycoprotein. These include erythromycin, clarithromycin [67], ketoconazole, sparfloxacin [68], almost all HIV-1 protease inhibitors [69], tenofovir disoproxil fumarate [70], posaconazole [71], and sofosbuvir [72]. Since drugs that have affinity for P-glycoprotein are not necessarily removed from the enterocyte by this efflux pump, anti-infectives may participate in, but are not necessarily influenced by, drug interactions involving P-glycoprotein. This concept is illustrated by an in vitro investigation of ketoconazole and erythromycin [73]. Both drugs demonstrate significant affinity for P-glycoprotein. However, in combination with verapamil (a classic P-glycoprotein inhibitor), significantly decreased P-glycoprotein-mediated efflux occurred only with erythromycin. Therefore, although ketoconazole exhibits binding affinity for P-glycoprotein, it can be concluded that P-glycoprotein does not contribute significantly to the process of first-pass effect of ketoconazole.

2.3 Drug Interactions Affecting Distribution

2.3.1 Protein Binding and Displacement

Drug interactions affecting distribution are in general those that alter protein binding (Table 2.1). Initially, the importance of drug displacement interactions has been overestimated, with the extrapolation of data from in vitro investigations without consideration for subsequent physiologic phenomena. The lack of well-designed studies has prevented precise quantification of the influence of protein binding on (anti-infective) therapeutic efficacy in vivo. The main reason for the general lack of clinical relevance of protein displacement effects is that redistribution and excretion of drugs generally occurs quickly after displacement, and hence the effects of any transient rise in unbound concentration of the object drug are rarely clinically important [74].

Albumin constitutes the main protein fraction (~5%) in blood plasma. As albumin contains both basic and acidic groups, it can bind basic as well as acidic drugs. Acidic drugs (e.g., penicillins, sulfonamides, doxycycline, and clindamycin) are strongly bound to albumin at a small number of binding sites, and basic drugs (e.g., erythromycin) are weakly bound to albumin at a larger number of sites [75, 76]. Basic drugs such as most HIV protease inhibitors [77] may also preferentially bind to α-1-acid glycoprotein.

Depending on relative plasma concentrations and protein-binding affinities, one drug may displace another with temporary clinically significant results. This interaction is much more likely to occur with drugs that are at least 80–90% bound to plasma proteins, with small changes in protein binding leading to large relative changes in free drug concentration. Drugs that are poorly bound to plasma proteins may also be displaced, but the relative increase in free drug concentration is generally of less consequence. When a protein displacement interaction occurs, the increased free drug in plasma quickly distributes throughout the body and will localize in tissues if the volume of distribution is large. An increase in unbound drug concentrations at metabolism and elimination sites will also lead to increased rates of elimination.

Generally, interactions between basic drugs and albumin are not clinically significant. In subjects with normal concentrations of albumin and anti-infective concentrations of less than 100 µg/mL, the degree of protein binding will be relatively constant. At higher anti-infective concentrations, available binding sites may theoretically become saturated, and the extent of binding subsequently decreased. Clinically significant displacement interactions for α-1-acid glycoprotein have not been described.

Before it is concluded that protein displacement interactions are never clinically relevant, one should keep this mechanism in mind in case unexpected acute toxicity occurs when (novel) drugs with high protein binding are combined. One such example is the recent occurrence of severe symptomatic bradycardia when sofosbuvir-containing HCV therapy was initiated in patients concomitantly taking amiodarone. Although the mechanism of this interaction has not yet been fully discovered, protein-binding displacement of amiodarone by anti-HCV agents is one of the hypotheses [78].

In summary, drug interactions involving albumin binding displacement may potentially be clinically significant if the compound is greater than 80% protein bound, has a high hepatic extraction ratio, a narrow therapeutic index, and a small volume of distribution. Although temporary increase in drug concentrations may be clinically significant with such drugs as warfarin and phenytoin, mean steady-state free drug concentrations will remain unaltered [79–82].

2.4 Drug Interactions Affecting Drug Metabolism

The principal site of drug metabolism is the liver. Metabolism generally converts lipophilic compounds into ionized metabolites for renal elimination. Drug-metabolizing activity can be classified according to nonsynthetic (Phase I) and synthetic (Phase II) reactions. Phase I reactions include oxidation, reduction, and hydrolysis and occur in the membrane of hepatocyte endoplasmic reticula. Phase II reactions result in conjugation (i.e., glucuronidation, sulfation) and occur in the cytosol of the hepatocyte.

2.4.1 Phase I Drug Metabolism

The majority of oxidative reactions are catalyzed by a superfamily of mixed-function mono-oxygenases called the cytochrome P450 enzyme system. Although cytochrome P450 (CYP) isozymes are located in numerous tissues throughout the body, the liver is the largest source of CYP protein. Many significant pharmacokinetic drug interactions involve the hepatic cytochrome P450 isozymes (Table 2.2).

Nomenclature for this superfamily is based on amino acid sequence homology and groups enzymes and genes into families and subfamilies [58, 83]. To designate the cytochrome P450 enzymes, the "CYP" prefix is used. All isozymes having at least 40% amino acid sequence homology are members of an enzyme family, as designated by an Arabic number (e.g., CYP3). All isozymes that have at least 55% amino acid sequence homology are members of an enzyme subfamily, as designated by a capital letter (e.g., CYP3A). An Arabic number is used to represent an individual enzyme (e.g., CYP3A4). Italicized nomenclature represents the gene coding for a specific enzyme (e.g., *CYP3A4*).

To date, at least 14 human families, 22 human subfamilies, and 36 human CYP enzymes have been identified [1, 84]. However, the CYP1, 2, and 3 families account for 70% of the total hepatic P450 content [85, 86]. Approximately 95% of all therapeutic drug oxidation can be accounted for by the activities of CYP1A2, CYP2C8/9, CYP2C19, CYP2D6, CYP2E1, and CYP3A4/5. Drug interactions involving these isozymes result from enzyme inhibition or induction, although genetic polymorphisms can attenuate these interactions.

2.4.1.1 Genetic Polymorphisms

Polymorphisms are generated by nonrandom genetic mutations that occur in at least 1% of a population and give rise to distinct subgroups within that population that differ in their ability to metabolize xenobiotics [87, 88]. Clinically significant

Table 2.2 Potential mechanisms of drug interactions involving metabolism

Phase I (nonsynthetic)
Genetic polymorphisms
Inhibition of activity
Suppression of activity
Induction of activity
Phase II (synthetic)
Genetic polymorphisms
Inhibition of activity
Induction of activity

polymorphisms in CYP enzymes have been documented for CYP2B6, CYP2D6, CYP2C9, and CYP2C19 [87, 89, 90]. Extensive or rapid metabolizers (generally the largest proportion of a population) have heterozygous or homozygous dominant alleles, poor metabolizers possess variant homozygous autosomal recessive alleles, and ultraextensive metabolizers exhibit gene amplification of autosomal dominant alleles.

Poor-metabolizer phenotypes can be at high risk for toxicity from drugs that require CYP inactivation and at high risk for therapeutic inefficacy from prodrugs that need CYP activation.

Two recent examples of the importance of genetic polymorphisms in evaluating the outcome of drug–drug interactions with anti-infectives are related to efavirenz (CYP2B6) and voriconazole (CYP2C19, CYP3A). The non-nucleoside reverse transcriptase inhibitor efavirenz is primarily metabolized by CYP2B6, but many patients possess a 516G > T variant in this enzyme (defined as CYP2B6*6 haplotype) that has almost no enzyme activity [91]. The prevalence of this polymorphism varies among ethnic groups: African Americans and sub-Saharan Africans, 45%; Hispanics and Caucasians, 21–27%; and Japanese and Asians, 18% [91]. Not only do these patients have a higher risk of discontinuation of efavirenz because of adverse effects (associated with higher efavirenz plasma concentrations), but they are also less prone to a drug–drug interaction with the enzyme inducer rifampin [92]. This has led to unexpected clinical observations of patients on efavirenz treated with rifampin that need a *lower* dose of efavirenz; there was no drug interaction, but the genetic polymorphism in CYP2B6 determined the therapeutic dose of efavirenz in such an individual [93].

The antifungal agent voriconazole is extensively metabolized by CYP2C19 and to a lesser extent by CYP2C9 and CYP3A. The antiretroviral combination atazanavir/ritonavir is an inhibitor of CYP3A but also an in vivo inducer of CYP2C19 and CYP2C9. It has been demonstrated [94] that when atazanavir/ritonavir is added to voriconazole in CYP2C19 extensive metabolizers, a moderate (10–40%) *reduction* in voriconazole exposure can be seen; this is explained by CYP2C9/19 induction by ritonavir. However, when atazanavir/ritonavir is added to voriconazole in CYP2C19 poor metabolizers, the net effect is 4.4–7.7-fold *increase* in voriconazole exposure. Here, atazanavir-/ritonavir-mediated CYP3A inhibition becomes dominant in the absence of CYP2C19 activity. Ideally, drug–drug interactions with (anti-infective) agents that are metabolized by polymorphic CYP enzymes should be studied in both extensive and poor metabolizers.

2.4.1.2 Mechanisms of Inhibition

Enzyme inhibition can result in sudden catastrophic drug interactions. Several mechanisms of inhibition exist, and many drugs can interact by multiple mechanisms [85, 86].

Reversible inhibition is the most common mechanism. Reversible inhibition occurs when compounds quickly form weak bonds with CYP isozymes without permanently disabling them [95]. This can occur both competitively (competition for the same binding site between inhibitor and substrate) and noncompetitively (inhibitor binds at a site on the enzyme distinct from the substrate).The magnitude of this type of inhibition depends both on the affinity of substrate and inhibitor for the enzyme, and on the concentration of the inhibitor at the enzyme site. Affinity is represented by an inhibitor constant (K_i), which is the concentration of inhibitor required to decrease the maximal rate of the reaction to half of the uninhibited value [96, 97]. For example, potent reversible CYP3A inhibitors generally have K_i values below 1 μM (e.g., ketoconazole, itraconazole, ritonavir, and cobicistat), although drugs with K_i values in the low micromolar range can also demonstrate competitive inhibition (e.g., erythromycin). Compounds with K_i's greater than 100 μM for the CYP3A subfamily tend not to produce clinically significant inhibition [98, 99].

CYP inhibition can also occur as a result of a slowly reversible reaction. When an inhibitor binds to a CYP isozyme and undergoes oxidation to a nitrosoalkane species, it can form a slowly reversible complex with the reduced heme in the CYP isozyme. This interaction has been documented between the macrolide antibiotics and CYP3A [100] and explains why clinically significant interactions (i.e., erythromycin and terfenadine) can occur with compounds that have modest K_i values.

A second, distinct type of enzyme inhibition is called mechanism-based inhibition (or suicide inhibition). This type of interaction is usually *irreversible* and generally occurs with the CYP-mediated formation of a reactive metabolite [95, 101]. This metabolite can covalently and irreversibly bind to the catalytic site residue and permanently inactivate the enzyme for subsequent reactions. The extent of the clinical importance of this reaction depends on the total amount of CYP isozyme present, the total amount of inhibitor to which the isozyme is exposed, and the rate of new isozyme synthesis. Examples of anti-infectives that display mechanism-based enzyme inhibition include isoniazid, ritonavir, and also macrolide antibiotics (which thus combine different mechanisms of enzyme inhibition).

2.4.1.3 Mechanisms of Suppression of Inflammation-Induced Enzyme Inhibition

As early as the 1960s, inflammation and infection were demonstrated to decrease Phase I metabolism of drugs and toxins in animals, thereby modulating pharmacologic and toxicologic effects. One of the earliest reports of infection altering human drug-metabolizing enzyme activity occurred a decade later, with quinidine concentrations consistently elevated in subjects experimentally infected with plasmodium falciparum malaria [102]. Since that time, numerous reports have described alterations in drug metabolism with viral and bacterial infections [103–105], in addition to complex events such as surgery and bone marrow transplantation.

The effects of inflammation and infection on CYP activity are ascribed to stimulation of the cellular immune response [104]. Although many different mediators may be involved, there has been particular focus on the major proinflammatory cytokines interleukin (IL)-1, IL-6, and tumor necrosis factor (TNF)-α. Generally, IL-1, IL-6, and TNFα demonstrate a suppressive effect on CYP isozymes by decreasing mRNA up to 80%. However correlations between mRNA, enzyme protein content, and enzyme activity are incomplete both within and between investigations [104].

A number of clinical investigations have also documented decreased drug-metabolizing enzyme activity during the administration of therapeutic interferons and interleukins. These studies demonstrate variable and conflicting results with respect to the magnitude of drug–cytokine interactions. With the increasing use of cytokines as therapeutic agents for a variety of disease states, further investigation is required to elucidate the mechanisms of drug–cytokine interactions in order to optimize anti-infective therapeutic regimens.

2.4.1.4   Mechanisms of Induction

An increase in cytochrome P450 activity through induction is less of an immediate concern than inhibition, since induction occurs gradually rather than rapidly and generally leads to compromised therapeutic goals rather than profound toxicity. Since the time course of enzyme induction is determined by the half-life of the substrate as well as the rate of isozyme turnover [99], it is often difficult to predict this time course specifically [106, 107]. Clinically significant induction results from $a > 50$-fold increase in the number of enzyme molecules. This generally occurs through an increase in P450 synthesis by either receptor-mediated transcriptional activation or mRNA stabilization. However, protein stabilization leading to decreased rates of P450 degradation has also been observed. It should be noted that enzyme induction also persists for days to weeks after stopping the inducing drug.

Induction of the CYP1 family by cigarette smoke, charcoal-broiled foods, indoles (found in broccoli, cauliflower, cabbage, Brussels sprouts, kale, watercress), and omeprazole occurs primarily by substrate binding to the aryl hydrocarbon receptor (AhR/dioxin receptor). This complex subsequently binds with a receptor nuclear translocator, enters the hepatocyte nucleus, and binds with regulatory DNA sequences to enhance gene transcription and stabilize mRNA.

The CYP2 and CYP3 families are induced by a variety of structurally diverse compounds. Activation of CYP2C genes is regulated by constitutive androstane receptor (CAR) and pregnane X receptor (PXR) in addition to multiple co-activators [98, 108–110]. Both PXR and CAR can regulate CYP2B6 and CYP3A expression; however, induction by efavirenz and nevirapine of these enzymes is mediated by specifically activating CAR [111]. PXR is activated by a range of drugs known to induce CYP3A4/5 expression (e.g., rifampicin, clotrimazole) [98]. PXR is expressed most abundantly in the liver, but is also present in the small intestine and colon. Transcriptional factors not directly activated by xenobiotics have also been shown to be critical for enzyme induction.

CYP3A can also be induced by posttranscriptional message stabilization and protein stabilization with the following anti-infectives: macrolides, imidazole antifungal agents, and rifampicin. A proposed mechanism for posttranscriptional protein stabilization is proteasome inhibition by NF kappaB activation [112], and message stabilization may involve a similar phosphorylation process.

2.4.2 Phase II Drug Metabolism

The term "Phase II" metabolism was developed originally to represent synthetic reactions occurring after "Phase I" processes. It is now known that many xenobiotics do not require Phase I metabolism before undergoing conjugation reactions [113]. The group of Phase II isozymes consists of UDP-glucuronosyltransferases, sulfotransferases, acetyltransferases, glutathione S-transferase, and methyltransferases. Many of these families of enzymes are still growing in complexity, and drug interactions involving these isozymes continue to be under investigation [114–118].

2.4.2.1 Genetic Polymorphisms

Many of the Phase II enzymes exhibit polymorphism [119, 120]. Although these polymorphisms have been implicated in selected anti-infective-associated adverse drug reactions (e.g., dapsone, isoniazid, sulfonamides [121]), influences of these polymorphisms on anti-infective drug interactions have not been documented.

2.4.2.2 Inhibition

Phase II drug-metabolizing enzymes do not currently appear to play as prominent a role in clinical drug interactions with anti-infectives as the cytochrome P450 enzyme system. This may be due to the large capacity of the conjugation system, in which only profound disturbances result in clinically significant alterations in drug pharmacokinetics.

UDP-glucuronosyltransferase represents the most common conjugation reaction in drug metabolism. Many drugs have been characterized as competitive inhibitors of UDP-glucuronosyltransferases [122], but the roles of these interactions in practical drug metabolism issues are currently only partly explored.

The HIV protease inhibitor atazanavir is a strong inhibitor of UGT1A1 [117]; the pharmacokinetic booster ritonavir is a moderate inducer of UGT1A1. When atazanavir is combined with ritonavir, the net inhibition effect is smaller than when atazanavir is given unboosted [117]. The HIV integrase inhibitors raltegravir and dolutegravir are UGT1A1 substrates, and their metabolism is thus inhibited by atazanavir [123, 124]. A well-known characteristic of UGT1A1 inhibitors is that hyperbilirubinemia occurs as bilirubin is an endogenous substrate of UGT1A1.

2.4.2.3 Induction

Far less is known about the potential for induction of Phase II enzymes than the cytochrome P450 enzyme system. The UDP-glucuronosyltransferases can be induced, but the clinical significance of this is not fully understood. However, the increased clearance of zidovudine that has been documented with the coadministration of rifampicin suggests that induction of these enzymes may be clinically significant [125]. Glutathione S-transferase is also known to be inducible, although these activities rarely exceed two- to threefold times baseline and are not involved in anti-infective metabolism [126]. Another example involves the induction of the sulfotransferase enzyme. Exposure to moxifloxacin is decreased by circa 30% upon coadministration of rifampicin [127, 128]. As moxifloxacin does not undergo Phase I metabolism, this interaction is probably due to induction of sulfation (and possibly glucuronidation) of moxifloxacin by rifampicin [128].

2.5   Drug Interactions Affecting Excretion

Renal elimination of drugs involves glomerular filtration, tubular secretion, and tubular reabsorption. Five mechanisms of drug–drug interactions can occur at the site of renal elimination. The most common mechanisms are discussed below (Table 2.3).

2.5.1   Glomerular Filtration

Rates of glomerular filtration can be affected by changes in renal blood flow, cardiac output, and extent of protein binding. With highly protein-bound drugs (e.g., >80%), a significant increase in the unbound fraction can lead to an increase in glomerular filtration and subsequent increased drug elimination [129, 130]. Conversely, if saturation of tubular secretion transporters occurs, and renal elimination is at a maximal, elimination rates may decrease significantly with increased free drug.

2.5.2   Tubular Secretion

The most common renal drug interactions occur at the transport site of tubular secretion. Many organic anionic and cationic drugs and metabolites compete with each other for secretion, as they share the same proximal tubular active

Table 2.3 Potential mechanisms of drug interactions involving excretion

Glomerular filtration
Tubular secretion
Tubular reabsorption

transport system [54, 131]. A classic example of this interaction, used long ago intentionally for therapeutic benefit, is the combination of probenecid and penicillin to decrease the secretion of penicillin and increase its serum concentrations [132]. Examples of other anti-infectives that may exhibit interactions by this mechanism include the sulfonamides, penicillins, and zidovudine. Also a range of antiretrovirals are subjected to tubular secretion and/or interact with the renal transport system [130].

P-glycoprotein has been identified in the apical membrane of the proximal tubule and can transport a large variety of drugs into the lumen [54]. A number of experimental drug interaction investigations have implicated the inhibition of renal p-glycoprotein to result in an increase in plasma drug concentrations. Quinolones [133], macrolides [134], and azole antifungals [135] demonstrate affinity for renal p-glycoprotein and can potentially be subjected to or cause significant drug interactions.

Besides p-glycoprotein many other renal transporters have been identified in the last 20 years [54]. For more detailed description, see Chap. 3.

2.5.3 Tubular Reabsorption

Reabsorption of drugs from the tubular lumen involves both passive diffusion and active transport processes. Only nonionized compounds are passively reabsorbed from the renal tubule, and thus manipulating urinary pH can alter the reabsorption of weak organic acids and bases. Renal clearance of weak organic bases ($pK_a = 7–10$) is increased with urine acidification (i.e., by salicylates and ascorbic acid) and decreased with urine alkalinization (i.e., by antacids, calcium carbonate, thiazide diuretics, and sodium bicarbonate). Likewise, renal elimination of weak organic acids ($pK_a = 3–7$; nitrofurantoin, sulfonamides, aminoglycosides, and vancomycin) is increased with urine alkalinization and decreased with urine acidification. Generally, these interactions are not clinically significant, since few drugs can have altered urinary excretion to a large enough extent to affect plasma half-life. The role of active transport reabsorption in anti-infective drug interactions remains largely unknown.

2.6 Pharmacodynamic Drug Interactions

Drug interactions are not limited to mechanisms of absorption, distribution, metabolism, and elimination, but can also result from pharmacodynamic interactions. Pharmacodynamic interactions may occur at the intended site of biological activity, i.e., on the same receptors or physiological systems, and they occur irrespective of drug concentrations in the blood or plasma. This type of interaction is fairly common, but is not always recognized or denoted as an interaction. For example, many antibiotics and antiviral drugs are applied in combination for their additive or synergistic effect to achieve improved efficacy or prevent the emergence of resistance.

Pharmacodynamic interactions may also have detrimental effects. Examples of such interactions include the potential for seizures with quinolones when combined with NSAIDs or other medications that lower seizure thresholds and the increased risk of serotonin syndrome after coadministration of linezolid with other medications with serotonergic activity such as antidepressants and opioids [136]. Another example is QT-interval prolongation by combination of anti-infectives and other agents including macrolides, quinolones, antimalarials, and azole antifungals [137]. A third example is the overlapping adverse effect profiles of antiretroviral and anti-TB drugs. Understanding drug mechanisms and side-effect profiles of the antimicrobial agent and concomitant therapy can prevent these complications.

2.7 Significance of Drug Interactions

Many drug interactions are primarily assessed in vitro (see Sect. 2.8 Preclinical Methods for Predicting Drug Interactions). However, absolute in vitro/in vivo correlations are infrequent. Even when assessed in a clinical trial, not all statistically significant drug interactions are of clinical significance. For example, interactions that involve drugs with wide therapeutic indices that demonstrate even more than 20% changes in exposure when combined with a second agent will most likely be of little, if any, clinical significance.

The greatest risk of documented clinically significant pharmacokinetic drug interactions involving anti-infective-induced altered protein binding, drug-metabolizing enzyme inhibition, and altered renal elimination include combinations of anti-infectives with anticoagulants, antidepressants, and cardiovascular agents. The most clinically significant anti-infective drug interactions involving enzyme induction are subtherapeutic concentrations resulting from the combination of rifampicin with various co-medications including anticoagulants, immunosuppressants, antiretrovirals, and oral contraceptives [125, 138, 139].

Conversely, the reduction of AUC and/or C_{max} of anti-infectives by other drugs or environmental influences can result in a much greater chance of failure of therapy and possibly an increase in the development of resistance. This now also includes the novel class of direct-acting antivirals against HCV where resistance may develop associated with low plasma concentrations of these agents [140, 141].

Again, not all pharmacokinetic drug interactions involving anti-infectives are detrimental, however. Ketoconazole has been used for a number of years to inhibit the metabolism of oral cyclosporine by approximately 80%, thereby reducing the cost of therapy as well as the rates of rejection and infection. As mentioned previously, the administration of ritonavir or cobicistat to enhance the oral absorption of antiretrovirals is a well-known component of potent antiretroviral combination regimens [142].

Beneficial and detrimental pharmacodynamic antimicrobial drug interactions also exist. The use of lower concentrations of two synergistic antibacterials to reduce the toxicity of each while having the same pharmacologic effect has been advocated, although the clinical data supporting superior efficacy is weak. Synergistic combinations of antimicrobials may produce better results in the treatment of *Pseudomonas aeruginosa* and *Enterococcus* species. Clinical data are largely lacking for detrimental effects of potentially antagonistic combinations of antimicrobials (e.g., a bacteriostatic drug combined with a bactericidal agent). However, these combinations are best avoided unless clinically warranted for the treatment of multiple pathogens.

2.8 Preclinical Methods for Predicting Drug Interactions

Although understanding and anticipating pharmacokinetic drug interactions are important components of rational therapeutics, there is a limit to the number and scope of clinical studies that can reasonably be performed. The development of human in vitro models allows information to be obtained without the expense and potential risks involved in conducting human trials. However, scaling of in vitro data to the clinical situation is not always accurate, and the results of these methods may not be definitive. A primary focus of preclinical screening methods for assessing drug–drug interactions is the identification of isozymes responsible for the metabolism of these compounds and the relative contribution of an inhibited pathway to a compound's overall elimination.

To account for variability in individual enzyme expression, positive controls for inhibition and induction should always be used (e.g., troleandomycin or ketoconazole for CYP3A inhibition, quinidine for CYP2D6 inhibition, and rifampicin for CYP3A induction). Modern technology has allowed in vitro screening techniques to become widely available, and much of these data are currently included in package inserts.

In addition, there is now guidance from FDA on how to select in vitro and in vivo systems for evaluating drug–drug interactions [143]. The following briefly summarizes the strengths and weaknesses of currently available in vitro human methodologies for assessing cytochrome P450 drug interactions and predicting their clinical significance (Table 2.4).

2.8.1 Purified P450 Isozymes

In an attempt to identify specific isozymes responsible for the metabolism of compounds, investigators have tried to isolate human cytochrome P450 enzymes and purify them from hepatic tissue. However, only small amounts of protein can be

Table 2.4 Preclinical methods for predicting drug interactions

	Advantages	Disadvantages
Purified P450 isozymes	Isozyme substrate identification Isozyme inhibitor identification Isozyme specificity	Limited protein yield Certain subfamilies undifferentiated Quality of purification affects result
Recombinant P450 isozymes	Isozyme substrate identification Isozyme inhibitor identification Isozyme specificity	Artificial system Results require confirmation
Human microsomes	Isozyme substrate identification Isozyme inhibitor identification Relative isozyme metabolic contribution Individual variability overcome by pooling Relatively low cost	Genetic/phenotypic variability Lack cellular machinery for induction/suppression
Immortalized cell lines	Ability to identify induction Method/system validation	P450 activity loss Important cellular processes may be lost
Liver slices	Relatively simple preparation Maintains hepatocyte ultrastructure Ability to identify metabolites inhibitors	Short-lived system Genetic/phenotypic variability Tissue-media distribution equilibrium not always achieved
Hepatocyte cultures	Phase I and II activity Physiologic processes maintained Better clinical extrapolation Ability to identify inhibition, induction and suppression	Genetic/phenotypic variability Requires fresh hepatic tissue Culture methods can be complex

isolated at any one time, and specific isozymes from certain subfamilies often cannot be separated (e.g., CYP2C9 vs CYP2C19 vs CYP2C10). To ensure correct interpretation of the results obtained from this method, it is most critical to examine the isozyme purification methods and quality control procedures. This method has now been superceded by the use of recombinant human cytochrome P450 isozymes.

2.8.2 Recombinant Human P450 Isozymes

Complementary DNA expression has been used to produce recombinant human cytochrome P450 isozymes in yeast, insects, bacteria, and mammalian cells, to be used in in vitro interaction experiments [115, 144]. An advantage of these systems is the ability to identify specific isozymes of a subfamily that are responsible for the metabolism of a compound and to confirm interaction of a compound with suspected isozyme-selective inhibitors. However, this remains an artificial system, and discrepancies can exist between results obtained by complementary DNA methods and other in vitro systems.

2.8.3 Microsomes

Microsomes isolated from human hepatocytes have become the "gold standard" of in vitro experimentation for drug interactions [145–147]. Microsomes are isolated membranes of hepatocyte endoplasmic reticula and contain the cytochromes P450 in proportion to their in vivo representation. Given the large interindividual variability in CYP expression, using microsomes from a single individual may produce distorted results. To circumvent this, pooling microsomes from multiple sources in order to obtain an average representation of activity is advocated. Human microsomes are widely available at relatively low cost, but they can only be used to determine direct inhibition of metabolism. Investigations of drug–drug interactions involving induction or suppression of CYP isozymes require intact cellular machinery [110, 148].

2.8.4 Immortalized Cell Lines

An ideal in vitro model for studying drug–drug interactions involving inhibition, suppression, and induction would be a validated, immortalized, readily available cell line, the results from which could be extrapolated directly to the clinical environment. However, no such model currently exists. All available immortalized human cell lines do not maintain a full complement of cytochrome P450 enzyme activities, nor do they maintain other potentially important physiologic processes, including membrane transporters. One commonly used immortalized cell line is derived from a human hepatoma (HepG2 cells). This model has been investigated for CYP1A1 induction, but does not significantly express other cytochrome P450s [149, 150].

2.8.5 Liver Slices

Human liver slices have been used with moderate success in determining the hepatic metabolism of certain compounds. Liver slices are relatively easy to prepare, and they maintain the hepatic ultrastructure [151–154]. However, up to half of constitutive (baseline) cytochrome P450 activity is lost within the first 24 h after isolation, and all constitutive cytochrome P450 activity is lost by 96 h. This makes investigations of induction and suppression of drug-metabolizing enzyme activity difficult. In addition, a distribution equilibrium is not achieved between all hepatocytes within the slice and the incubation media, resulting in decreased rates of metabolism compared to a hepatocyte monolayer culture system.

2.8.6 Human Hepatocyte Cultures

Primary human hepatocyte culture systems are ideal for studying drug interactions, as they maintain both Phase I and Phase II activity and form and maintain physiologic processes such as biliary canaliculi and transporters [153, 155]. Determining drug interactions in this system often allows for the closest prediction of potential drug interactions. Although this system does not mimic the pharmacokinetic alterations in drug concentrations seen clinically, it does allow quantitation of "best" and "worst" scenarios that may be extrapolated to the clinical setting. Inhibition, suppression, and induction interactions can all be performed with this model. Although maintaining constitutive levels of cytochrome P450 activity has been challenging, currently available enriched media and improved culture conditions allow for maintenance of control activity for at least 72–96 h after isolation. Challenges encountered with this system are primarily in obtaining fresh hepatic tissue for digestion and the specialized technique of perfusion for isolation of the hepatocytes. In addition, with the wide variability in enzyme activity seen clinically, investigations in a limited number of hepatocyte preparations will not be able to definitively reflect the occurrence of drug interactions in an entire population, but only suggest the potential for interactions to occur. These limitations (availability and reproducibility) can be partially overcome with cryopreserved human hepatocytes.

2.9 In Vitro/In Vivo Scaling of Drug Interactions

Extrapolating in vitro results to an in vivo situation is often complicated. The process of using in vitro models to predict drug interactions in vivo, preferably in humans, is still under development, and extensive validation of this approach is needed. In vitro models predictive of drug interactions are essential for rapid, cost-effective screening of pharmaceutical compounds and are important for reducing risks to patient safety. Currently these models are constructed from a combination of laboratory and theoretical components [150, 156–158]. In addition, preclinical screening of promising compounds frequently include the study of nonhuman mammalian species, although interspecies differences in expression and regulation of transporters and enzymes are well documented [159–161]. These differences limit the translation of preclinical animal data to the human situation.

Ideally, in a valid model, the clinical decrease in clearance caused by coadministration of an inhibitor would be specifically predicted by the decrease in reaction velocity (e.g., formation rate of a metabolite) for the same compound in vitro when the inhibitor is present in the same concentration. However, presently available models contain a number of weaknesses and assumptions that make scaling of in vitro data to the clinical situation complicated and not always accurate. Poor predictions occur with compounds that have flow-dependent hepatic clearance, with mechanism-based inhibition, and with compounds that concurrently induce and inhibit enzyme activity. In addition, inhibitor and substrate plasma concentrations are not always proportional to the inhibitor and substrate concentrations to which the enzyme is exposed in vitro. For example, supratherapeutic, as opposed to clinically relevant, concentrations of inhibitors and substrates may be utilized. Furthermore, experimental conditions such as enzyme protein concentration and buffers can critically affect specific results and confound in vitro/in vivo correlations [158]. For example, in vitro and cell culture models can demonstrate extensive partitioning of lipophilic compounds into cells, with uptake not restricted by plasma protein binding.

In order to establish the feasibility of in vitro to in vivo scaling, most currently reported predictions of inhibitory drug interactions are retrospective. Presently available methods allow a general assessment of what may occur (i.e., an unlikely interaction versus a probable interaction). However, to be most useful, in vitro data should not only indicate the possibility of an interaction but also predict its magnitude and clinical importance. Until such a time, the clinical study remains the ultimate means by which a drug interaction and its importance can be assessed.

2.10 Overview of Clinical Methods for Predicting Drug Interactions

The primary cause of clinically significant drug interactions is the involvement of drug-metabolizing enzymes. An overview of relevant substrates, inhibitors, and inducers of CYP450 enzymes is given in Table 2.5. Because great variability exists in drug-metabolizing enzyme activity among subjects, and drug interactions may not achieve clinical significance in all patients, interactions may be better clinically predicted by the knowledge of *individual* patient isozyme activities. However, there is currently a need for the development of reliable, accurate, and noninvasive methods to monitor drug-metabolizing enzyme expression in humans in order to guide drug dosage, reduce toxicity, and predict potential drug interactions.

Genotyping involves identification of mutant genes causing poor or ultra-extensive metabolizer activity. Genotyping has been demonstrated to predict the clinical outcome of drug interactions involving both Phase I and Phase II metabolism. However, drug-metabolizing enzyme activity can be exquisitely sensitive to other non-genetic factors, i.e., environmental and physiologic influences. Therefore, genotyping allows for the determination of an individual's genetic predisposition to a specific enzyme activity, but may not reflect true phenotype at any one point in time.

Phenotyping for drug-metabolizing enzymes or transporters is defined as measuring its actual in vivo activity in an individual [162]. This is performed by administration of a selective substrate ("probe") for this enzyme and subsequent

Table 2.5 Major cytochrome P450 substrates, inhibitors, and inducers

1A2	2B6	2C8	2C9	2C19	2D6	2E1	3A4/5
Substrates							
Caffeine	Bupropion	Repaglinide	Tolbutamide	Omeprazole	Tamoxifen	Halothane	Midazolam
Clozapine	Efavirenz	Paclitaxel	Ibuprofen	Phenytoin	Metoprolol	Acetaminophen	Cyclosporine
Theophylline	Methadone	Torsemide	Naproxen	Clopidogrel	Amphetamine	Ethanol	Tacrolimus
	Cyclophosphamide		Etravirine	Cyclophosphamide	Chlorpromazine	Theophylline	Protease
			S-warfarin	R-warfarin	Dextromethorphan		Inhibitors
			Losartan	Etravirine	Fluoxetine		Cisapride
					Paroxetine		Terfenadine
					Promethazine		Diltiazem
							Verapamil
							Simvastatin
							Fentanyl
							Sildenafil
							Amiodarone
							Etravirine
Inhibitors							
Ciprofloxacin	Ticlopidine	Gemfibrozil	Fluconazole	Omeprazole	Bupropion	Disulfiram	Ketoconazole
Fluvoxamine	Thiotepa	Trimethoprim	Amiodarone	Lansoprazole	Fluoxetine	Ethanol	Ritonavir
		Montelukast	Probenecid	Chloramphenicol	Paroxetine		Saquinavir
		Sorafenib	Isoniazid	Indomethacin	Terbinafine		Nelfinavir
		Quercetin	Sorafenib	Etravirine			Clarithromycin
			Etravirine				Aprepitant
							Grapefruit

Inducers

Nafcillin Cruciferous Vegetables Chargrilled Meat Tobacco	Rifampin Phenobarbital Phenytoin	Rifampin	Rifampin Secobarbital Carbamazepine	Rifampicin Carbamazepine Prednisone	Rifampin Dexamethasone	Ethanol Isoniazid	Rifampin Efavirenz Nevirapine Etravirine Carbamazepine Phenytoin St. John's Wort

For a more comprehensive list see http://medicine.iupui.edu/clinpharm/ddis/table.asp

determination of appropriate pharmacokinetic parameters. The metric used may be systemic clearance of a drug eliminated exclusively by the respective enzyme, partial clearance for a metabolic pathway, or absorption rate in the case of a transporter. Other parameters such as single-point concentrations or ratios of metabolite over parent concentrations in plasma, saliva, and/or urine are also often used [162, 163]. Specific methods have been developed to phenotype CYP1A2, CYP2C9, CYP2C19, CYP2D6, CYP2E1, CYP3A, and N-acetyltransferase activities [162]. Multiple substrates can be studied in a combination using a "cocktail approach," which involves the administration of more than one probe drug simultaneously [162, 164].

Phenotyping offers the primary advantage of quantitating enzyme activity and accounts for combined genetic, environmental, and endogenous influences on drug-metabolizing or drug-transporting enzyme activity. However, a number of currently available phenotyping methods are invasive (requiring pharmacokinetic sampling of blood) and impractical (requiring multiple samples), and analytical methods are not readily available. With a simplification of phenotyping methods, and an increase in the availability of analytical procedures [163], it may be possible to use these methods to determine correlations between enzyme activity and the risk of significant drug interactions in individual patients.

More details can be found in Chap. 23.

The practice of therapeutic drug monitoring (TDM), i.e., the measurement of drug concentrations and subsequent individualization of doses, is also a means to detect and monitor drug interactions in clinical practice. Currently, TDM is available for a range of antibiotics, among others for HIV drugs [165], anti-TB agents [166], antifungals [167], aminoglycosides [168], and vancomycin [169].

2.11 Conclusions and Future Directions

It is difficult to assess the true incidence and clinical significance of drug interactions. Understanding the mechanisms underlying drug interactions is important for the prediction and avoidance of drug toxicity when initiating combination therapy. Although multiple in vitro methods are currently in use to assess drug interactions, not all have allowed the prediction of clinically significant events. As drug interactions most commonly result from influences on drug-metabolizing enzymes, future research defining the origins of enzyme activity variability and characterizing individual patient activity will certainly improve our ability to predict these interactions and improve drug therapy.

Acknowledgments We would like to thank Kevin C. Brown and Angela D. M. Kashuba for providing us with the text files of this chapter in the previous edition.

References

1. Guengerich FP, Waterman MR, Egli M (2016) Recent structural insights into cytochrome P450 function. Trends Pharmacol Sci 37(8):625–640
2. Ong CE, Pan Y, Mak JW, Ismail R (2013) In vitro approaches to investigate cytochrome P450 activities: update on current status and their applicability. Expert Opin Drug Metab Toxicol 9(9):1097–1113
3. Rendic S, Guengerich FP (2015) Survey of human oxidoreductases and cytochrome P450 enzymes involved in the metabolism of xenobiotic and natural chemicals. Chem Res Toxicol 28(1):38–42
4. Ogawa R, Echizen H (2011) Clinically significant drug interactions with antacids: an update. Drugs 71(14):1839–1864
5. Ogawa R, Echizen H (2010) Drug-drug interaction profiles of proton pump inhibitors. Clin Pharmacokinet 49(8):509–533
6. Zhang L, Wu F, Lee SC, Zhao H, Zhang L (2014) pH-dependent drug-drug interactions for weak base drugs: potential implications for new drug development. Clin Pharmacol Ther 96(2):266–277
7. Terrault NA, Zeuzem S, Di Bisceglie AM, Lim JK, Pockros PJ, Frazier LM et al (2016) Effectiveness of Ledipasvir-Sofosbuvir combination in patients with hepatitis C virus infection and factors associated of sustained virologic response. Gastroenterology 151:1131
8. Falcon RW, Kakuda TN (2008) Drug interactions between HIV protease inhibitors and acid-reducing agents. Clin Pharmacokinet 47(2):75–89
9. Crauwels H, van Heeswijk RP, Stevens M, Buelens A, Vanveggel S, Boven K et al (2013) Clinical perspective on drug-drug interactions with the non-nucleoside reverse transcriptase inhibitor rilpivirine. AIDS Rev 15(2):87–101
10. Burger DM, Hugen PW, Kroon FP, Groeneveld P, Brinkman K, Foudraine NA et al (1998) Pharmacokinetic interaction between the proton pump inhibitor omeprazole and the HIV protease inhibitor indinavir. AIDS (London, England) 12(15):2080–2082
11. Chin TW, Loeb M, Fong IW (1995) Effects of an acidic beverage (Coca-Cola) on absorption of ketoconazole. Antimicrob Agents Chemother 39(8):1671–1675
12. Kanda Y, Kami M, Matsuyama T, Mitani K, Chiba S, Yazaki Y et al (1998) Plasma concentration of itraconazole in patients receiving chemotherapy for hematological malignancies: the effect of famotidine on the absorption of itraconazole. Hematol Oncol 16(1):33–37
13. Jaruratanasirikul S, Sriwiriyajan S (1998) Effect of omeprazole on the pharmacokinetics of itraconazole. Eur J Clin Pharmacol 54(2):159–161
14. Jaruratanasirikul S, Kleepkaew A (1997) Influence of an acidic beverage (Coca-Cola) on the absorption of itraconazole. Eur J Clin Pharmacol 52(3):235–237
15. Moreno F, Hardin TC, Rinaldi MG, Graybill JR (1993) Itraconazole-didanosine excipient interaction. JAMA 269(12):1508
16. Alffenaar JW, van Assen S, van der Werf TS, Kosterink JG, Uges DR (2009) Omeprazole significantly reduces posaconazole serum trough level. Clin Infect Dis Off Publ Infect Dis Soc Am 48(6):839
17. Krishna G, Moton A, Ma L, Medlock MM, McLeod J (2009) Pharmacokinetics and absorption of posaconazole oral suspension under various gastric conditions in healthy volunteers. Antimicrob Agents Chemother 53(3):958–966
18. Horowitz HW, Jorde UP, Wormser GP (1992) Drug interactions in use of dapsone for pneumocystis carinii prophylaxis. Lancet (London, England) 339(8795):747
19. Metroka CE, McMechan MF, Andrada R, Laubenstein LJ, Jacobus DP (1991) Failure of prophylaxis with dapsone in patients taking dideoxyinosine. N Engl J Med 325(10):737
20. Kraft WK, Chang PS, van Iersel ML, Waskin H, Krishna G, Kersemaekers WM (2014) Posaconazole tablet pharmacokinetics: lack of effect of concomitant medications altering gastric pH and gastric motility in healthy subjects. Antimicrob Agents Chemother 58(7):4020–4025

21. Johnson MD, Hamilton CD, Drew RH, Sanders LL, Pennick GJ, Perfect JR (2003) A randomized comparative study to determine the effect of omeprazole on the peak serum concentration of itraconazole oral solution. J Antimicrob Chemother 51(2):453–457

22. Knupp CA, Barbhaiya RH (1997) A multiple-dose pharmacokinetic interaction study between didanosine (Videx) and ciprofloxacin (Cipro) in male subjects seropositive for HIV but asymptomatic. Biopharm Drug Dispos 18(1):65–77

23. Polk RE (1989) Drug-drug interactions with ciprofloxacin and other fluoroquinolones. Am J Med 87(5A):76S–81S

24. Sahai J, Gallicano K, Oliveras L, Khaliq S, Hawley-Foss N, Garber G (1993) Cations in the didanosine tablet reduce ciprofloxacin bioavailability. Clin Pharmacol Ther 53(3):292–297

25. Campbell NR, Hasinoff BB (1991) Iron supplements: a common cause of drug interactions. Br J Clin Pharmacol 31(3):251–255

26. Moss DM, Siccardi M, Murphy M, Piperakis MM, Khoo SH, Back DJ et al (2012) Divalent metals and pH alter raltegravir disposition in vitro. Antimicrob Agents Chemother 56(6):3020–3026

27. Ramanathan S, Mathias A, Wei X, Shen G, Koziara J, Cheng A et al (2013) Pharmacokinetics of once-daily boosted elvitegravir when administered in combination with acid-reducing agents. J Acquir Immune Defic Syndr 64(1):45–50

28. Song I, Borland J, Arya N, Wynne B, Piscitelli S (2015) Pharmacokinetics of dolutegravir when administered with mineral supplements in healthy adult subjects. J Clin Pharmacol 55(5):490–496

29. Yuk JH, Nightingale CH, Quintiliani R, Yeston NS, Orlando R 3rd, Dobkin ED et al (1990) Absorption of ciprofloxacin administered through a nasogastric or a nasoduodenal tube in volunteers and patients receiving enteral nutrition. Diagn Microbiol Infect Dis 13(2):99–102

30. Yuk JH, Nightingale CH, Sweeney KR, Quintiliani R, Lettieri JT, Frost RW (1989) Relative bioavailability in healthy volunteers of ciprofloxacin administered through a nasogastric tube with and without enteral feeding. Antimicrob Agents Chemother 33(7):1118–1120

31. Greiff JM1, Rowbotham D (1994) Pharmacokinetic drug interactions with gastrointestinal motility modifying agents. 27(6):447—61

32. van Waterschoot RA, Schinkel AH (2011) A critical analysis of the interplay between cytochrome P450 3A and P-glycoprotein: recent insights from knockout and transgenic mice. Pharmacol Rev 63(2):390–410

33. Knight B, Troutman M, Thakker DR (2006) Deconvoluting the effects of P-glycoprotein on intestinal CYP3A: a major challenge. Curr Opin Pharmacol 6(5):528–532

34. Bertz RJ, Granneman GR (1997) Use of in vitro and in vivo data to estimate the likelihood of metabolic pharmacokinetic interactions. Clin Pharmacokinet 32(3):210–258

35. Peters SA et al. (2016) Predicting drug extraction in the Human Gut Wall: Assessing contributions from drug metabolizing enzymes and transporter proteins using preclinical models. Clin Pharmacokinet 55:673–696

36. Thummel KE, Wilkinson GR (1998) In vitro and in vivo drug interactions involving human CYP3A. Annu Rev Pharmacol Toxicol 38:389–430

37. Komura H, Iwaki M (2011) In vitro and in vivo small intestinal metabolism of CYP3A and UGT substrates in preclinical animals species and humans: species differences. Drug Metab Rev 43(4):476–498

38. Thelen K, Dressman JB (2009) Cytochrome P450-mediated metabolism in the human gut wall. J Pharm Pharmacol 61(5):541–558

39. Zanger UM, Schwab M (2013) Cytochrome P450 enzymes in drug metabolism: regulation of gene expression, enzyme activities, and impact of genetic variation. Pharmacol Ther 138(1):103–141

40. Thummel KE, O'Shea D, Paine MF, Shen DD, Kunze KL, Perkins JD et al (1996) Oral first-pass elimination of midazolam involves both gastrointestinal and hepatic CYP3A-mediated metabolism. Clin Pharmacol Ther 59(5):491–502

41. Hanley MJ, Cancalon P, Widmer WW, Greenblatt DJ (2011) The effect of grapefruit juice on drug disposition. Expert Opin Drug Metab Toxicol 7(3):267–286

42. Seden K, Dickinson L, Khoo S, Back D (2010) Grapefruit-drug interactions. Drugs 70(18):2373–2407

43. Kupferschmidt HH, Fattinger KE, Ha HR, Follath F, Krahenbuhl S (1998) Grapefruit juice enhances the bioavailability of the HIV protease inhibitor saquinavir in man. Br J Clin Pharmacol 45(4):355–359

44. Kawakami M, Suzuki K, Ishizuka T, Hidaka T, Matsuki Y, Nakamura H (1998) Effect of grapefruit juice on pharmacokinetics of itraconazole in healthy subjects. Int J Clin Pharmacol Ther 36(6):306–308

45. Cheng KL, Nafziger AN, Peloquin CA, Amsden GW (1998) Effect of grapefruit juice on clarithromycin pharmacokinetics. Antimicrob Agents Chemother 42(4):927–929

46. Kempf DJ, Marsh KC, Kumar G, Rodrigues AD, Denissen JF, McDonald E et al (1997) Pharmacokinetic enhancement of inhibitors of the human immunodeficiency virus protease by coadministration with ritonavir. Antimicrob Agents Chemother 41(3):654–660

47. Hsu A, Granneman GR, Bertz RJ (1998) Ritonavir. Clinical pharmacokinetics and interactions with other anti-HIV agents. Clin Pharmacokinet 35(4):275–291

48. Hill A, van der Lugt J, Sawyer W, Boffito M (2009) How much ritonavir is needed to boost protease inhibitors? Systematic review of 17 dose-ranging pharmacokinetic trials. AIDS (London, England) 23(17):2237–2245

49. Menon RM, Klein CE, Podsadecki TJ, Chiu YL, Dutta S, Awni WM (2016) Pharmacokinetics and tolerability of paritaprevir, a direct acting antiviral agent for hepatitis C virus treatment, with and without ritonavir in healthy volunteers. Br J Clin Pharmacol 81(5):929–940

50. Brayer SW, Reddy KR (2015) Ritonavir-boosted protease inhibitor based therapy: a new strategy in chronic hepatitis C therapy. Expert Rev Gastroenterol Hepatol 9(5):547–558

51. Shah BM, Schafer JJ, Priano J, Squires KE (2013) Cobicistat: a new boost for the treatment of human immunodeficiency virus infection. Pharmacotherapy 33(10):1107–1116

52. Nathan B, Bayley J, Waters L, Post FA (2013) Cobicistat: a novel pharmacoenhancer for co-formulation with HIV protease and integrase inhibitors. Infectious diseases and therapy 2(2):111–122

53. Silva R, Vilas-Boas V, Carmo H, Dinis-Oliveira RJ, Carvalho F, de Lourdes BM et al (2015) Modulation of P-glycoprotein efflux pump: induction and activation as a therapeutic strategy. Pharmacol Ther 149:1–123

54. Konig J, Muller F, Fromm MF (2013) Transporters and drug-drug interactions: important determinants of drug disposition and effects. Pharmacol Rev 65(3):944–966

55. Estudante M, Morais JG, Soveral G, Benet LZ (2013) Intestinal drug transporters: an overview. Adv Drug Deliv Rev 65(10):1340–1356

56. Mulgaonkar A, Venitz J, Sweet DH (2012) Fluoroquinolone disposition: identification of the contribution of renal secretory and reabsorptive drug transporters. Expert Opin Drug Metab Toxicol 8(5):553–569

57. Pal D, Kwatra D, Minocha M, Paturi DK, Budda B, Mitra AK (2011) Efflux transporters- and cytochrome P-450-mediated interactions between drugs of abuse and antiretrovirals. Life Sci 88(21–22):959–971

58. Sevrioukova IF, Poulos TL (2013) Understanding the mechanism of cytochrome P450 3A4: recent advances and remaining problems. Dalton Trans (Cambridge, England : 2003) 42(9):3116–3126

59. Feng B, Varma MV, Costales C, Zhang H, Tremaine L (2014) In vitro and in vivo approaches to characterize transporter-mediated disposition in drug discovery. Expert Opin Drug Discovery 9(8):873–890

60. Giacomini KM, Huang SM, Tweedie DJ, Benet LZ, Brouwer KL, Chu X et al (2010) Membrane transporters in drug development. Nat Rev Drug Discov 9(3):215–236

61. Salama NN, Eddington ND, Fasano A (2006) Tight junction modulation and its relationship to drug delivery. Adv Drug Deliv Rev 58(1):15–28

62. Assimakopoulos SF, Dimitropoulou D, Marangos M, Gogos CA (2014) Intestinal barrier dysfunction in HIV infection: pathophysiology, clinical implications and potential therapies. Infection 42(6):951–959

63. Zakeri-Milani P, Valizadeh H (2014) Intestinal transporters: enhanced absorption through P-glycoprotein-related drug interactions. Expert Opin Drug Metab Toxicol 10(6):859–871

64. Glaeser H (2011) Importance of P-glycoprotein for drug-drug interactions. Handb Exp Pharmacol 201:285–297

65. Lown KS, Fontana RJ, Schmiedlin-Ren P et al (1995) Interindividual variation in intestinal mdr1:lack of short diet effects. Gastroenterology 108:A737

66. Lamba J, Strom S, Venkataramanan R, Thummel KE, Lin YS, Liu W et al (2006) MDR1 genotype is associated with hepatic cytochrome P450 3A4 basal and induction phenotype. Clin Pharmacol Ther 79(4):325–338

67. Wakasugi H, Yano I, Ito T, Hashida T, Futami T, Nohara R et al (1998) Effect of clarithromycin on renal excretion of digoxin: interaction with P-glycoprotein. Clin Pharmacol Ther 64(1):123–128

68. Cormet-Boyaka E, Huneau JF, Mordrelle A, Boyaka PN, Carbon C, Rubinstein E et al (1998) Secretion of sparfloxacin from the human intestinal Caco-2 cell line is altered by P-glycoprotein inhibitors. Antimicrob Agents Chemother 42(10):2607–2611

69. Srinivas RV, Middlemas D, Flynn P, Fridland A (1998) Human immunodeficiency virus protease inhibitors serve as substrates for multidrug transporter proteins MDR1 and MRP1 but retain antiviral efficacy in cell lines expressing these transporters. Antimicrob Agents Chemother 42(12):3157–3162

70. Neumanova Z, Cerveny L, Ceckova M, Staud F (2014) Interactions of tenofovir and tenofovir disoproxil fumarate with drug efflux transporters ABCB1, ABCG2, and ABCC2; role in transport across the placenta. AIDS (London, England) 28(1):9–17

71. Lempers VJ, van den Heuvel JJ, Russel FG, Aarnoutse RE, Burger DM, Bruggemann RJ et al (2016) Inhibitory potential of antifungal drugs on ATP-binding cassette transporters P-glycoprotein, MRP1 to MRP5, BCRP, and BSEP. Antimicrob Agents Chemother 60(6):3372–3379

72. Kirby BJ, Symonds WT, Kearney BP, Mathias AA (2015) Pharmacokinetic, pharmacodynamic, and drug-interaction profile of the hepatitis C virus NS5B polymerase inhibitor Sofosbuvir. Clin Pharmacokinet 54(7):677–690

73. Takano M, Hasegawa R, Fukuda T, Yumoto R, Nagai J, Murakami T (1998) Interaction with P-glycoprotein and transport of erythromycin, midazolam and ketoconazole in Caco-2 cells. Eur J Pharmacol 358(3):289–294

74. Sansom LN, Evans AM (1995) What is the true clinical significance of plasma protein binding displacement interactions? Drug Saf 12(4):227–233

75. Onufrak NJ, Forrest A, Gonzalez D (2016) Pharmacokinetic and pharmacodynamic principles of anti-infective dosing. Clin Ther 38:1930

76. Barbour A, Scaglione F, Derendorf H (2010) Class-dependent relevance of tissue distribution in the interpretation of anti-infective pharmacokinetic/pharmacodynamic indices. Int J Antimicrob Agents 35(5):431–438

77. Boffito M, Back DJ, Blaschke TF, Rowland M, Bertz RJ, Gerber JG et al (2003) Protein binding in antiretroviral therapies. AIDS Res Hum Retrovir 19(9):825–835

78. Back DJ, Burger DM (2015) Interaction between amiodarone and sofosbuvir-based treatment for hepatitis C virus infection: potential mechanisms and lessons to be learned. Gastroenterology 149(6):1315–1317

79. Rolan PE (1994) Plasma protein binding displacement interactions--why are they still regarded as clinically important? Br J Clin Pharmacol 37(2):125–128

80. Schmidt S, Gonzalez D, Derendorf H (2010) Significance of protein binding in pharmacokinetics and pharmacodynamics. J Pharm Sci 99(3):1107–1122

81. Benet LZ, Hoener BA (2002) Changes in plasma protein binding have little clinical relevance. Clin Pharmacol Ther 71(3):115–121

82. Schmidt S, Barbour A, Sahre M, Rand KH, Derendorf H (2008) PK/PD: new insights for antibacterial and antiviral applications. Curr Opin Pharmacol 8(5):549–556

83. Nelson DR (2006) Cytochrome P450 nomenclature, 2004. Methods Mol Biol 320:1–10

84. Martiny VY, Miteva MA (2013) Advances in molecular modeling of human cytochrome P450 polymorphism. J Mol Biol 425(21):3978–3992
85. Murray M (1997) Drug-mediated inactivation of cytochrome P450. Clin Exp Pharmacol Physiol 24(7):465–470
86. Vanden Bossche H, Koymans L, Moereels H (1995) P450 inhibitors of use in medical treatment: focus on mechanisms of action. Pharmacol Ther 67(1):79–100
87. Ma Q, Lu AY (2011) Pharmacogenetics pharmacogenomics, and individualized medicine. Pharmacol Rev 63(2):437–459
88. Michaud V, Bar-Magen T, Turgeon J, Flockhart D, Desta Z, Wainberg MA (2012) The dual role of pharmacogenetics in HIV treatment: mutations and polymorphisms regulating antiretroviral drug resistance and disposition. Pharmacol Rev 64(3):803–833
89. Backman JT, Filppula AM, Niemi M, Neuvonen PJ (2016) Role of cytochrome P450 2C8 in drug metabolism and interactions. Pharmacol Rev 68(1):168–241
90. Werk AN, Cascorbi I (2014) Functional gene variants of CYP3A4. Clin Pharmacol Ther 96(3):340–348
91. Naidoo P, Chetty VV, Chetty M (2014) Impact of CYP polymorphisms, ethnicity and sex differences in metabolism on dosing strategies: the case of efavirenz. Eur J Clin Pharmacol 70(4):379–389
92. Semvua HH, Mtabho CM, Fillekes Q, van den Boogaard J, Kisonga RM, Mleoh L et al (2013) Efavirenz, tenofovir and emtricitabine combined with first-line tuberculosis treatment in tuberculosis-HIV-coinfected Tanzanian patients: a pharmacokinetic and safety study. Antivir Ther 18(1):105–113
93. van Luin M, Brouwer AM, van der Ven A, de Lange W, van Schaik RH, Burger DM (2009) Efavirenz dose reduction to 200 mg once daily in a patient treated with rifampicin. AIDS (London, England) 23(6):742–744
94. Zhu L, Bruggemann RJ, Uy J, Colbers A, Hruska MW, Chung E et al (2016) CYP2C19 genotype-dependent pharmacokinetic drug interaction between voriconazole and ritonavir-boosted atazanavir in healthy subjects. J Clin Pharmacol
95. Kamel A, Harriman S (2013) Inhibition of cytochrome P450 enzymes and biochemical aspects of mechanism-based inactivation (MBI). Drug Discov Today Technol 10(1):e177–e189
96. Greenblatt DJ (2014) In vitro prediction of clinical drug interactions with CYP3A substrates: we are not there yet. Clin Pharmacol Ther 95(2):133–135
97. Brown HS, Galetin A, Hallifax D, Houston JB (2006) Prediction of in vivo drug-drug interactions from in vitro data : factors affecting prototypic drug-drug interactions involving CYP2C9, CYP2D6 and CYP3A4. Clin Pharmacokinet 45(10):1035–1050
98. Hisaka A, Ohno Y, Yamamoto T, Suzuki H (2010) Prediction of pharmacokinetic drug-drug interaction caused by changes in cytochrome P450 activity using in vivo information. Pharmacol Ther 125(2):230–248
99. Pelkonen O, Turpeinen M, Hakkola J, Honkakoski P, Hukkanen J, Raunio H (2008) Inhibition and induction of human cytochrome P450 enzymes: current status. Arch Toxicol 82(10):667–715
100. Ke AB, Zamek-Gliszczynski MJ, Higgins JW, Hall SD (2014) Itraconazole and clarithromycin as ketoconazole alternatives for clinical CYP3A inhibition studies. Clin Pharmacol Ther 95(5):473–476
101. Zhou S, Yung Chan S, Cher Goh B, Chan E, Duan W, Huang M et al (2005) Mechanism-based inhibition of cytochrome P450 3A4 by therapeutic drugs. Clin Pharmacokinet 44(3):279–304
102. Trenholme GM, Williams RL, Rieckmann KH, Frischer H, Carson PE (1976) Quinine disposition during malaria and during induced fever. Clin Pharmacol Ther 19(4):459–467
103. Lee JI, Zhang L, Men AY, Kenna LA, Huang SM (2010) CYP-mediated therapeutic protein-drug interactions: clinical findings, proposed mechanisms and regulatory implications. Clin Pharmacokinet 49(5):295–310
104. Morgan ET (2009) Impact of infectious and inflammatory disease on cytochrome P450-mediated drug metabolism and pharmacokinetics. Clin Pharmacol Ther 85(4):434–438

105. Aitken AE, Richardson TA, Morgan ET (2006) Regulation of drug-metabolizing enzymes and transporters in inflammation. Annu Rev Pharmacol Toxicol 46:123–149

106. Zhu M, Kaul S, Nandy P, Grasela DM, Pfister M (2009) Model-based approach to characterize efavirenz autoinduction and concurrent enzyme induction with carbamazepine. Antimicrob Agents Chemother 53(6):2346–2353

107. Smythe W, Khandelwal A, Merle C, Rustomjee R, Gninafon M, Bocar Lo M et al (2012) A semimechanistic pharmacokinetic-enzyme turnover model for rifampin autoinduction in adult tuberculosis patients. Antimicrob Agents Chemother 56(4):2091–2098

108. Rana R, Chen Y, Ferguson SS, Kissling GE, Surapureddi S, Goldstein JA (2010) Hepatocyte nuclear factor 4{alpha} regulates rifampicin-mediated induction of CYP2C genes in primary cultures of human hepatocytes. Drug Metab Dispos 38(4):591–599

109. Fahmi OA, Ripp SL (2010) Evaluation of models for predicting drug-drug interactions due to induction. Expert Opin Drug Metab Toxicol 6(11):1399–1416

110. Sinz M, Wallace G, Sahi J (2008) Current industrial practices in assessing CYP450 enzyme induction: preclinical and clinical. AAPS J 10(2):391–400

111. Faucette SR, Zhang TC, Moore R, Sueyoshi T, Omiecinski CJ, LeCluyse EL et al (2007) Relative activation of human pregnane X receptor versus constitutive androstane receptor defines distinct classes of CYP2B6 and CYP3A4 inducers. J Pharmacol Exp Ther 320(1):72–80

112. Zangar RC, Bollinger N, Verma S, Karin NJ, Lu Y (2008) The nuclear factor-kappa B pathway regulates cytochrome P450 3A4 protein stability. Mol Pharmacol 73(6):1652–1658

113. Wu B, Kulkarni K, Basu S, Zhang S, Hu M (2011) First-pass metabolism via UDP-glucuronosyltransferase: a barrier to oral bioavailability of phenolics. J Pharm Sci 100(9):3655–3681

114. Devineni D, Vaccaro N, Murphy J, Curtin C, Mamidi RN, Weiner S et al (2015) Effects of rifampin, cyclosporine A, and probenecid on the pharmacokinetic profile of canagliflozin, a sodium glucose co-transporter 2 inhibitor, in healthy participants. Int J Clin Pharmacol Ther 53(2):115–128

115. Walsky RL, Bauman JN, Bourcier K, Giddens G, Lapham K, Negahban A et al (2012) Optimized assays for human UDP-glucuronosyltransferase (UGT) activities: altered alamethicin concentration and utility to screen for UGT inhibitors. Drug Metab Dispos 40(5):1051–1065

116. Belanger AS, Caron P, Harvey M, Zimmerman PA, Mehlotra RK, Guillemette C (2009) Glucuronidation of the antiretroviral drug efavirenz by UGT2B7 and an in vitro investigation of drug-drug interaction with zidovudine. Drug Metab Dispos 37(9):1793–1796

117. Burger DM, Huisman A, Van Ewijk N, Neisingh H, Van Uden P, Rongen GA et al (2008) The effect of atazanavir and atazanavir/ritonavir on UDP-glucuronosyltransferase using lamotrigine as a phenotypic probe. Clin Pharmacol Ther 84(6):698–703

118. van der Lee MJ, Dawood L, ter Hofstede HJ, de Graaff-Teulen MJ, van Ewijk-Beneken Kolmer EW, Caliskan-Yassen N et al (2006) Lopinavir/ritonavir reduces lamotrigine plasma concentrations in healthy subjects. Clin Pharmacol Ther 80(2):159–168

119. Sim SC, Kacevska M, Ingelman-Sundberg M (2013) Pharmacogenomics of drug-metabolizing enzymes: a recent update on clinical implications and endogenous effects. Pharmacogenomics J 13(1):1–11

120. Yiannakopoulou E (2013) Pharmacogenomics of phase II metabolizing enzymes and drug transporters: clinical implications. Pharmacogenomics J 13(2):105–109

121. Perwitasari DA, Atthobari J, Wilffert B (2015) Pharmacogenetics of isoniazid-induced hepatotoxicity. Drug Metab Rev 47(2):222–228

122. Knights KM, Rowland A, Miners JO (2013) Renal drug metabolism in humans: the potential for drug-endobiotic interactions involving cytochrome P450 (CYP) and UDP-glucuronosyltransferase (UGT). Br J Clin Pharmacol 76(4):587–602

123. Jansen A, Colbers EP, van der Ven AJ, Richter C, Rockstroh JK, Wasmuth JC et al (2013) Pharmacokinetics of the combination raltegravir/atazanavir in HIV-1-infected patients. HIV Med 14(7):449–452

124. Bollen P, Reiss P, Schapiro J, Burger D (2015) Clinical pharmacokinetics and pharmacody-namics of dolutegravir used as a single tablet regimen for the treatment of HIV-1 infection. Expert Opin Drug Saf 14(9):1457–1472

125. Regazzi M, Carvalho AC, Villani P, Matteelli A (2014) Treatment optimization in patients co-infected with HIV and mycobacterium tuberculosis infections: focus on drug-drug inter-actions with rifamycins. Clin Pharmacokinet 53(6):489–507

126. Higgins LG, Hayes JD (2011) Mechanisms of induction of cytosolic and microsomal glu-tathione transferase (GST) genes by xenobiotics and pro-inflammatory agents. Drug Metab Rev 43(2):92–137

127. Nijland HM, Ruslami R, Suroto AJ, Burger DM, Alisjahbana B, van Crevel R et al (2007) Rifampicin reduces plasma concentrations of moxifloxacin in patients with tuberculosis. Clin Infect Dis Off Publ Infect Dis Soc Am 45(8):1001–1007

128. Weiner M, Burman W, Luo CC, Peloquin CA, Engle M, Goldberg S et al (2007) Effects of rifampin and multidrug resistance gene polymorphism on concentrations of moxifloxacin. Antimicrob Agents Chemother 51(8):2861–2866

129. Feng B, LaPerle JL, Chang G, Varma MV (2010) Renal clearance in drug discovery and development: molecular descriptors, drug transporters and disease state. Expert Opin Drug Metab Toxicol 6(8):939–952

130. Yombi JC, Pozniak A, Boffito M, Jones R, Khoo S, Levy J et al (2014) Antiretrovirals and the kidney in current clinical practice: renal pharmacokinetics, alterations of renal function and renal toxicity. AIDS (London, England) 28(5):621–632

131. Morrissey KM, Stocker SL, Wittwer MB, Xu L, Giacomini KM (2013) Renal transporters in drug development. Annu Rev Pharmacol Toxicol 53:503–529

132. Kampmann J, Hansen JM, Siersboek-Nielsen K, Laursen H (1972) Effect of some drugs on penicillin half-life in blood. Clin Pharmacol Ther 13(4):516–519

133. Douros A, Grabowski K, Stahlmann R (2015) Safety issues and drug-drug interactions with commonly used quinolones. Expert Opin Drug Metab Toxicol 11(1):25–39

134. Cohen O, Locketz G, Hershko AY, Gorshtein A, Levy Y (2015) Colchicine-clarithromycin-induced rhabdomyolysis in familial mediterranean fever patients under treatment for helico-bacter pylori. Rheumatol Int 35(11):1937–1941

135. Bruggemann RJ, Alffenaar JW, Blijlevens NM, Billaud EM, Kosterink JG, Verweij PE et al (2009) Clinical relevance of the pharmacokinetic interactions of azole antifungal drugs with other coadministered agents. Clin Infect Dis Off Publ Infect Dis Soc Am 48(10):1441–1458

136. Woytowish MR, Maynor LM (2013) Clinical relevance of linezolid-associated serotonin tox-icity. Ann Pharmacother 47(3):388–397

137. Owens RC Jr, Nolin TD (2006) Antimicrobial-associated QT interval prolongation: pointes of interest. Clin Infect Dis Off Publ Infect Dis Soc Am 43(12):1603–1611

138. Chen J, Raymond K (2006) Roles of rifampicin in drug-drug interactions: underlying molec-ular mechanisms involving the nuclear pregnane X receptor. Ann Clin Microbiol Antimicrob 5:3

139. Semvua HH, Kibiki GS, Kisanga ER, Boeree MJ, Burger DM, Aarnoutse R (2015) Pharmacological interactions between rifampicin and antiretroviral drugs: challenges and research priorities for resource-limited settings. Ther Drug Monit 37(1):22–32

140. Langness JA, Everson GT (2016) Viral hepatitis: drug-drug interactions in HCV treatment--the good, the bad and the ugly. Nat Rev Gastroenterol Hepatol 13(4):194–195

141. Burger D, Back D, Buggisch P, Buti M, Craxi A, Foster G et al (2013) Clinical management of drug-drug interactions in HCV therapy: challenges and solutions. J Hepatol 58(4):792–800

142. Larson KB, Wang K, Delille C, Otofokun I, Acosta EP (2014) Pharmacokinetic enhancers in HIV therapeutics. Clin Pharmacokinet 53(10):865–872

143. FDA. Guidance for Industry. Drug interaction studies — study design, data analysis, impli-cations for dosing, and labeling recommendations 2012 [September 13, 2016]. Available from: http://www.fda.gov/Drugs/GuidanceComplianceRegulatoryInformation/Guidances/default.htm

144. Stringer RA, Strain-Damerell C, Nicklin P, Houston JB (2009) Evaluation of recombinant cytochrome P450 enzymes as an in vitro system for metabolic clearance predictions. Drug Metab Dispos Biol Fate Chem 37(5):1025–1034

145. Spaggiari D, Geiser L, Daali Y, Rudaz S (2014) A cocktail approach for assessing the in vitro activity of human cytochrome P450s: an overview of current methodologies. J Pharm Biomed Anal 101:221–237

146. Khojasteh SC, Prabhu S, Kenny JR, Halladay JS, Lu AY (2011) Chemical inhibitors of cytochrome P450 isoforms in human liver microsomes: a re-evaluation of P450 isoform selectivity. Eur J Drug Metab Pharmacokinet 36(1):1–16

147. Parkinson A, Kazmi F, Buckley DB, Yerino P, Ogilvie BW, Paris BL (2010) System-dependent outcomes during the evaluation of drug candidates as inhibitors of cytochrome P450 (CYP) and uridine diphosphate glucuronosyltransferase (UGT) enzymes: human hepatocytes versus liver microsomes versus recombinant enzymes. Drug Metab Pharmacokinet 25(1):16–27

148. Donato MT, Lahoz A, Castell JV, Gomez-Lechon MJ (2008) Cell lines: a tool for in vitro drug metabolism studies. Curr Drug Metab 9(1):1–11

149. Alqahtani S, Mohamed LA, Kaddoumi A (2013) Experimental models for predicting drug absorption and metabolism. Expert Opin Drug Metab Toxicol 9(10):1241–1254

150. Scotcher D, Jones C, Posada M, Rostami-Hodjegan A, Galetin A (2016) Key to opening kidney for in vitro-in vivo extrapolation entrance in health and disease: part I: in vitro systems and physiological data. AAPS J

151. Graaf IA, Groothuis GM, Olinga P (2007) Precision-cut tissue slices as a tool to predict metabolism of novel drugs. Expert Opin Drug Metab Toxicol 3(6):879–898

152. de Graaf IA, Olinga P, de Jager MH, Merema MT, de Kanter R, van de Kerkhof EG et al (2010) Preparation and incubation of precision-cut liver and intestinal slices for application in drug metabolism and toxicity studies. Nat Protoc 5(9):1540–1551

153. Godoy P, Hewitt NJ, Albrecht U, Andersen ME, Ansari N, Bhattacharya S et al (2013) Recent advances in 2D and 3D in vitro systems using primary hepatocytes, alternative hepatocyte sources and non-parenchymal liver cells and their use in investigating mechanisms of hepatotoxicity, cell signaling and ADME. Arch Toxicol 87(8):1315–1530

154. Olinga P, Schuppan D (2013) Precision-cut liver slices: a tool to model the liver ex vivo. J Hepatol 58(6):1252–1253

155. Lash LH, Putt DA, Cai H (2008) Drug metabolism enzyme expression and activity in primary cultures of human proximal tubular cells. Toxicology 244(1):56–65

156. Peters SA, Jones CR, Ungell AL, Hatley OJ (2016) Predicting drug extraction in the human gut wall: assessing contributions from drug metabolizing enzymes and transporter proteins using preclinical models. Clin Pharmacokinet 55(6):673–696

157. Moss DM, Marzolini C, Rajoli RK, Siccardi M (2015) Applications of physiologically based pharmacokinetic modeling for the optimization of anti-infective therapies. Expert Opin Drug Metab Toxicol 11(8):1203–1217

158. Rostami-Hodjegan A (2012) Physiologically based pharmacokinetics joined with in vitro-in vivo extrapolation of ADME: a marriage under the arch of systems pharmacology. Clin Pharmacol Ther 92(1):50–61

159. Chu X, Bleasby K, Evers R (2013) Species differences in drug transporters and implications for translating preclinical findings to humans. Expert Opin Drug Metab Toxicol 9(3):237–252

160. Graham MJ, Lake BG (2008) Induction of drug metabolism: species differences and toxicological relevance. Toxicology 254(3):184–191

161. Martignoni M, Groothuis GM, de Kanter R (2006) Species differences between mouse, rat, dog, monkey and human CYP-mediated drug metabolism, inhibition and induction. Expert Opin Drug Metab Toxicol 2(6):875–894

162. Fuhr U, Jetter A, Kirchheiner J (2007) Appropriate phenotyping procedures for drug metabolizing enzymes and transporters in humans and their simultaneous use in the "cocktail" approach. Clin Pharmacol Ther 81(2):270–283

163. De Kesel PM, Lambert WE, Stove CP (2016) Alternative sampling strategies for cytochrome P450 phenotyping. Clin Pharmacokinet 55(2):169–184

164. Breimer DD, Schellens JH (1990) A 'cocktail' strategy to assess in vivo oxidative drug metabolism in humans. Trends Pharmacol Sci 11(6):223–225
165. Pretorius E, Klinker H, Rosenkranz B (2011) The role of therapeutic drug monitoring in the management of patients with human immunodeficiency virus infection. Ther Drug Monit 33(3):265–274
166. Alsultan A, Peloquin CA (2014) Therapeutic drug monitoring in the treatment of tuberculosis: an update. Drugs 74(8):839–854
167. Bruggemann RJ, Aarnoutse RE (2015) Fundament and prerequisites for the application of an antifungal TDM service. Curr Fungal Infect Rep 9(2):122–129
168. Croes S, Koop AH, van Gils SA, Neef C (2012) Efficacy, nephrotoxicity and ototoxicity of aminoglycosides, mathematically modelled for modelling-supported therapeutic drug monitoring. Eur J Pharm Sci Off J Eur Fed Pharm Sci 45(1–2):90–100
169. Ye ZK, Li C, Zhai SD (2014) Guidelines for therapeutic drug monitoring of vancomycin: a systematic review. PLoS One 9(6):e99044

Chapter 3
Mechanisms of Drug Interactions II: Transport Proteins

Darren Michael Moss, Marco Siccardi, and Catia Marzolini

3.1 Introduction

The movement of drugs across biological membranes was once thought to proceed by simple diffusion depending on their lipophilic properties. However, due to significant advances in molecular biology and biotechnology, a wide variety of drug uptake and efflux transporters have been identified and characterized over the last 20 years. Major membrane transporters have been classified into the solute carrier (SLC) transporter family and the ATP-binding cassette (ABC) transporter family as designated by the Human Genome Organization (HUGO) Gene Nomenclature Committee (http://www.genenames.org). With exception of the multi-antimicrobial extrusion protein/multidrug and toxic compound extrusion transporter (MATE), the SLC transporter family is mainly characterized by uptake transporters which transfer substrates, either by facilitated diffusion down the electrochemical gradient or by secondary active transport against a diffusion gradient coupled to the symport or antiport of inorganic or small organic ions to provide the driving force [1]. The SLC

D.M. Moss
Department of Molecular and Clinical Pharmacology, University of Liverpool, Liverpool, UK

School of Pharmacy, Keele University, Keele, UK

M. Siccardi
Department of Molecular and Clinical Pharmacology, University of Liverpool, Liverpool, UK

C. Marzolini (✉)
Division of Infectious Diseases and Hospital Epidemiology, University Hospital Basel
and University of Basel, Basel, Switzerland
e-mail: catia.marzolini@usb.ch

© Springer International Publishing AG 2018
M.P. Pai et al. (eds.), *Drug Interactions in Infectious Diseases: Mechanisms and Models of Drug Interactions*, Infectious Disease,
https://doi.org/10.1007/978-3-319-72422-5_3

transporter family comprises various members of the organic anion-transporting polypeptide (OATP) family, organic cation transporter (OCT), organic anion transporter (OAT), organic cation/carnitine transporter (OCTN), peptide transporter (PEPT), concentrative nucleoside transporter (CNT), equilibrative nucleoside transporter (ENT) and MATE [2]. The ABC transporter family is primarily characterized by efflux transporters that function to export drugs out of a cell against a concentration gradient and are driven by the hydrolysis of adenosine triphosphate (ATP) as an energy source. Members of the ABC transporter family are the multidrug resistance protein (MDR), multidrug resistance-related protein (MRP) family, bile salt export pump (BSEP) and the breast cancer resistance protein (BCRP) [2]. SLC and ABC transporters are involved in the transport of a broad range of drugs in clinical use and share a wide distribution in the body, notably in key organs for drug disposition such as the intestine, liver and kidney. Tables 3.1 and 3.2 list uptake and efflux transporters considered to be relevant for the disposition of anti-infective agents, their tissue distribution as well as selected drug substrates [2, 3].

The role of uptake and efflux transporters in the drug disposition process, including particular emphasis on their documented or potential role in clinically relevant drug-drug interactions involving anti-infective medications, will be discussed in the following sections.

3.2   Transporter Effect on Drug Disposition

Most of the SLC and ABC transporters are found at either the apical or basolateral membrane of transporting epithelia (Fig. 3.1). Depending on their function and localization, these transporters will facilitate the entry or the removal of a drug substrate into a given organ. The net pharmacokinetic effect of active transport processes mostly results from the involvement of several transporters that may not always belong to the same family. For example, the transport pathway for the renal elimination of the nucleotide analogue tenofovir involves the uptake from the blood into the renal proximal tubular cells mediated by OAT1/3 and the efflux into urine by MRP4 [6]. Since the coordinated expression and function of transporters are critical in determining the extent and direction of drug movement, modulation of their activity (i.e. inhibition or induction) will directly impact the absorption, distribution, metabolism and excretion of a drug substrate. Of note, the expression and function of drug transporters can be dramatically influenced by disease pathology. A full review of this topic lies outside the scope of this chapter, although several excellent reviews exist which summarize the influence of specific disease states on transporter expression and function [7, 8].

Table 3.1 Uptake transporters involved in the disposition of anti-infective agents

Gene	Transporter	Location	Selected drug substrates	Influence on drug disposition
SLC01A2	**OATP1A2**	Brain, kidney, intestine	Fexofenadine, methotrexate, digoxin, statins, doxorubicin, **levofloxacin, darunavir, lopinavir, saquinavir**	oral absorption, renal excretion, CNS distribution
SLC01B1	**OATP1B1**	Liver	Statins, methotrexate, paclitaxel, repaglinide, fexofenadine, bosentan, olmesartan, valsartan, torasemide, **rifampin, benzylpenicillin, capsofungin, lopinavir, darunavir, saquinavir, maraviroc, tenofovir alafenamide, paritepravir, grazoprevir, simeprevir, velpatasvir**	hepatic uptake role in clinically relevant DDI
SLC01B3	**OATP1B3**	Liver	Statins, methotrexate, docetaxel, paclitaxel, fexofenadine, bosentan, olmesartan, telmisartan, valsartan, digoxin, enalapril, **erythromycin, rifampin, paritaprevir, grazoprevir, simeprevir, velpatasvir**	hepatic uptake role in clinically relevant DDI
SLC02B1	**OATP2B1**	Liver, intestine, placenta	Statins, fexofenadine, **benzylpenicillin**	hepatic uptake, distribution
SLC22A1	**OCT1**	Liver, intestine	Quinidine, cisplatin, imatinib, sorafenib, oxaliplatin, metformin, cimetidine, famotidine, ranitidine, **aciclovir, ganciclovir, lamivudine**	hepatic uptake role in clinically relevant DDI
SLC22A2	**OCT2**	Kidney, brain (choroid plexus)	Metformin, ranitidine, amiloride, cisplatin, oxaliplatin, varenicline, **lamivudine**	CNS distribution, renal excretion role in clinically relevant DDI
SLC22A6	**OAT1**	Kidney	Indomethacine, methotrexate, **adefovir, cidofovir, aciclovir, ganciclovir, didanosine, tenofovir, zidovudine, raltegravir**	renal excretion role in clinically relevant DDI
SLC22A7	**OAT2**	Liver, kidney	5-fluorouracil, methotrexate, paclitaxel, valproic acid, **tetracycline, zidovudine**	hepatic upatke, renal excretion

(continued)

Table 3.1 (continued)

Gene	Transporter	Location	Selected drug substrates	Influence on drug disposition
SLC22A8	**OAT3**	Kidney, brain	Cimetidine,ranitidine, methotrexate, furosemide, sitagliptin, **cidofovir, aciclovir, valaciclovir, amoxicillin, cefazolin, cefotaxime, meropenem, tetracycline, tenofovir**	renal excretion role in clinically relevant DDI
SLC22A11	**OAT4**	Kidney, placenta	Methotrexate, **tetracycline, zidovudine**	renal excretion
SLC15A1	**PepT1**	Intestine, kidney	Enalapril, captopril, **amoxicillin, ampicillin, cefaclor, valaciclovir**	oral absorption, renal excretion role in clinically relevant DDI
SLC15A2	**PepT2**	Kidney	Enalapril, captopril, **amoxicillin, valaciclovir**	renal excretion
SLC29A1	**ENT1**	Ubiquitous	Gemcitabine, cytarabine, **ribavirin**	distribution
SLC29A2	**ENT2**	Ubiquitous	**Didanosine, zalcitabine, zidovudine**	distribution
SLC47A1	**MATE1**	Liver, kidney, adrenal gland, skeletal muscle	Cimetidine, metformin, cisplatin, **aciclovir, ganciclovir, fluoroquinolones, emtricitabine, lamivudine**	biliary excretion, renal excretion
SLC47A2	**MATE2-K**	Kidney	Cimetidine, metformin, cisplatin, oxaliplatin, **aciclovir, ganciclovir**	renal excretion

Anti-infective agents are highlighted in bold. DDI, drug-drug interactions

3.2.1 Intestinal Absorption

The small intestine not only can limit the absorption of drugs through intestinal metabolism [9] but also through active drug transport back into the lumen by efflux transporters located at the apical brush border membrane of enterocytes such as MDR1 (i.e. P-glycoprotein) or BCRP [10]. Conversely, uptake transporters such as PEPT1 or OATPs will facilitate the intestinal drug absorption across the brush border membrane [11, 12] (Fig. 3.1). Consequently, modification of the expression or function of uptake or efflux transporters in the gastrointestinal tract will impact the bioavailability of orally administered drug substrates. However, it should be noted that the transport capacity can be saturated by the high concentration of drugs present in the intestinal lumen. Thus, the relative contribution of intestinal P-glycoprotein to the overall drug absorption is unlikely to be quantitatively important because its transport activity is easily saturated for most drugs at clinically relevant doses [13]. Nevertheless, some drugs administered at high doses are still influenced by the

Table 3.2 Efflux transporters involved in the disposition of anti-infective agents

Gene	Transporter	Location	Selected drug substrates	Influence on drug disposition
ABCB1	**MDR1/ P-gp**	Kidney, liver, brain, intestine, placenta, testes, lymphocyte	Anticancer agents, antihypertensive agents, antiarrhythmics, antihistamines, immunosuppressants, antidepressants, antiepileptics, **antifungals, HIV protease inhibitors, dolutegravir, maraviroc, tenofovir disoproxil fumarate, tenofovir alafenamide, direct acting antivirals against HCV**	oral absorption, biliary excretion, renal excretion, CNS distribution role in clinically relevant DDI
ABCC1	**MRP1**	Many tissues, testes, lymphocyte	Daunorubicin, doxorubicin, epirubicin, etoposide, methotrexate, vincristine, **HIV protease inhibitors**	distribution
ABCC2	**MRP2**	Liver, kidney, intestine	Methotrexate, vinblastine, etoposide, vincristine, valsartan, olmesartan, pravastatin, **HIV protease inhibitors**	oral absorption, biliary excretion renal excretion role in clinically relevant DDI
ABCC4	**MRP4**	Kidney, liver, brain	Methotrexate, topotecan, furosemide, adefovir, **tenofovir**, abacavir	distribution, renal excretion.
ABCC5	**MRP5**	Many tissues	Adefovir, **lamivudine**	distribution
ABCG2	**BCRP**	Intestine, liver, brain, kidney, placenta	Mitoxantrone, doxorubicine, topotecan, methotrexate, imatinib, irinotecan, rosuvastatin, **abacavir, tenofovir alafenamid**e, **sofosbuvir, dasabuvir, ombitasvir, boceprevir, paritaprevir, simeprevir, ledipasvir, velpatasvir**	oral absorption, biliary excretion role in clinically relevant DDI

Anti-infective agents are highlighted in bold
DDI drug-drug interactions

effects of intestinal P-glycoprotein. Typically, such drugs are poorly water soluble, dissolve slowly and are large in size (>800 Da), e.g. cyclosporine and saquinavir [13]. In general, transporter-mediated drug-drug interactions at the level of intestinal absorption are more likely to be clinically relevant for drugs with a narrow therapeutic index and characterized by an exclusive transporter-mediated disposition profile, e.g. digoxin [14].

3.2.2 Hepatobiliary Elimination

The hepatic elimination of drugs includes several steps: extraction of drugs from the portal blood into the hepatocytes which is often mediated by SLC transporters expressed on the sinusoidal (basolateral) membrane, hepatic metabolism mediated

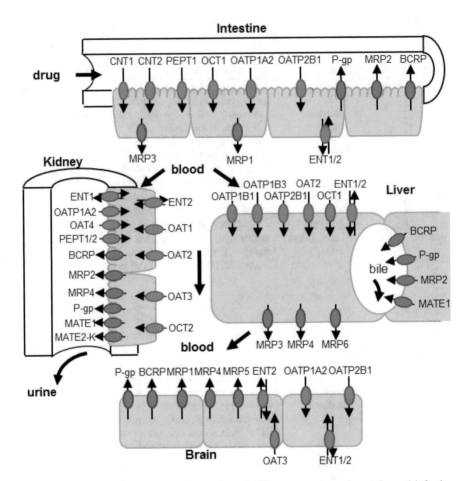

Fig. 3.1 Schematic diagram depicting uptake and efflux transporters relevant for anti-infective drug disposition and their localization in the human intestinal epithelia, hepatocyte, kidney proximal tubule and brain capillary endothelial cell. Uptake transporters are represented in blue and efflux transporters in red. *BCRP* breast cancer resistance protein, *CNT* concentrative nucleoside transporter, *ENT* equilibrative nucleoside transporter, *MATE* multidrug and toxin extrusion protein, *MDR* multidrug resistance protein, *MRP* multidrug resistance associated protein, *OAT* organic anion transporter, *OATP* organic anion-transporting polypeptide, *OCT* organic cation transporter, *PEPT* peptide transport protein (From Refs. [3–5])

by phase I (cytochrome P450) and phase II (glucuronidation) enzymes and either secretion of the drug back into the circulation for subsequent renal elimination mediated by ABC transporters expressed on the sinusoidal membrane or secretion into the bile via the efflux transporters expressed on the canalicular (apical) membrane of the hepatocyte [15] (Fig. 3.1). The cooperation of sinusoidal uptake and canalicular efflux transporters allows the directional transport across the hepatocytes. Members of the SLC family are considered to be of particular importance for hepatic drug elimination and drug pharmacokinetics. Specifically, these transporters regulate the amount of drug available for metabolism by liver enzymes and the subsequent biliary excretion of drugs [16]. Of interest, saturation of hepatic uptake by OATP1B1/3 may occur which leads to nonlinear pharmacokinetics of substrate drugs such as observed for simeprevir, a drug against hepatitis C virus (HCV) [17].

Inhibition of these hepatic uptake transporters will increase the systemic exposure of a drug substrate and potentially lead to side effects [18]. For instance, the mean AUC of several statins, not significantly metabolized by drug-metabolizing enzymes, increased four- to tenfold (i.e. fluvastatin (fourfold), pitavastatin (fivefold), pravastatin (tenfold), rosuvastatin (sevenfold)) in the presence of cyclosporine, an OATP1B1/3 inhibitor [19]. On the other hand, inhibition of canalicular efflux transporters will impact the biliary clearance of a drug substrate.

3.2.3   Renal Excretion

Renal elimination involves both passive and active processes: glomerular filtration and transporter-mediated secretion and reabsorption of drugs. Renal transporters, located mainly in the proximal tubular cells, play a role in several steps: drug uptake into the proximal tubular cell, drug efflux into the glomerular filtrate, reabsorption of the drug from the filtrate and drug efflux back into the blood (Fig. 3.1). Overall, renal excretion results from a coordinated function of uptake and efflux transporters located at the basolateral and apical membranes of proximal tubular cells. Members of OAT and OCT families present at the basolateral membrane are characterized by a high clearance capacity and are considered as major renal transporters for the uptake of organic anions and cations, respectively. As a result, highly efficient uptake of certain drugs in the cell can result in accumulation which can cause nephrotoxicity. For instance, preclinical experiments have shown that both cidofovir and adefovir are taken up by OAT1, which contributes to increased cytotoxicity [20, 21]. The large number of efflux transporters expressed at the brush border membrane emphasizes the importance of rapid efflux of potentially toxic compounds into urine. The competitive inhibition of proximal tubular secretion is one of the most common types of drug-drug interaction at the renal level. A decrease in renal secretion can lead to an increase in systemic drug exposure. However, competitive inhibition of renal secretion will result in clinically relevant drug-drug interactions only if the affected drug is actively secreted in the kidney and if the transporter-mediated renal clearance accounts for the majority of the total clearance of the affected drug. In addition, the concentration of the fraction unbound in plasma for the interacting drug must be high enough to produce a pronounced effect. The potential for a significant drug interaction is likely to be small if the concentration of the interacting agent is < Ki (i.e. Michaelis-Menten inhibitory constant), unless the drug has a narrow therapeutic window [22]. For example, interactions with fatal complications have been reported after concomitant administration of tenofovir with didanosine [23] and methotrexate with nonsteroidal anti-inflammatory drugs (NSAID) [24]. These interactions result from interactions on basolateral renal tubular uptake transporters (OAT1 and/or OAT3) as well as on the efflux transporters (MRP2 and/or MRP4) [6, 25].

3.2.4   Tissue Distribution

In organs such as the brain, transporter expression is critical for the brain homeo-stasis by limiting the entry of potentially harmful endogenous and exogenous substances. The blood-brain barrier consists of complex tight junctions of the brain capillary endothelial cells that express various transport proteins [26] (Fig. 3.1). The mechanism of blood-brain barrier transport has been divided into three separate processes: blood uptake of drugs and nutrients into the brain, efflux of compounds to prevent entry into the brain and uptake of metabolites, neu-rotransmitters and neurotoxins from the brain into the blood [27, 28]. Successful treatment of certain CNS infections requires adequate brain penetration of anti-infective medications. Thus, drug transporters may act as major barrier to current and effective drug treatment. For instance, the critical role of P-glycoprotein in restricting brain uptake of HIV protease inhibitors [29] or abacavir [30] has been demonstrated in animal studies. Measurements of antiretroviral concentrations in the cerebrospinal fluid indicate indeed that most drugs have very low brain pen-etration [31, 32].

Expression of several transporters detected in lymphocytes may also have an impact on HIV therapy [33, 34]. Specifically, drug transporters are believed to have a role in limiting drug uptake into lymphocytes. For instance, BCRP has been impli-cated in conferring cellular resistance to zidovudine and lamivudine by limiting their entry in lymphocytes [35]. Similarly, several studies have shown that P-glycoprotein, MRP, BCRP and OATP limit intracellular levels of various HIV protease inhibitors in lymphocytes. Therefore the effectiveness of antiretroviral therapy may be compromised since HIV virus replicates and is primarily contained within CD4+ cells [36–38].

3.2.5   Impact of Genetic Polymorphisms on Drug Disposition

As highlighted previously, alterations in uptake or efflux transporter function will directly impact the disposition of a drug substrate. Impaired transport function may result from genetic variations in the gene encoding the transporter protein. Numerous transporters relevant to anti-infective disposition show potentially function-altering mutations in the population, for example, in the influx transporters OATP1A2, OATP2B1, OATP1B3, OCT1, OCT2 and OCT3 and the efflux transporter P-gp [39]. For instance, the variant SLCO1B1*5, which is characterized by a nucleotide change from T to C in position 521 (521 T > C) of the SLCO1B1 gene encoding human OATP1B1, is associated with a reduced in vitro transport activity [40]. HIV infected patients carrying this genetic variant were shown to have higher plasma concentrations of the HIV protease inhibitor lopinavir when compared to noncarriers [41, 42].

The disposition of tenofovir has also been shown to be influenced by genetic variants in MRP2 ($-24C > T$) and MRP4 ($3436A > G$ and $3463A > G$) [43–46]. Interestingly, the risk for tenofovir-induced proximal tubulopathy has been associated with homozygosity for the C allele at position -24 in MRP2 [47]. However, the mechanism by which MRP2 influences the risk of kidney tubular dysfunction is not well understood as in vitro studies have shown that tenofovir is not a substrate for human MRP2 [6, 48]. More detailed information regarding genetic variations in drug transporters and their effect on the pharmacokinetic of drugs in clinical use can be found in the following references [6, 48–53]. In addition to genetic variations, modulation of transporter function may result from the inhibiting or inducing properties of a drug substrate, thereby influencing the transport kinetics of a simultaneously administered drug.

3.3 Transporter-Mediated Drug-Drug Interactions

Drug-drug interactions observed in the clinic can be linked with drug transporters. In this regard, in vitro transporter-expressing systems have been particularly useful in understanding and predicting transporter-mediated drug-drug interactions. Table 3.3 compiles the substrate and inhibiting properties of selected anti-infective agents obtained in vitro for the major drug transporters with a documented role in drug disposition [6, 20, 30, 35, 38, 41, 48, 54–121]. In vitro approaches are now commonly used as a critical first step for the assessment of drug interaction potential and to support subsequent in vivo studies which help define the clinical relevance. For that purpose, the Food and Drug Administration (FDA) and the European Medicines Agency (EMA) have released guidance documents about the conduct and decision-making criteria of in vitro transporter assay (see Table 3.4) [151, 152].

This section will discuss the mechanisms of transporter-mediated interactions and describe examples of clinically relevant drug-drug interactions involving anti-infective agents (Table 3.5) [56, 60, 65, 66, 83, 93, 105, 106, 122–150].

3.3.1 Mechanisms of Inhibition and Induction of Transporters

Transporter-mediated drug interactions in the clinic may be either inhibitory, inductive or both and may involve influx or efflux transporters. Transporters can be inhibited in a competitive or non-competitive manner similarly to drug-metabolizing enzymes. Competitive inhibition occurs when two substrates compete at the same binding site where only one substrate can bind. For non-competitive inhibition, two substrates will simultaneously bind at two different sites which might inhibit the subsequent translocation process. Induction of drug transporters and drug-metabolizing enzymes occurs indirectly, i.e. through the interaction with nuclear receptors

Table 3.3 Selected anti-infective agents and their substrate and/or inhibitor properties for key transporters in drug disposition

	OATP 1A2	OATP 1B1	OATP 1B3	OATP 2B1	OAT 1	OAT 2	OAT 3	OAT 4	OCT 1	OCT 2	MATE 1	MATE 2K	Pgp	BCRP	MRP 1	MRP 2	MRP 4	MRP 5
Antibiotics																		
Amoxicillin																		
Benzylpenicillin						x	x											
Cefaclor						x												
Cefazolin					x	x	x											
Clarithromycin		x*	x*										x*					
Erythromycin		x	x		x								x					
Levofloxacin											x	x						
Meropenem					x		x											
Rifampin	x	x*	x*	x														
Tetracycline					x	x		x										
Trimethoprim									x	x*	x*	x						
Antifungals																		
Capsofungin		x												x				
Itraconazole									x				x*	x				
Ketoconazole	x	x					x		x	x	x		x*	x				
Antivirals																		
Aciclovir					x		x		x									
Adefovir					x												x	
Cidofovir					x												x	
Ganciclovir					x	x	x		x									
Oseltamivir																		
Valaciclovir																		
HCV direct acting antivirals																		
Boceprevir	x*	x*																
Grazoprevir														x*				
Paritaprevir	x*	x*											x*	x*				
Simeprevir	x*	x*	x										x*	x*		x		
Telaprevir	x	x							x	x	x		x*					
Daclatasvir	x*	x*							x				x*	x*				
Elbasvir													x*	x*				
Ledipasvir	x	x											x*	x*				
Ombitasvir																		
Velpatasvir	x*	x*											x*	x*				
Dasabuvir													x*	x*				
Sofosbuvir													x	x				
Antiretrovirals																		
Atazanavir		x*	x*	x									x*	x	x	x		
Darunavir		x*	x*	x					x				x*			x		
Indinavir	x	x	x	x					x	x			x*		x	x		
Lopinavir		x*	x*	x									x*	x		x		
Ritonavir	x	x	x	x					x	x	x		x*	x	x	x		
Saquinavir	x	x	x	x					x	x			x*	x	x	x		
Dolutegravir										x*	x							
Elvitegravir/cobicistat	x*	x*							x	x	x		x*	x*				
Raltegravir					x		x											
Maraviroc																		
Efavirenz				x					x						x	x		
Etravirine														x				
Nevirapine															x	x		
Rilpivirine									x	x	x							
Abacavir									x	x				x				
Emtricitabine									x	x					x	x		
Lamivudine									x	x								
Tenofovir									x	x			#					
Zidovudine					x	x	x	x	x	x				x				

This table compiles data obtained from in vitro transport/inhibition studies using human isoforms (From Refs. [6, 20, 30, 38, 41, 48, 54–121]). In vitro data using rodent isoforms or data resulting from knockout animal model are not included in this table. Substrates are annotated by grey squares and inhibitors by a cross. Clinically significant inhibitions as observed in clinical studies are marked * (see Table 3.5). # denotes that the prodrugs of tenofovir (tenofovir disoproxil fumarate and tenofovir alafenamide) are substrates of P-glycoprotein, whereas tenofovir is not

Table 3.4 Recommendations for drug transporter testing as outlined in the European Medicines Agency guidelines and the Food and Drug Administration's draft guidance on drug interaction studies [151, 152]

		Inhibition studies		Substrate studies	
	Transporter	EMA	FDA	EMA	FDA
Efflux	P-gp	Yes	Yes	Consider	Yes
	BCRP	Yes	Yes	Consider	Yes
	BSEP	Preferred	Consider	Consider	Consider
	MRPs	No	Consider	Consider	Consider
Uptake	OAT1	Yes	Yes	Consider	If >25% active renal secretion
	OAT3	Yes	Yes	Consider	If >25% active renal secretion
	OATP1B1	Yes	Yes	If >25% clearance is hepatic	If >25% clearance is hepatic or biliary
	OATP1B3	Yes	Yes	If >25% clearance is hepatic	If >25% clearance is hepatic or biliary
	OCT1	Consider	No	Consider	No
	OCT2	Yes	Yes	Consider	If >25% active renal secretion
	MATE1	Consider	Consider	Consider	Consider
	MATE2K	Consider	Consider	Consider	Consider

such as the pregnane X receptor (PXR) or constitutive androstane receptor (CAR) [2, 153]. These nuclear receptors share a common signalling pathway, which involve ligand (e.g. rifampin) binding to the receptor, heterodimerization with the 9-cis-retinoic acid receptor (RXR), binding of the heterodimer to response elements of target genes (i.e. drug transporters, drug-metabolizing enzymes) and subsequent initiation of the gene transcription [154].

3.3.2 Interplay Between Drug Transporters and Drug-Metabolizing Enzymes

Drug transporters and drug-metabolizing enzymes often share overlapping tissue expression and substrate specificities. For instance, many P-glycoprotein substrates and inhibitors are also substrates and inhibitors of cytochrome P450 3A4 (CYP3A4) (i.e. erythromycin, itraconazole, HIV protease inhibitors) [155]. This overlap between P-glycoprotein (as well as other transporters) and CYP3A4 is thought to result from their common mechanism of regulation. Drug interactions involving compounds that inhibit both drug-metabolizing and transporter pathways may result in profound interactions. For instance, the antidiabetic drug repaglinide is a substrate for CYP2C8, CYP3A4 and OATP1B1. Gemfibrozil and its metabolite are both inhibitors of CYP2C8 and OATP1B1 [156]. Co-administration of gemfibrozil

Table 3.5 Clinical examples of transporter-mediated drug-drug interactions

Drug	Inhibitor/inducer	PK effect/toxicity[a]	Putative mechanism	Reference
Atorvastatin	Rifampin (sd)	AUC ↑ 682%	Inhibition of OATP1B1	[122]
Atorvastatin	Rifampin	AUC ↓ 80%	Induction of P-gp, CYP3A4	[123]
Bosentan	Rifampin	Ctrough ↑ 5 fold	Inhibition of OATP1B1/3 initial phase	[124]
Bosentan	Rifampin	AUC ↓ 58%, Cmax ↓ 53%	Induction of CYP3A4 at steady-state	[124]
Grazoprevir	Rifampin (sd)	AUC ↑ 735%, Cmax ↑ 552%	Inhibition of OATP1B1	[106]
Grazoprevir	Rifampin	AUC ↓ 7%, Cmin ↓ 90%	Induction of CYPs/P-gp at steady-state	[106]
Velpatasvir	Rifampin (sd)	AUC ↑ 47%, Cmax ↑ 28%	Inhibition of OATP1B1	[83]
Velpatasvir	Rifampin	AUC ↓ 82%, Cmax ↓ 71%	Induction of CYPs/P-gp at steady-state	[83]
Bosentan	Lopinavir/ ritonavir	AUC ↑ 422%, Cmax ↑ 377%, ↑ AE	Inhibition of OATP1B1/3, CYP3A4	[125]
Grazoprevir	Atazanavir/ ritonavir	AUC ↑ 900%, Cmax ↑ 500%	Inhibition of OATP1B1/3, CYP3A4	[106]
Grazoprevir	Darunavir/ ritonavir	AUC ↑ 600%, Cmax ↑ 400%	Inhibition of OATP1B1/3, CYP3A4	[106]
Grazoprevir	Lopinavir/ ritonavir	AUC ↑ 1100%, Cmax ↑ 600%	Inhibition of OATP1B1/3, CYP3A4	[106]
Velpatasvir	Ciclosporine	AUC ↑ 103%, Cmax ↑ 56%	Inhibition of OATP1B1/3, P-gp	[83]
Rosuvastatin	Atazanavir/ ritonavir	AUC ↑ 213%, Cmax ↑ 600%	Inhibition of OATP1B1 and BCRP	[126]
Rosuvastatin	Lopinavir/ ritonavir	AUC ↑ 107%, Cmax ↑ 365%	Inhibition of OATP1B1 and BCRP	[127]
Rosuvastatin	Elvitegravir/ cobicistat	AUC ↑ 38%, Cmax ↑ 89%	Inhibition of OATP1B1 and BCRP	[60]
Rosuvastatin	Daclatasvir	AUC ↑ 58%, Cmax ↑104%	Inhibition of OATP1B1 and BCRP	[66]
Rosuvastatin	Simeprevir	AUC ↑ 181%, Cmax ↑217%	Inhibition of OATP1B1 and BCRP	[93]
Rosuvastatin	Velpatasvir	AUC ↑ 160%, Cmax ↑ 170%	Inhibition of OATP1B1 and BCRP	[83]

(continued)

Table 3.5 (continued)

Drug	Inhibitor/inducer	PK effect/toxicity[a]	Putative mechanism	Reference
Rosuvastatin	Elbasvir/ grazoprevir	AUC ↑ 126%, Cmax ↑ 449%	Inhibition of BCRP	[106]
Pravastatin	Clarithromycin	AUC ↑ 110%, Cmax ↑ 128%	Inhibition of OATP1B1/3	[128]
Pravastatin	Boceprevir	AUC ↑ 63%, Cmax ↑ 49%	Inhibition of OATP1B1/3	[129]
Pravastatin	Paritaprevir/r	AUC ↑ 76%, Cmax ↑ 43%	Inhibition of OATP1B1/3	[56]
Pravastatin	Velpatasvir	AUC ↑ 35%, Cmax ↑ 28%	Inhibition of OATP1B1/3	[83]
Lamivudine	Trimethoprim	AUC ↑ 43%, CL ↓ 35%	Inhibition of OCT1/2	[130]
Memantine	Trimethoprim	Myoclonus, delirium	Inhibition of OCT2	[131]
Metformin	Dolutegravir (qd/ bid)	AUC ↑ 79/145%, Cmax ↑ 66/111%	Inhibition of OCT2	[105]
Metformin	Trimethoprim	AUC ↑ 37%, Cmax ↑ 38%, CL ↓ 20%	Inhibition of OCT2 and MATE1	[132]
Metformin	Cephalexin	AUC ↑ 24%, Cmax ↑ 34%, CL ↓ 14%	Inhibition of MATE1	[133]
Metformin	Pyrimethamine	AUC ↑ 39%, Cmax ↑ 42%, CL ↓ 35%	Inhibition of MATE1 and MATE2-K	[134]
Ciprofloxacin	Probenecid	CL ↓ 65%	Inhibition of OAT3 and/or OCT2	[135]
Cidofovir	Probenecid	CL ↓ 32%	Inhibition of OAT1/3	[136]
Zalcitabine	Probenecid	AUC ↑ 54%, CL ↓ 42%	Inhibition of OAT1	[137]
Flucloxacillin	Piperacillin	CL ↓ 58%	Inhibition of OAT1	[138]
Digoxin	Ritonavir	AUC ↑ 80%, CL ↓ 35%	Inhibition of P-gp	[139]
Grazoprevir	Ketoconazole	AUC ↑ 200%, Cmax ↑ 13%	Inhibition of P-gp, CYP3A4	[106]
Tacrolimus	Darunavir/ ritonavir	Increase in Ctrough	Inhibition of P-gp, CYP3A4	[140]
Sirolimus	Clarithromycin	Increase in Ctrough	Inhibition of P-gp, CYP3A4	[141]
Digoxin	Clarithromycin	CL ↓ 50%	Inhibition of P-gp	[142]
Colchicine	Clarithromycin	Colchicine intoxication	Inhibition of P-gp	[143]
Digoxin	Itraconazole	AUC ↑ 50%, CL ↓ 20%	Inhibition of P-gp	[144]
Digoxin	Daclatasvir	AUC ↑ 27%, Cmax ↑ 65%	Inhibition of P-gp	[66]
Digoxin	Elbasvir	AUC ↑ 11%, Cmax ↑ 47%	Inhibition of P-gp	[106]
Digoxin	Paritaprevir/r	AUC ↑ 35%, Cmax ↑ 61%	Inhibition of P-gp	[56]
Digoxin	Simeprevir	AUC ↑ 39%, Cmax ↑ 31%	Inhibition of P-gp	[93]

(continued)

Table 3.5 (continued)

Drug	Inhibitor/inducer	PK effect/toxicity[a]	Putative mechanism	Reference
Digoxin	Velpatasvir	AUC ↑ 34%, Cmax ↑ 88%	Inhibition of P-gp	[83]
Digoxin	Telaprevir	AUC ↑ 85%, Cmax ↑ 50%	Inhibition of P-gp	[65]
Tenofovir (DF)[b]	Ledipasvir/ sofosbuvir	AUC ↑ 98%, Cmax ↑ 79%	Inhibition of P-gp and possibly BCRP	[145]
Digoxin	Rifampin	AUC ↓ 30%, Cmax ↓ 58%	Induction of P-gp	[146]
Elbasvir	Efavirenz	AUC ↓ 54%, Cmax ↓ 45%	Induction of P-gp, CYP3A4	[106]
Grazoprevir	Efavirenz	AUC ↓ 83%, Cmax ↓ 87%	Induction of P-gp, CYP3A4	[106]
Indinavir	St John's wort	AUC ↓ 57%, Ctrough ↓ 81%	Induction of P-gp, CYP3A4	[147]
Mycophenolic acid	Rifampin	AC AUC ↓, metabolites AUC ↑	Induction of UGT, inhibition MRP2	[148]
Tenofovir	Diclofenac	Nephrotoxicity	Inhibition of MRP4	[149, 150]

[a]Percent change refers to the difference between the area under the curve (AUC), maximal concentration (C_{max}), concentration just before the next dose (C_{trough}) or renal clearance (CL) in the presence and the absence of the interacting drug, b = tenofovir derived from the prodrug tenofovir disoproxil fumarate when given in an efavirenz based regimen

AC active compound, *UGT* uridine diphosphate-glucuronosyltransferase, *AE* adverse effect, *bid* twice daily, *qd* once daily, *sd* single dose

and repaglinide caused a profound increase in repaglinide area under the curve (AUC) (eightfold), whereas co-administration of repaglinide with itraconazole, a CYP3A4 inhibitor, caused a modest change in repaglinide AUC (1.4-fold) [157]. Interestingly, the simultaneous administration of these three drugs led to a major increase in repaglinide AUC (19-fold). Thus, the interplay between drug-metabolizing enzymes and transporter proteins must be considered when evaluating any drug-drug interaction potential.

Emerging evidence suggests that drug transporters such as OATP may indirectly regulate the expression of drug disposition genes through modulation of the intracellular concentrations of PXR or CAR ligands [158]. This concept evolves from previous in vitro observations which suggest that OATP1B1 is a major determinant of PXR activation via rifampin [110]. This interplay can result in time-dependent drug-drug interactions. For instance, a single dose of rifampin co-administered with atorvastatin resulted in a sevenfold increase in atorvastatin AUC [122], whereas the treatment with rifampin over 5 days decreased the AUC of atorvastatin by 80% [123]. An initial increase in drug exposure followed by a decrease is also observed for grazoprevir and velpatasvir, when co-administered with rifampin [83, 106] (Table 3.5). The increase in atorvastatin, grazoprevir or velpatasvir levels after a single dose of rifampin most likely results from OATP1B1/3 inhibition by rifampin [114]. Conversely, the subsequent decrease in drug exposure upon continued dosing reflects the time-dependent induction of drug-metabolizing enzymes by rifampin.

Thus, as illustrated in these examples, the interplay between drug transporters and drug-metabolizing enzymes further complicates the prediction of the effect of a drug interaction [159].

3.3.3 Role of OATP in Drug-Drug Interactions

The organic anion-transporting polypeptides (OATP) represent an important family of uptake carriers mediating the transport of relatively large (molecular weight > 400–500 Da) and hydrophobic organic anions. Typical endogenous and exogenous substrates include bile salts, thyroid hormones as well as numerous drugs in clinical use such as statins, rifampin and protease inhibitors [2, 19, 160] (Tables 3.1 and 3.3). OATPs are expressed in several tissues [11, 161–165] (Fig. 3.1). As major hepatic uptake transporters, OATP1B1/3 regulate the amount of drugs available for phase I/II metabolism or biliary excretion. Several clinically relevant drug-drug interactions involving OATPs have been reported in the literature. A few examples are highlighted below (Table 3.5).

Macrolides are well known to cause drug-drug interactions via the inhibition of drug-metabolizing enzymes. For instance, clarithromycin was shown to increase the AUC of several concomitantly administered statins (i.e. simvastatin (tenfold), atorvastatin (fourfold) and pravastatin (twofold)) [128]. For simvastatin and atorvastatin, this increase can be explained by the inhibition of CYP3A4 as both drugs are substrates of this enzyme. In contrast, pravastatin is mainly eliminated as unchanged drug. In vitro experiments have revealed that clarithromycin but also erythromycin inhibited OATP1B1-mediated pravastatin uptake [102]. Thus, inhibition of OATP1B1 probably explains the observed clinical drug interaction. This same mechanism of interaction explains the increase in pravastatin exposure when co-administered with direct-acting antivirals against HCV (Table 3.5). The co-administration of the HIV protease inhibitor lopinavir boosted with ritonavir (i.e. lopinavir/r) and rosuvastatin surprisingly led to increased plasma concentrations of the statin (i.e. AUC and Cmax increased 107% and 365%, respectively) [127]. Similarly, rosuvastatin is mainly excreted as unchanged drug [166]. In vitro experiments have shown that rosuvastatin is a substrate of the hepatic transporter OATP1B1 [167] and the intestinal efflux transporter BCRP [168]. Of interest, lopinavir and ritonavir were shown to inhibit BCRP and OATP1B1 [54, 68, 117]. As a consequence, more rosuvastatin will be absorbed at the intestinal level and less rosuvastatin will enter the hepatocyte for elimination thereby explaining the increased plasma concentrations of the statin [127]. This same mechanism of interaction explains the interaction between atazanavir/r, elvitegravir/cobicistat, daclatasvir, simeprevir or velpatasvir and rosuvastatin [60, 66, 83, 93, 126]. Of note, this interaction could possibly negatively impact the pharmacodynamic effect of the statin, by inhibiting its entry into the liver, which is the site of action and elimination. Consequently, the lipid-lowering effect of the statin may be attenuated despite the increase in plasma concentration and potential associated risk of myotoxicity. Bosentan is used for the treatment of pulmonary arterial hypertension. This drug is metabolized by CYP2C9 and CYP3A4 [169] and is a substrate of OATP1B1 and OATP1B3 [170]. The co-administration of bosentan and lopinavir/r, an inhibitor of CYP3A4 and OATP1B1/3 [54], resulted in a marked increase in bosentan exposure (AUC and Cmax increased 422% and 377%, respectively) as well as an increase in adverse events of bosentan [125]. Inhibition of CYP3A4 and OATP1B1/3 by boosted protease inhibitors explains also the profound increase in grazoprevir exposure [106]. Interestingly, the co-administration of bosentan, grazoprevir or velpatasvir and rifampin resulted in a time-dependent interaction with an initial increase in bosentan, grazoprevir or velpatasvir exposure followed by a decrease in exposure at steady state (Table 3.5). The inhibition of the OATP1B1/3-mediated transport of bosentan, grazoprevir or velpatasvir by rifampin [114] most likely explains the initial increase in exposure, whereas the CYP and P-gp inductive properties of rifampin resulted in the decrease in exposure of bosentan, grazoprevir or velpatasvir observed at steady state [83, 106, 124].

3.3.4   Role of OAT in Drug-Drug Interactions

The organic anion transporters (OAT) accept relatively small (molecular weight < 400–500 Da), hydrophilic organic anions. Their substrates include several drugs such as beta-lactams, nonsteroidal anti-inflammatory drugs (NSAID) and antiviral nucleoside analogues [2, 59, 171, 172] (Tables 3.1 and 3.3). OAT1, OAT3 and OAT4 are mainly expressed in the kidney, whereas OAT2 is predominantly expressed in the liver [173–176] (Fig. 3.1). OATs are considered as major excretory systems and have been involved in drug-drug interactions of clinical importance (Table 3.5).

Perhaps the most widely understood drug interaction, first noted six decades ago, is that of penicillin and probenecid, which results in elevated serum penicillin concentrations [177]. In vitro experiments have revealed that probenecid strongly inhibits human OAT1 and OAT3 [178]. Interactions between probenecid and beta-lactam antibiotics have been reported extensively [179, 180]. This beneficial interaction has been intentionally utilized to enhance the activity of antibiotics in treating infections. Probenecid has also been used deliberately to alter the renal clearance of concomitantly administered drugs in order to reduce their toxicity. For instance, cidofovir, a nucleoside analogue, is predominantly excreted in the urine as unchanged drug [181]. The nephrotoxicity related to this compound is due to its high concentration in the kidney as a result of rapid drug uptake at the basolateral membrane of tubular cells and slower efflux into the urine via transporters of the brush-boarder membrane [20, 21]. Co-administration of probenecid decreased cidofovir clearance which subsequently resulted in reduction of nephrotoxicity [136]. In vitro studies have shown that cidofovir is a substrate of OAT1 [21, 59]. Thus, inhibition of OAT1-mediated uptake of cidofovir by probenecid can prevent its nephrotoxicity. Similar findings have been observed when NSAID are co-administered with adefovir, another nucleoside analogue [182]. However, drug-drug interactions at the level of renal excretion may also have detrimental effects. For instance, the co-administration of tenofovir and diclofenac has been associated with tubular nephrotoxicity [149] due to inhibition of MRP4 by diclofenac [25] and thereby reduced renal excretion of tenofovir. All together these examples indicate that inhibition of basolateral or apical renal transporters will have different impact on intrarenal drug accumulation and thereby nephrotoxicity. Inhibition of basolateral uptake transporters will tend to reduce drug accumulation in the renal tubular cell and therefore be protective for the nephron. Conversely, inhibition of apical efflux transporters will diminish drug exit from renal tubular cells which increases drug accumulation in the tubular cell and thereby nephrotoxicity.

3.3.5 Role of OCT in Drug-Drug Interactions

The organic cation transporters (OCT) mediate the cellular uptake of small organic cations (molecular weight < 400 Da). Typical substrates are drugs such as metformin, cytostatic drugs or antiviral nucleoside analogues [2, 183, 184] (Tables 3.1 and 3.3). OCT1 is mostly expressed in the liver; low levels are also present in the apical surface of intestinal enterocytes [4], whereas OCT2 is most abundant in the kidney [185, 186] (Fig. 3.1). Both transporters have been implicated in drug-drug interactions (Table 3.5).

Trimethoprim is used in HIV patients with low CD4-cell counts for primary and secondary prophylaxis against *Pneumocystis jirovecii* infection. Co-administration of trimethoprim and lamivudine resulted in a 35% reduction in lamivudine renal clearance [130]. In vitro experiments have shown that lamivudine is transported by OCT1/2 and that trimethoprim inhibits these same transporters thereby explaining the reduced renal elimination of lamivudine [72]. The co-administration of trimethoprim with memantine, a drug prescribed for the treatment of Alzheimer's disease and primarily excreted unchanged in the kidney, led to the development of myoclonic activity and delirium [131]. Interestingly, these symptoms rapidly subsided after trimethoprim discontinuation. In vitro data have indicated that memantine is transported by OCT2 [187]. Thus, the observed adverse events are most likely attributed to the inhibition of memantine renal excretion by trimethoprim. The antidiabetic metformin is eliminated largely by renal secretion via OCT2 (uptake in the tubular cell) and MATE1(secretion in urine). The co-administration with the HIV integrase inhibitor dolutegravir, an inhibitor of OCT2, was shown to increase metformin exposure [105]. Thus dose adjustments of metformin might be needed when patients are starting or stopping dolutegravir.

3.3.6 Role of MATE in Drug-Drug Interactions

The multidrug and toxic compound extrusion (MATE) transporters expressed in humans include the solute carriers MATE1 and MATE2 and the splice variant MATE2K. All three isoforms are expressed on the apical membrane of renal proximal tubule cells where, in tandem with OCT2 expressed on the basolateral membrane, they facilitate the movement of substrates from the blood and into the lumen [188]. Additionally, MATE1 is expressed on the canalicular membrane of hepatocytes and is believed to work in combination with OCT1 in biliary excretion. MATE transporter substrates, which tend to overlap with OCT1 and OCT2 substrates, are typically hydrophilic, low-molecular-weight cations. However, MATE transporters show broader substrate specificity than for OCTs and include anionic and zwitterionic compounds [189]. MATE transporters are capable of transporting the antibacterials cephradine (MATE1 and MATE2K) and levofloxacin (MATE1), the antimalarial quinine (MATE1 and MATE2K) and the antivirals aciclovir (MATE1 and MATE2K), ganciclovir (MATE1 and MATE2K) and lamivudine (MATE1 and MATE2K) [89, 189]. Clinically relevant drug interactions involving MATE transporters have been hypothesized, although it is difficult to separate the possible actions of OCTs from these interactions. The pharmacokinetics of metformin was altered in human subjects (14% decrease in renal clearance, 24% and 34% increase in AUC and Cmax, respectively) when co-administered with the antibiotic cephalexin [133], and it has been hypothesized that cephalexin-mediated MATE inhibition is involved in this interaction [5]. The pharmacokinetics of metformin (35% decrease in renal clearance, 42% and 39% increase in Cmax and AUC, respectively) was influenced when co-administered with anti-protozoal pyrimethamine, which has been confirmed as an inhibitor of MATE transporters [134, 190]. Importantly, supporting the involvement of MATE transporters in these interactions, pyrimethamine was shown to be up to 200 times more potent in inhibiting MATE transporters than in inhibiting OCT in vitro. Co-administration with trimethoprim was also shown to increase metformin exposure as a result of OCT2 and MATE1 inhibition [132]. Finally, it is important to mention that the transporters MATE1 and OCT2 are involved in the renal excretion of creatinine [77]. Consequently, a small increase in serum creatinine with a related decrease in estimated glomerular filtration rate has been reported upon treatment with ritonavir- and cobicistat-based regimens (via inhibition of MATE1) as well as upon treatment with dolutegravir-based regimens (inhibition of OCT2). This effect has been shown to reflect mainly the inhibition of creatinine secretion by MATE1 or OCT2 rather than an actual impairment of the renal function [191, 192].

3.3.7   Role of ENT and CNT in Drug-Drug Interactions

Equilibrative nucleoside transporters (ENT) and concentrative nucleoside transporters (CNT) mediate the uptake of endogenous nucleosides and nucleoside-derived drugs across cellular barriers and are expressed to varying degrees in a wide variety of tissues. When assessed, nucleoside and nucleotide analogue antiviral drugs tend to be less efficiently transported by ENTs than the extent observed for endogenous compounds. ENT2 and ENT3 transport the antiviral nucleosides zidovudine, zalcitabine and didanosine, and to a lesser extent, ENT1 transports zalcitabine and didanosine, although the impact of ENT on the clinical pharmacokinetics of these drugs is not fully understood [193, 194]. Animal studies have suggested that ENT1 is involved in the absorption and distribution of the hepatitis C drug ribavirin [195]. Additionally to their affinity for ENTs, ribavirin and didanosine are also transported by CNT2, and zidovudine and zalcitabine are transported by CNT1 [194]. There is currently no requirement by the FDA or the EMA for investigating drug interactions involving ENT and CNT prior to drug approval.

3.3.8 Role of PEPT in Drug-Drug Interactions

The peptide transporters (PEPT) are responsible for the cellular uptake of several drugs such as angiotensin-converting enzyme inhibitors, beta-lactam antibiotics and antiviral drugs [2, 196, 197] (Tables 3.1 and 3.3). Interestingly, PEPT1 has been targeted as a way to improve oral drug absorption. For instance, the bioavailability of aciclovir was considerably enhanced after oral administration of its valine ester (i.e. valaciclovir), which is a PEPT1 substrate [198]. PEPT1 is primarily located in the intestine and kidney, whereas PEPT2 is mainly located in the kidney [64, 199] (Fig. 3.1). PEPT1 plays a major role in the intestinal absorption of beta-lactam antibiotics. In vitro data showed that the intestinal transport of 23 beta-lactam antibiotics and the bioavailability in humans both correlated with their affinity for PEPT1 [200]. Because of their role in facilitating oral absorption and renal reabsorption of several drugs in clinical use, these transporters may be subject to drug-drug interactions.

3.3.9 Role of P-gp in Drug-Drug Interactions

P-glycoprotein (P-gp), the encoded product of the human MDR1 gene, was first discovered for its role in mediating the multidrug resistance phenotype associated with certain cancers [201]. P-gp has a large substrate specificity and can recognize hundreds of compounds ranging from small molecules of 350 Da up to polypeptides of 4000 Da. Therapeutic compounds transported by P-gp include anticancer drugs, antihypertensive agents, antiarrhythmics, glucocorticoids, HIV protease inhibitors, antibiotics, antimycotics, immunosuppressive agents, antidepressants, neuroleptics, antiepileptics, antiacids, opioids and antiemetics [52] (Tables 3.2 and 3.3). As mentioned previously, many substrates of P-gp are also substrates of drug-metabolizing enzymes, which make it difficult to assess the extent of interactions associated with P-gp. P-gp is expressed in various tissues and serves as a permeation barrier in the gastrointestinal tract, brain, lymphocytes, placenta, testes and ovaries while contributing to the elimination of drugs in the liver and kidney [202–205] (Fig. 3.1). The anatomical localization coupled with the broad variety of drug substrates contributes to the significant role of P-gp in drug disposition. The effect of P-gp on the pharmacokinetics of substrate drugs has been demonstrated in several studies using mdr1a/1b knockout mice. Mice lacking mdr genes usually present with increased drug absorption, increased distribution in the brain and decreased drug elimination compared with wild-type mice [29, 206–208]. Interestingly, animal studies revealed that P-gp inhibition had a much greater impact on the tissue distribution of drug substrates than on their systemic exposure [209]. Thus, the potential risk of P-gp-mediated drug interactions might be underestimated if only plasma concentrations are monitored [210].

Several drug-drug interactions mediated by P-gp have been reported in the literature (Table 3.5). Digoxin, a widely prescribed agent for congestive heart failure, has a negligible metabolism and is primarily eliminated in the kidney through glomerular filtration and active secretion. In vitro and in vivo animal studies have clearly shown that digoxin is a high-affinity substrate for P-gp [211, 212]. Concomitant administration of ritonavir, a potent inhibitor of P-gp [213], was shown to substantially increase digoxin exposure and reduce its renal clearance (80% increase in AUC and 35% decrease in clearance) [139]. Increase in digoxin exposure or decrease in digoxin renal clearance has also been reported with the concomitant use of various direct-acting antivirals against HCV or clarithromycin, itraconazole and erythromycin (Table 3.5) [142, 144, 214]. All these drugs are inhibitors of P-gp-mediated digoxin transport [142, 215, 216]. Another clinically relevant interaction with digoxin involves the co-administration of rifampin. The oral bioavailability of digoxin was decreased by 30% during rifampin therapy due to induction of intestinal P-gp [146]. Similar interactions with rifampin have been reported for fexofenadine [217–219], cyclosporine [218] and tacrolimus [220]. Finally, of interest for HIV therapy, tenofovir disoproxil fumarate, the prodrug of tenofovir, is a substrate of P-gp. Thus, the co-administration with inhibitors of P-gp such as boosted protease inhibitors or ledipasvir/sofosbuvir has been shown to increase the absorption of tenofovir disoproxil fumarate, thereby resulting in higher systemic levels of tenofovir [145]. As an example, the co-administration of tenofovir disoproxil fumarate (as part of an efavirenz-based regimen) with ledipasvir/sofosbuvir increased tenofovir AUC and Cmax by 98% and 79%, respectively. Although this increase has been attributed to inhibition of intestinal P-gp, in vitro data seem to indicate that carboxylesterase inhibition by protease inhibitors could also contribute to this effect [221].

3.3.10   Role of BCRP in Drug-Drug Interactions

The breast cancer resistance protein (BCRP) was originally identified in a breast cancer cell line that exhibited resistance to anthracyclines [222]; therefore, anticancer drugs are among the first reported substrates [2, 223] (Table 3.2). Some nucleoside analogues have been shown to be transported by BCRP, whereas HIV protease inhibitors and several direct-acting antivirals against HCV are BCRP inhibitors [35, 56, 66, 68, 83, 93, 106, 117] (Table 3.5). BCRP is primarily expressed in the small intestine, the liver, the blood-brain barrier and the placenta [224, 225] (Fig. 3.1). The localization of BCRP suggests that this transporter as well as other transporters of the ABC family play a protective role in limiting oral bioavailability and transport across the blood-brain barrier or the placenta [226]. Drug-drug interactions possibly involving BCRP have been described for the combination of HIV protease inhibitors (i.e. atazanavir/r or lopinavir/r) or several direct-acting antivirals against HCV and rosuvastatin (see section "Role of OATP in Drug Interactions") (Table 3.5). However, BCRP interactions are difficult to investigate as BCRP and P-gp have extensive substrate overlap; therefore, one transporter may compensate when the other is inhibited.

3.3.11 Role of MRP in Drug-Drug Interactions

The multidrug resistance-related proteins (MRPs) also are known to confer multiple drug resistance to cancer cells [227]. Collectively, MRPs often share substrates in common with P-gp and are known to mediate the transport of numerous medications such as anticancer drugs, statins, nucleoside analogues or HIV protease inhibitors [2, 228] (Tables 3.2 and 3.3). These transporters are widely distributed in nearly all human tissues [229–234] (Fig. 3.1). In particular, MRP2 is localized at the canalicular membrane of the hepatocytes and is primarily responsible for hepatobiliary excretion of drugs. In the kidney, MRP2 and MRP4 are expressed at the apical membrane of the tubular cells where they facilitate the renal excretion of anionic compounds. A few examples of drug-drug interactions involving MRP are described below (Table 3.5).

Mycophenolate mofetil (MMF), an immunosuppressant used for organ transplant recipients, is de-esterified to form mycophenolic acid (MPA), which is the active compound. MPA is subsequently glucuronidated into phenyl (MPAG) and acyl (AcMPAG) glucuronide metabolites whose biliary excretion is mediated by MRP2 [235]. Following excretion into bile, these metabolites can be deconjugated back to MPA and reabsorbed via an enterohepatic cycling process [236]. MPA and AcMPAG are eliminated in the urine via OAT1/3 and possibly MRP2-mediated tubular secretion [237]. Interestingly, co-administration of rifampin and MMF required dose increase for a lung graft recipient [238]. The PK analysis of this interaction revealed a significant total MPA AUC decrease of 17.5% after rifampin co-administration, whereas MPAG and AcMPAG AUC increased by 34.4% and 193%, respectively [148]. This interaction likely resulted from the induction of MPA glucuronidation through rifampin-mediated PXR activation and possibly through inhibition of MRP2-mediated enterohepatic recirculation or renal excretion [148]. In vitro experiments have shown that rifampin is a substrate for MRP2 and thus could compete for this specific transporter [239]. Since MPA has a narrow therapeutic index, this interaction may lead to MPA underexposure and loss of clinical efficacy. In addition, increased plasma levels of toxic glucuronide metabolites may lead to side effects [148].

Adefovir and cidofovir are both inhibitors and substrates of MRP2 [81]. As mentioned previously, these compounds undergo renal tubular secretion and can cause nephrotoxicity resulting from accumulation in proximal renal tubules via OAT-mediated cellular uptake. Inhibition of MRP2 in renal cells may also contribute to adefovir and cidofovir nephrotoxicity by reduction in efflux. Similarly, the use of NSAID in HIV patients treated by tenofovir leads to the nucleotide analogue renal accumulation and subsequent nephrotoxicity through MRP4 inhibition [150].

3.3.12   Role of BSEP in Drug-Drug Interactions

Bile salt efflux pump (BSEP) is exclusively expressed on the apical membrane of hepatocytes and functions to remove bile salts from hepatocytes and into the bile [240]. Other than indirect evidence suggesting BSEP-mediated transport of the antifungal micafungin, investigations have not suggested that BSEP plays a major role in anti-infective drug elimination from the liver [241]. However, the high specificity of BSEP for bile salts and the lack of backup transporters on the apical surface indicate that the action of BSEP is a rate-limiting step in bile salt excretion. Bile salts are toxic to hepatocytes at high intracellular concentrations, and therefore any drug which inhibits BSEP activity has the potential to cause cholestasis and liver damage. A wide variety of drugs has been shown to inhibit BSEP activity, including the antitubercular drug rifampicin [242]. Prospective BSEP interaction testing is not endorsed by the International Transporter Consortium (ITC) or the FDA at this time, although the EMA guidelines recommend that investigating BSEP inhibition potential should be considered in drug development programmes.

3.4   Challenges in Predicting In Vivo Drug-Drug Interactions

Estimating the contribution of transporters to total tissue uptake and excretion is necessary for predicting their role in drug-drug interactions. Although remarkable advances have been made in the functional characterization of drug transporters, the quantitative evaluation of transporter-mediated drug interactions is difficult to predict. Unlike drug-metabolizing enzymes which are largely concentrated in the liver and intestine, drug transporters are expressed in various tissues with different functions (absorption, distribution and elimination). Therefore the influence of transporters on the disposition of a drug requires investigation of numerous transporters with different functions in both hepatic and extrahepatic tissues. Another difficulty relates to the overlapping substrate specificities and the considerable functional redundancy in transport proteins. Furthermore, the interplay between transporters and drug-metabolizing enzymes adds complexity in estimating the role of a single transporter in drug disposition. Other limitations include the lack of specific and potent inhibitors for individual transporters which precludes accurate extrapolations from in vitro inhibition studies [210] . Differences in tissue localization and in substrate recognition of transporters between humans and animals often complicate translation from preclinical findings to the clinic. Finally, drug interactions involving transporters at the level of absorption and elimination alter the plasma concentrations of drugs. In contrast, interactions occurring at the blood-brain barrier do not affect the drug exposure in the circulating blood but only the pharmacological and/or toxicological effect of the drug. Therefore, drug interactions studies that assess only plasma drug concentrations do not fully characterize the transport-mediated influence of one drug on another; thus changes in the tissue distribution of drugs should also be considered [210].

Physiologically based pharmacokinetic (PBPK) modelling exists as a potentially useful tool for extrapolating in vitro transporter data to in vivo prediction potential. PBPK modelling is a bottom-up technique which uses in vitro drug data (i.e. physicochemical characteristics, intrinsic clearance, cellular permeability, etc.) to simulate compound pharmacokinetics through mathematical descriptions of absorption, distribution and elimination [243]. This approach has been applied successfully to characterize drug-drug interactions for drug combinations used in the daily clinical practice but for which limited clinical data are available on drug-drug interactions as, for instance, the co-administration of efavirenz or boosted protease inhibitors and commonly prescribed antidepressants [244]. By taking into account the concurrent inhibitory and inductive effect of antiretroviral drugs on cytochromes, PBPK simulations showed that the magnitude of drug-drug interactions with antidepressants was overall weak to moderate. The modest magnitude has been attributed to the fact that antidepressants are substrates of multiple isoforms, and thus metabolism can still occur through cytochromes that are weakly impacted by efavirenz or boosted protease inhibitors. When compared to drugs displaying passively mediated pharmacokinetics, the prediction of pharmacokinetics for substrates of drug transporters is far less established. This is partly due to the lack of knowledge in important areas, such as the levels of transporter expression and activity in different tissues, the often incomplete data of substrate transporter affinity rates and the interplay of transporters with other transporters and enzymes. However, attempts have been made to create PBPK models which account for the activity and expression of specific transporters. A prediction strategy for novel OATP substrates was developed by inputting data from human hepatocyte models combined with available clinical plasma concentration-time data [245]. The study was able to more successfully predict the active liver uptake and efflux of pravastatin, cerivastatin, bosentan, fluvastatin, rosuvastatin, valsartan and repaglinide by OATP expressed in the liver. A PBPK model was recently developed to investigate the influence of cimetidine on metformin passive permeability and active transport in the kidney [246]. The model included the action of transporters MATE1, MATE2K, OCT1 and OCT2 and succeeded in simulating metformin AUC within 50% of observed data. Regarding the development of PBPK models for anti-infectives, information on transporter activity is often not included due to insufficient data. In several PBPK models, an "active transport" elimination factor has been included without considering the actions of individual transporters, such as for the elimination of darunavir via the liver [247] and nucleoside reverse transcriptase inhibitors tenofovir, emtricitabine and lamivudine via the kidneys [248].

3.5   Clinical Drug Interaction Studies with Transporters

The understanding of the rate-limiting step in the clearance of a given drug (i.e. transport vs metabolism) and how a potential co-administered drug can alter the clearance is critical for the correct prediction of the drug-drug interaction and subsequent design and data interpretation of clinical drug interaction studies. Clinical drug interaction studies are usually designed to assess the effect of a known inhibitor of a transporter on the disposition of a drug candidate or the effect of the drug candidate on the disposition of a known substrate of a transporter. The selection of either a substrate or an inhibitor of a given transporter has to be based on the substrate and inhibitor transporter selectivity properties. Other considerations should include the therapeutic window of the substrate drug and the maximum effect that would be expected if the clearance of the substrate drug was totally inhibited, or the therapeutic use. For instance, a substrate or inhibitor of a drug candidate should be selected based on their likelihood of being co-administered in a therapeutic setting. Prior to the licencing of a new drug, the FDA and EMA require that certain tests are performed which determine if a drug is a substrate or inhibitor of a selection of clinically relevant transporters (Table 3.4).

The ITC has published decision trees to help determine when to conduct in vivo human interaction studies based on in vitro evaluation of transporters [3]. The guidance from the transporter white paper published in 2010 is summarized below:

- If the drug candidate is a substrate of OATP, a clinical drug interaction study should be performed with OATP inhibitors such as rifampin or cyclosporine. If the drug candidate is an inhibitor of OATP, possible probe substrates for OATP include atorvastatin, pravastatin, pitavastatin or rosuvastatin.
- If the drug candidate is a substrate of OAT, then inhibition should be studied with probenecid. Multiple probe substrates for OAT can be used in clinical drug interaction studies including zidovudine, lamivudine, aciclovir, ciprofloxacin, tenofovir or methotrexate.
- If the drug candidate is a substrate of OCT2, a clinical drug interaction study should be performed with cimetidine. Possible probe substrates for OCT2 include metformin or varenicline.
- If the drug candidate is a dual substrate for P-gp and CYP3A4, then inhibition should be studied using an inhibitor that shows strong inhibition for both P-gp and CYP3A4, such as itraconazole, ketoconazole, ritonavir or cyclosporine. Possible probe substrates for P-gp include digoxin or loperamide.
- If the drug candidate is an inhibitor of BCRP, possible probes include sulphasalazine, rosuvastatin, pitavastatin, ciprofloxacin or dipyridamole.

A more recent update by the ITC includes additional recommendations, such as the assessment of BSEP in cases where bile-related hepatic toxicity is suspected or observed, the assessment of ENTs specifically in the field of anticancer nucleoside-based treatments and the assessment of MRP and MATE transporters in specific circumstances [5]. The FDA and EMA have updated their recommendations based on the advice of the ITC and now include MRPs, BSEP, MATE1 and MATE2K in their list of potential transporter investigations, although some (such as for the MRPs in the EMA guidelines) remain unrequired (Table 3.4).

It is important to note that some of these inhibitors or substrates may inhibit or be transported by multiple transporters or may also affect drug-metabolizing enzymes, therefore the clinical interaction data should be interpreted cautiously.

3.6 Summary

Whereas drug-drug interactions mediated via known drug-metabolizing enzymes have been established over several decades, SLC and ABC transporters are now becoming recognized as significant determinants of drug disposition and drug-drug interactions. The magnitude of transporter-mediated drug interactions is generally smaller when compared to cytochrome-mediated interactions, and therefore, to date, few clinically significant drug interactions have been demonstrated to be based on a single mechanism through transporter inhibition. In general, transporter-mediated drug interactions are likely to be most critical when the elimination of the affected drug or the distribution in a target tissue is characterized by an exclusive transporter-mediated disposition profile or when the involved drug exhibits a narrow therapeutic window of safety. Since the exposure of a majority of drugs in clinical use is defined by the interplay between enzymes and transporters, both pharmacological pathways need to be considered when evaluating the potential risk for drug-drug interactions. Finally, several issues remain to be addressed in order to better understand transporter-mediated drug-drug interactions. Efforts should be made to identify more specific inhibitors of drug transporters, to improve the ability to predict the magnitude of transporter-mediated drug-drug interactions based on in vitro data and to better understand changes in the drug transport-drug metabolism interplay during co-administration of drugs. Furthermore, future research should aim at better understanding the impact of transporter-mediated drug-drug interactions at the level of blood-tissue barriers (i.e. blood-brain barrier) or tumours.

References

1. Hagenbuch B (2010) Drug uptake systems in liver and kidney: a historic perspective. Clin Pharmacol Ther 87:39–47
2. Klaassen CD, Aleksunes LM (2010) Xenobiotic, bile acid, and cholesterol transporters: function and regulation. Pharmacol Rev 62:1–96
3. International, Transporter, and Consortium (2010) Membranes transporters in drug development. Nat Rev Drug Discov 9:215–236
4. Han TK, Everett RS, Proctor WR et al (2013) Organic cation transporter 1 (OCT1/mOct1) is localized in the apical membrane of Caco-2 cell monolayers and enterocytes. Mol Pharmacol 84:182–189

5. Hillgren KM, Keppler D, Zur AA et al (2013) Emerging transporters of clinical importance: an update from the international transporter consortium. Clin Pharmacol Ther 94:52–63
6. Ray AS, Cihlar T, Robinson KL et al (2006) Mechanism of active renal tubular efflux of tenofovir. Antimicrob Agents Chemother 50:3297–3304
7. Alam C, Whyte-Allman SK, Omeragic A, Bendayan R (2016) Role and modulation of drug transporters in HIV-1 therapy. Adv Drug Deliv Rev 103:121–143
8. Atilano-Roque A, Roda G, Fogueri U, Kiser JJ, Joy MS (2016) Effect of disease pathologies on transporter expression and function. J Clin Pharmacol 56(Suppl 7):S205–S221
9. Bonkovsky HL, Hauri HP, Marti U, Gasser R, Meyer UA (1985) Cytochrome P450 of small intestinal epithelial cells. Immunochemical characterization of the increase in cytochrome P450 caused by phenobarbital. Gastroenterology 88:458–467
10. Dietrich CG, Geier A, Oude Elferink RPABC (2003) Of oral bioavailability: transporters as gatekeepers in the gut. Gut 52:1788–1795
11. Glaeser H, Bailey DG, Dresser GK et al (2007) Intestinal drug transporter expression and the impact of grapefruit juice in humans. Clin Pharmacol Ther 81:362–370
12. Katsura T, Inui K (2003) Intestinal absorption of drugs mediated by drug transporters: mechanisms and regulation. Drug Metab Pharmacokinet 18:1–15
13. Lin JH, Yamazaki M (2003) Role of P-glycoprotein in pharmacokinetics: clinical implications. Clin Pharmacokinet 42:59–98
14. Lee CA, Cook JA, Reyner EL, Smith DA (2010) P-glycoprotein related drug interactions: clinical importance and a consideration of disease states. Expert Opin Drug Metab Toxicol 6:603–619
15. Nies AT, Schwab M, Keppler D (2008) Interplay of conjugating enzymes with OATP uptake transporters and ABCC/MRP efflux pumps in the elimination of drugs. Expert Opin Drug Metab Toxicol 4:545–568
16. Smith NF, Figg WD, Sparreboom A (2005) Role of the liver-specific transporters OATP1B1 and OATP1B3 in governing drug elimination. Expert Opin Drug Metab Toxicol 1:429–445
17. Snoeys J, Beumont M, Monshouwer M, Ouwerkerk-Mahadevan S (2016) Mechanistic understanding of the nonlinear pharmacokinetics and intersubject variability of simeprevir: a PBPK-guided drug development approach. Clin Pharmacol Ther 99:224–234
18. Kusuhara H, Sugiyama Y (2010) Pharmacokinetic modeling of the hepatobiliary transport mediated by cooperation of uptake and efflux transporters. Drug Metab Rev 42:539–550
19. Kalliokoski A, Niemi M (2009) Impact of OATP transporters on pharmacokinetics. Br J Pharmacol 158:693–705
20. Cihlar T, Ho ES, Lin DC, Mulato AS (2001) Human renal organic anion transporter 1 (hOAT1) and its role in the nephrotoxicity of antiviral nucleotide analogs. Nucleosides Nucleotides Nucleic Acids 20:641–648
21. Ho ES, Lin DC, Mendel DB, Cihlar T (2000) Cytotoxicity of antiviral nucleotides adefovir and cidofovir is induced by the expression of human renal organic anion transporter 1. J Am Soc Nephrol 11:383–393
22. Li M, Anderson GD, Wang J (2006) Drug-drug interactions involving membrane transporters in the human kidney. Expert Opin Drug Metab Toxicol 2:505–532
23. Murphy MD, O'Hearn M, Chou S (2003) Fatal lactic acidosis and acute renal failure after addition of tenofovir to an antiretroviral regimen containing didanosine. Clin Infect Dis 36:1082–1085
24. Maiche AG (1986) Acute renal failure due to concomitant action of methotrexate and indomethacin. Lancet 1:1390
25. El-Sheikh AA, van den Heuvel JJ, Koenderink JB, Russel FG (2007) Interaction of nonsteroidal anti-inflammatory drugs with multidrug resistance protein (MRP) 2/ABCC2- and MRP4/ABCC4-mediated methotrexate transport. J Pharmacol Exp Ther 320:229–235
26. Urquhart BL, Kim RB (2009) Blood-brain barrier transporters and response to CNS-active drugs. Eur J Clin Pharmacol 65:1063–1070
27. Ohtsuki S (2004) New aspects of the blood-brain barrier transporters; its physiological roles in the central nervous system. Biol Pharm Bull 27:1489–1496

28. Ohtsuki S, Terasaki T (2007) Contribution of carrier-mediated transport systems to the blood-brain barrier as a supporting and protecting interface for the brain; importance for CNS drug discovery and development. Pharm Res 24:1745–1758
29. Kim RB, Fromm MF, Wandel C et al (1998) The drug transporter P-glycoprotein limits oral absorption and brain entry of HIV-1 protease inhibitors. J Clin Invest 101:289–294
30. Shaik N, Giri N, Pan G, Elmquist WF (2007) P-glycoprotein-mediated active efflux of the anti-HIV1 nucleoside abacavir limits cellular accumulation and brain distribution. Drug Metab Dispos 35:2076–2085
31. Best BM, Letendre SL, Brigid E et al (2009) Low atazanavir concentrations in cerebrospinal fluid. AIDS 23:83–87
32. Marra CM, Zhao Y, Clifford DB et al (2009) Impact of combination antiretroviral therapy on cerebrospinal fluid HIV RNA and neurocognitive performance. AIDS 23:1359–1366
33. Minuesa G, Purcet S, Erkizia I et al (2008) Expression and functionality of anti-human immunodeficiency virus and anticancer drug uptake transporters in immune cells. J Pharmacol Exp Ther 324:558–567
34. Turriziani O, Gianotti N, Falasca F et al (2008) Expression levels of MDR1, MRP1, MRP4, and MRP5 in peripheral blood mononuclear cells from HIV infected patients failing antiretroviral therapy. J Med Virol 80:766–771
35. Wang X, Furukawa T, Nitanda T et al (2003) Breast cancer resistance protein (BCRP/ABCG2) induces cellular resistance to HIV-1 nucleoside reverse transcriptase inhibitors. Mol Pharmacol 63:65–72
36. Janneh O, Anwar T, Jungbauer C et al (2009) P-glycoprotein, multidrug resistance-associated proteins and human organic anion transporting polypeptide influence the intracellular accumulation of atazanavir. Antivir Ther 14:965–974
37. Janneh O, Hartkoorn RC, Jones E et al (2008) Cultured CD4T cells and primary human lymphocytes express hOATPs: intracellular accumulation of saquinavir and lopinavir. Br J Pharmacol 155:875–883
38. Janneh O, Owen A, Chandler B et al (2005) Modulation of the intracellular accumulation of saquinavir in peripheral blood mononuclear cells by inhibitors of MRP1, MRP2, P-gp and BCRP. AIDS 19:2097–2102
39. Kerb R (2006) Implications of genetic polymorphisms in drug transporters for pharmacotherapy. Cancer Lett 234:4–33
40. Tirona RG, Leake BF, Merino G, Kim RB (2001) Polymorphisms in OATP-C: identification of multiple allelic variants associated with altered transport activity among European- and African-Americans. J Biol Chem 276:35669–35675
41. Hartkoorn RC, Kwan WS, Shallcross V et al (2010) HIV protease inhibitors are substrates for OATP1A2, OATP1B1 and OATP1B3 and lopinavir plasma concentrations are influenced by SLCO1B1 polymorphisms. Pharmacogenet Genomics 20:112–120
42. Lubomirov R, di Iulio J, Fayet A et al (2010) ADME pharmacogenetics: investigation of the pharmacokinetics of the antiretroviral agent lopinavir coformulated with ritonavir. Pharmacogenet Genomics 20:217–230
43. Kiser JJ, Aquilante CL, Anderson PL et al (2008) Clinical and genetic determinants of intracellular tenofovir diphosphate concentrations in HIV-infected patients. J Acquir Immune Defic Syndr 47:298–303
44. Kiser JJ, Carten ML, Aquilante CL et al (2008) The effect of lopinavir/ritonavir on the renal clearance of tenofovir in HIV-infected patients. Clin Pharmacol Ther 83:265–272
45. Manosuthi W, Sukasem C, Thongyen S, Nilkamhang S, Sungkanuparph S (2014) ABCC2*1C and plasma tenofovir concentration are correlated to decreased glomerular filtration rate in patients receiving a tenofovir-containing antiretroviral regimen. J Antimicrob Chemother 69:2195–2201
46. Nishijima T, Komatsu H, Higasa K et al (2012) Single nucleotide polymorphisms in ABCC2 associate with tenofovir-induced kidney tubular dysfunction in Japanese patients with HIV-1 infection: a pharmacogenetic study. Clin Infect Dis 55:1558–1567

47. Rodriguez-Novoa S, Labarga P, Soriano V et al (2009) Predictors of kidney tubular dysfunction in HIV-infected patients treated with tenofovir: a pharmacogenetic study. Clin Infect Dis 48:e108–e116

48. Imaoka T, Kusuhara H, Adachi M et al (2007) Functional involvement of multidrug resistance-associated protein 4 (MRP4/ABCC4) in the renal elimination of the antiviral drugs adefovir and tenofovir. Mol Pharmacol 71:619–627

49. Choi MK, Song IS (2008) Organic cation transporters and their pharmacokinetic and pharmacodynamic consequences. Drug Metab Pharmacokinet 23:243–253

50. Ieiri I, Higuchi S, Sugiyama Y (2009) Genetic polymorphisms of uptake (OATP1B1, 1B3) and efflux (MRP2, BCRP) transporters: implications for inter-individual differences in the pharmacokinetics and pharmacodynamics of statins and other clinically relevant drugs. Expert Opin Drug Metab Toxicol 5:703–729

51. Maeda K, Sugiyama Y (2008) Impact of genetic polymorphisms of transporters on the pharmacokinetic, pharmacodynamic and toxicological properties of anionic drugs. Drug Metab Pharmacokinet 23:223–235

52. Marzolini C, Paus E, Buclin T, Kim RB (2004) Polymorphisms in human MDR1 (P-glycoprotein): recent advances and clinical relevance. Clin Pharmacol Ther 75:13–33

53. Marzolini C, Tirona RG, Kim RB (2004) Pharmacogenomics of the OATP and OAT families. Pharmacogenomics 5:273–282

54. Annaert P, Ye ZW, Stieger B, Augustijns P (2010) Interaction of HIV protease inhibitors with OATP1B1, 1B3, and 2B1. Xenobiotica 40:163–176

55. Babu E, Takeda M, Narikawa S et al (2002) Human organic anion transporters mediate the transport of tetracycline. Jpn J Pharmacol 88:69–76

56. Badri PS, Dutta S, Wang H et al (2015) Drug interactions with the direct-acting antiviral combination of Ombitasvir and Paritaprevir-ritonavir. Antimicrob Agents Chemother 60:105–114

57. Chu X, Cai X, Cui D et al (2013) Vitro assessment of drug-drug interaction potential of boceprevir associated with drug metabolizing enzymes and transporters. Drug Metab Dispos 41:668–681

58. Ci L, Kusuhara H, Adachi M et al (2007) Involvement of MRP4 (ABCC4) in the luminal efflux of ceftizoxime and cefazolin in the kidney. Mol Pharmacol 71:1591–1597

59. Cihlar T, Lin DC, Pritchard JB et al (1999) The antiviral nucleotide analogs cidofovir and adefovir are novel substrates for human and rat renal organic anion transporter 1. Mol Pharmacol 56:570–580

60. Custodio JM, Wang H, Hao J et al (2014) Pharmacokinetics of cobicistat boosted-elvitegravir administered in combination with rosuvastatin. J Clin Pharmacol 54:649–656

61. Fujita Y, Noguchi K, Suzuki T, Katayama K, Sugimoto Y (2013) Biochemical interaction of anti-HCV telaprevir with the ABC transporters P-glycoprotein and breast cancer resistance protein. BMC Res Notes 6:445

62. Furihata T, Fu Z, Suzuki Y et al (2015) Differential inhibition features of direct-acting anti-hepatitis C virus agents against human organic anion transporting polypeptide 2B1. Int J Antimicrob Agents 46:381–388

63. Furihata T, Matsumoto S, Fu Z et al (2014) Different interaction profiles of direct-acting anti-hepatitis C virus agents with human organic anion transporting polypeptides. Antimicrob Agents Chemother 58:4555–4564

64. Ganapathy ME, Brandsch M, Prasad PD, Ganapathy V, Leibach FH (1995) Differential recognition of beta -lactam antibiotics by intestinal and renal peptide transporters, PEPT 1 and PEPT 2. J Biol Chem 270:25672–25677

65. Garg V, Chandorkar G, Farmer HF et al (2012) Effect of telaprevir on the pharmacokinetics of midazolam and digoxin. J Clin Pharmacol 52:1566–1573

66. Garimella T, You X, Wang R et al (2016) A review of Daclatasvir drug-drug interactions. Adv Ther 33:1867–1884

67. Gupta A, Unadkat JD, Mao Q (2007) Interactions of azole antifungal agents with the human breast cancer resistance protein (BCRP). J Pharm Sci 96:3226–3235

68. Gupta A, Zhang Y, Unadkat JD, Mao QHIV (2004) Protease inhibitors are inhibitors but not substrates of the human breast cancer resistance protein (BCRP/ABCG2). J Pharmacol Exp Ther 310:334–341
69. Hill G, Cihlar T, Oo C et al (2002) The anti-influenza drug oseltamivir exhibits low potential to induce pharmacokinetic drug interactions via renal secretion-correlation of in vivo and in vitro studies. Drug Metab Dispos 30:13–19
70. Hoque MT, Kis O, De Rosa MF, Bendayan R (2015) Raltegravir permeability across blood-tissue barriers and the potential role of drug efflux transporters. Antimicrob Agents Chemother 59:2572–2582
71. Hyland R, Dickins M, Collins C, Jones H, Jones B (2008) Maraviroc: in vitro assessment of drug-drug interaction potential. Br J Clin Pharmacol 66:498–507
72. Jung N, Lehmann C, Rubbert A et al (2008) Relevance of the organic cation transporters 1 and 2 for antiretroviral drug therapy in human immunodeficiency virus infection. Drug Metab Dispos 36:1616–1623
73. Kis O, Robillard K, Chan GN, Bendayan R (2009) The complexities of antiretroviral drug-drug interactions: role of ABC and SLC transporters. Trends Pharmacol Sci 31:22–35
74. Kobayashi Y, Sakai R, Ohshiro N et al (2005) Possible involvement of organic anion transporter 2 on the interaction of theophylline with erythromycin in the human liver. Drug Metab Dispos 33:619–622
75. Kunze A, Huwyler J, Camenisch G, Gutmann H (2012) Interaction of the antiviral drug telaprevir with renal and hepatic drug transporters. Biochem Pharmacol 84:1096–1102
76. Lempers VJ, van den Heuvel JJ, Russel FG et al (2016) Inhibitory potential of antifungal drugs on ATP-binding cassette transporters P-glycoprotein, MRP1 to MRP5, BCRP, and BSEP. Antimicrob Agents Chemother 60:3372–3379
77. Lepist EI, Zhang X, Hao J et al (2014) Contribution of the organic anion transporter OAT2 to the renal active tubular secretion of creatinine and mechanism for serum creatinine elevations caused by cobicistat. Kidney Int 86:350–357
78. Li M, Andrew MA, Wang J et al (2009) Effects of cranberry juice on pharmacokinetics of beta-lactam antibiotics following oral administration. Antimicrob Agents Chemother 53:2725–2732
79. Maeda T, Takahashi K, Ohtsu N et al (2007) Identification of influx transporter for the quinolone antibacterial agent levofloxacin. Mol Pharm 4:85–94
80. Marzolini C, Gibbons S, Khoo S, Back D (2016) Cobicistat versus ritonavir boosting and differences in the drug-drug interaction profiles with comedications. J Antimicrob Chemother 71:1755–1758
81. Miller DS (2001) Nucleoside phosphonate interactions with multiple organic anion transporters in renal proximal tubule. J Pharmacol Exp Ther 299:567–574
82. Minuesa G, Volk C, Molina-Arcas M et al (2009) Transport of lamivudine [(−)-beta-L-2′,3′-dideoxy-3′-thiacytidine] and high-affinity interaction of nucleoside reverse transcriptase inhibitors with human organic cation transporters 1, 2, and 3. J Pharmacol Exp Ther 329:252–261
83. Mogalian E, German P, Kearney BP et al (2016) Use of multiple probes to assess transporter- and cytochrome P450-mediated drug-drug interaction potential of the Pangenotypic HCV NS5A inhibitor Velpatasvir. Clin Pharmacokinet 55:605–613
84. Morimoto K, Nakakariya M, Shirasaka Y et al (2008) Oseltamivir (Tamiflu) efflux transport at the blood-brain barrier via P-glycoprotein. Drug Metab Dispos 36:6–9
85. Moss DM, Kwan WS, Liptrott NJ et al (2011) Raltegravir is a substrate for SLC22A6: a putative mechanism for the interaction between raltegravir and tenofovir. Antimicrob Agents Chemother 55:879–887
86. Moss DM, Liptrott NJ, Curley P et al (2013) Rilpivirine inhibits drug transporters ABCB1, SLC22A1, and SLC22A2 in vitro. Antimicrob Agents Chemother 57:5612–5618
87. Moss DM, Liptrott NJ, Siccardi M, Owen A (2015) Interactions of antiretroviral drugs with the SLC22A1 (OCT1) drug transporter. Front Pharmacol 6:78

88. Moss DM, Siccardi M, Murphy M et al (2012) Divalent metals and pH alter raltegravir disposition in vitro. Antimicrob Agents Chemother 56:3020–3026

89. Muller F, Konig J, Hoier E, Mandery K, Fromm MF (2013) Role of organic cation transporter OCT2 and multidrug and toxin extrusion proteins MATE1 and MATE2-K for transport and drug interactions of the antiviral lamivudine. Biochem Pharmacol 86:808–815

90. Nies AT, Damme K, Schaeffeler E, Schwab M (2012) Multidrug and toxin extrusion proteins as transporters of antimicrobial drugs. Expert Opin Drug Metab Toxicol 8:1565–1577

91. Ogihara T, Kano T, Wagatsuma T et al (2009) Oseltamivir (tamiflu) is a substrate of peptide transporter 1. Drug Metab Dispos 37:1676–1681

92. Okuda M, Kimura N, Inui K (2006) Interactions of fluoroquinolone antibacterials, DX-619 and levofloxacin, with creatinine transport by renal organic cation transporter hOCT2. Drug Metab Pharmacokinet 21:432–436

93. Ouwerkerk-Mahadevan S, Snoeys J, Peeters M, Beumont-Mauviel M, Simion A (2016) Drug-drug interactions with the NS3/4A protease inhibitor Simeprevir. Clin Pharmacokinet 55:197–208

94. Polli JW, Jarrett JL, Studenberg SD et al (1999) Role of P-glycoprotein on the CNS disposition of amprenavir (141W94), an HIV protease inhibitor. Pharm Res 16:1206–1212

95. Reese MJ, Savina PM, Generaux GT et al (2013) Vitro investigations into the roles of drug transporters and metabolizing enzymes in the disposition and drug interactions of dolutegravir, a HIV integrase inhibitor. Drug Metab Dispos 41:353–361

96. Reznicek J, Ceckova M, Cerveny L, Muller F, Staud F (2017) Emtricitabine is a substrate of MATE1 but not of OCT1, OCT2, P-gp, BCRP or MRP2 transporters. Xenobiotica 47:77–85

97. Rizk ML, Houle R, Chan GH et al (2014) Raltegravir has a low propensity to cause clinical drug interactions through inhibition of major drug transporters: an in vitro evaluation. Antimicrob Agents Chemother 58:1294–1301

98. Sakurai Y, Motohashi H, Ueo H et al (2004) Expression levels of renal organic anion transporters (OATs) and their correlation with anionic drug excretion in patients with renal diseases. Pharm Res 21:61–67

99. Sala-Rabanal M, Loo DD, Hirayama BA, Turk E, Wright EM (2006) Molecular interactions between dipeptides, drugs and the human intestinal H+ −oligopeptide cotransporter hPEPT1. J Physiol 574:149–166

100. Sandhu P, Lee W, Xu X et al (2005) Hepatic uptake of the novel antifungal agent caspofungin. Drug Metab Dispos 33:676–682

101. Schuetz JD, Connelly MC, Sun D et al (1999) MRP4: a previously unidentified factor in resistance to nucleoside-based antiviral drugs. Nat Med 5:1048–1051

102. Seithel A, Eberl S, Singer K et al (2007) The influence of macrolide antibiotics on the uptake of organic anions and drugs mediated by OATP1B1 and OATP1B3. Drug Metab Dispos 35:779–786

103. Shibayama T, Sugiyama D, Kamiyama E et al (2007) Characterization of CS-023 (RO4908463), a novel parenteral carbapenem antibiotic, and meropenem as substrates of human renal transporters. Drug Metab Pharmacokinet 22:41–47

104. Siccardi M, D'Avolio A, Nozza S et al (2010) Maraviroc is a substrate for OATP1B1 in vitro and maraviroc plasma concentrations are influenced by SLCO1B1 521 T>C polymorphism. Pharmacogenet Genomics 20:759–765

105. Song IH, Zong J, Borland J et al (2016) The effect of Dolutegravir on the pharmacokinetics of metformin in healthy subjects. J Acquir Immune Defic Syndr 72:400–407

106. Sulejmani N, Jafri SM, Gordon SC (2016) Pharmacodynamics and pharmacokinetics of elbasvir and grazoprevir in the treatment of hepatitis C. Expert Opin Drug Metab Toxicol 12:353–361

107. Susanto M, Benet LZ (2002) Can the enhanced renal clearance of antibiotics in cystic fibrosis patients be explained by P-glycoprotein transport? Pharm Res 19:457–462

108. Takeda M, Khamdang S, Narikawa S et al (2002) Characterization of methotrexate transport and its drug interactions with human organic anion transporters. J Pharmacol Exp Ther 302:666–671

109. Takeda M, Khamdang S, Narikawa S et al (2002) Human organic anion transporters and human organic cation transporters mediate renal antiviral transport. J Pharmacol Exp Ther 300:918–924

110. Tirona RG, Leake BF, Wolkoff AW, Kim RB (2003) Human organic anion transporting polypeptide-C (SLC21A6) is a major determinant of rifampin-mediated pregnane X receptor activation. J Pharmacol Exp Ther 304:223–228

111. Ueo H, Motohashi H, Katsura T, Inui K (2005) Human organic anion transporter hOAT3 is a potent transporter of cephalosporin antibiotics, in comparison with hOAT1. Biochem Pharmacol 70:1104–1113

112. Uwai Y, Ida H, Tsuji Y, Katsura T, Inui K (2007) Renal transport of adefovir, cidofovir, and tenofovir by SLC22A family members (hOAT1, hOAT3, and hOCT2). Pharm Res 24:811–815

113. VanWert AL, Gionfriddo MR, Sweet DH (2010) Organic anion transporters: discovery, pharmacology, regulation and roles in pathophysiology. Biopharm Drug Dispos 31:1–71

114. Vavricka SR, Van Montfoort J, Ha HR, Meier PJ, Fattinger K (2002) Interactions of rifamycin SV and rifampicin with organic anion uptake systems of human liver. Hepatology 36:164–172

115. Vermeer LM, Isringhausen CD, Ogilvie BW, Buckley DB (2016) Evaluation of ketoconazole and its alternative clinical CYP3A4/5 inhibitors as inhibitors of drug transporters: the in vitro effects of ketoconazole, ritonavir, clarithromycin, and Itraconazole on 13 clinically-relevant drug transporters. Drug Metab Dispos 44:453–459

116. Weiss J, Becker JP, Haefeli WE (2014) Telaprevir is a substrate and moderate inhibitor of P-glycoprotein, a strong inductor of ABCG2, but not an activator of PXR in vitro. Int J Antimicrob Agents 43:184–188

117. Weiss J, Rose J, Storch CH et al (2007) Modulation of human BCRP (ABCG2) activity by anti-HIV drugs. J Antimicrob Chemother 59:238–245

118. Weiss J, Theile D, Ketabi-Kiyanvash N, Lindenmaier H, Haefeli WE (2007) Inhibition of MRP1/ABCC1, MRP2/ABCC2, and MRP3/ABCC3 by nucleoside, nucleotide, and non-nucleoside reverse transcriptase inhibitors. Drug Metab Dispos 35:340–344

119. Yamashita F, Ohtani H, Koyabu N et al (2006) Inhibitory effects of angiotensin II receptor antagonists and leukotriene receptor antagonists on the transport of human organic anion transporter 4. J Pharm Pharmacol 58:1499–1505

120. Ye ZW, Camus S, Augustijns P, Annaert P (2010) Interaction of eight HIV protease inhibitors with the canalicular efflux transporter ABCC2 (MRP2) in sandwich-cultured rat and human hepatocytes. Biopharm Drug Dispos 31:178–188

121. Zembruski NC, Haefeli WE, Weiss J (2011) Interaction potential of etravirine with drug transporters assessed in vitro. Antimicrob Agents Chemother 55:1282–1284

122. Lau YY, Huang Y, Frassetto L, Benet LZ (2007) Effect of OATP1B transporter inhibition on the pharmacokinetics of atorvastatin in healthy volunteers. Clin Pharmacol Ther 81:194–204

123. Backman JT, Luurila H, Neuvonen M, Neuvonen PJ (2005) Rifampin markedly decreases and gemfibrozil increases the plasma concentrations of atorvastatin and its metabolites. Clin Pharmacol Ther 78:154–167

124. van Giersbergen PL, Treiber A, Schneiter R, Dietrich H, Dingemanse J (2007) Inhibitory and inductive effects of rifampin on the pharmacokinetics of bosentan in healthy subjects. Clin Pharmacol Ther 81:414–419

125. Dingemanse J, van Giersbergen PL, Patat A, Nilsson PN (2010) Mutual pharmacokinetic interactions between bosentan and lopinavir/ritonavir in healthy participants. Antivir Ther 15:157–163

126. Busti AJ, Bain AM, Hall RG 2nd et al (2008) Effects of atazanavir/ritonavir or fosamprenavir/ritonavir on the pharmacokinetics of rosuvastatin. J Cardiovasc Pharmacol 51:605–610

127. Kiser JJ, Gerber JG, Predhomme JA et al (2008) Drug/drug interaction between lopinavir/ritonavir and rosuvastatin in healthy volunteers. J Acquir Immune Defic Syndr 47:570–578

128. Jacobson TA (2004) Comparative pharmacokinetic interaction profiles of pravastatin, simvastatin, and atorvastatin when coadministered with cytochrome P450 inhibitors. Am J Cardiol 94:1140–1146

129. Hulskotte EG, Feng HP, Xuan F et al (2013) Pharmacokinetic evaluation of the interaction between hepatitis C virus protease inhibitor boceprevir and 3-hydroxy-3-methylglutaryl coenzyme a reductase inhibitors atorvastatin and pravastatin. Antimicrob Agents Chemother 57:2582–2588

130. Moore KH, Yuen GJ, Raasch RH et al (1996) Pharmacokinetics of lamivudine administered alone and with trimethoprim-sulfamethoxazole. Clin Pharmacol Ther 59:550–558

131. Moellentin D, Picone C, Leadbetter E (2008) Memantine-induced myoclonus and delirium exacerbated by trimethoprim. Ann Pharmacother 42:443–447

132. Grun B, Kiessling MK, Burhenne J et al (2013) Trimethoprim-metformin interaction and its genetic modulation by OCT2 and MATE1 transporters. Br J Clin Pharmacol 76:787–796

133. Jayasagar G, Krishna Kumar M, Chandrasekhar K, Madhusudan Rao C, Madhusudan Rao Y (2002) Effect of cephalexin on the pharmacokinetics of metformin in healthy human volunteers. Drug Metabol Drug Interact 19:41–48

134. Kusuhara H, Ito S, Kumagai Y et al (2011) Effects of a MATE protein inhibitor, pyrimethamine, on the renal elimination of metformin at oral microdose and at therapeutic dose in healthy subjects. Clin Pharmacol Ther 89:837–844

135. Landersdorfer CB, Kirkpatrick CM, Kinzig M et al (2010) Competitive inhibition of renal tubular secretion of ciprofloxacin and metabolite by probenecid. Br J Clin Pharmacol 69:167–178

136. Cundy KC (1999) Clinical pharmacokinetics of the antiviral nucleotide analogues cidofovir and adefovir. Clin Pharmacokinet 36:127–143

137. Massarella JW, Nazareno LA, Passe S, Min B (1996) The effect of probenecid on the pharmacokinetics of zalcitabine in HIV-positive patients. Pharm Res 13:449–452

138. Landersdorfer CB, Kirkpatrick CM, Kinzig M et al (2008) Inhibition of flucloxacillin tubular renal secretion by piperacillin. Br J Clin Pharmacol 66:648–659

139. Ding R, Tayrouz Y, Riedel KD et al (2004) Substantial pharmacokinetic interaction between digoxin and ritonavir in healthy volunteers. Clin Pharmacol Ther 76:73–84

140. Mertz D, Battegay M, Marzolini C, Mayr M (2009) Drug-drug interaction in a kidney transplant recipient receiving HIV salvage therapy and tacrolimus. Am J Kidney Dis 54:e1–e4

141. Capone D, Palmiero G, Gentile A et al (2007) A pharmacokinetic interaction between clarithromycin and sirolimus in kidney transplant recipient. Curr Drug Metab 8:379–381

142. Wakasugi H, Yano I, Ito T et al (1998) Effect of clarithromycin on renal excretion of digoxin: interaction with P-glycoprotein. Clin Pharmacol Ther 64:123–128

143. Rollot F, Pajot O, Chauvelot-Moachon L et al (2004) Acute colchicine intoxication during clarithromycin administration. Ann Pharmacother 38:2074–2077

144. Jalava KM, Partanen J, Neuvonen PJ (1997) Itraconazole decreases renal clearance of digoxin. Ther Drug Monit 19:609–613

145. Mathias A (2015) Drug interactions between the anti-HCV regimen ledipasvir/sofosbuvir and antiretrovirals. In: International Workshop on Clinical Pharmacology of HIV & Hepatitis Therapy, Washington, DC

146. Greiner B, Eichelbaum M, Fritz P et al (1999) The role of intestinal P-glycoprotein in the interaction of digoxin and rifampin. J Clin Invest 104:147–153

147. Piscitelli SC, Burstein AH, Chaitt D, Alfaro RM, Falloon J (2000) Indinavir concentrations and St John's wort. Lancet 355:547–548

148. Naesens M, Kuypers DR, Streit F et al (2006) Rifampin induces alterations in mycophenolic acid glucuronidation and elimination: implications for drug exposure in renal allograft recipients. Clin Pharmacol Ther 80:509–521

149. Bickel M, Khaykin P, Stephan C et al (2013) Acute kidney injury caused by tenofovir disoproxil fumarate and diclofenac co-administration. HIV Med 14:633–638

150. Morelle J, Labriola L, Lambert M et al (2009) Tenofovir-related acute kidney injury and proximal tubule dysfunction precipitated by diclofenac: a case of drug-drug interaction. Clin Nephrol 71:567–570

151. Food and Drug Administration (2012) Guidance for Industry: drug interactions studies- study design, data analysis, and implications for dosing, and labeling recommendations

152. European Medicine Agency (2012) Guideline on the investigation of drug interactions
153. Urquhart BL, Tirona RG, Kim RB (2007) Nuclear receptors and the regulation of drug-metabolizing enzymes and drug transporters: implications for interindividual variability in response to drugs. J Clin Pharmacol 47:566–578
154. Mangelsdorf DJ, Thummel C, Beato M et al (1995) The nuclear receptor superfamily: the second decade. Cell 83:835–839
155. Kim RB, Wandel C, Leake B et al (1999) Interrelationship between substrates and inhibitors of human CYP3A and P-glycoprotein. Pharm Res 16:408–414
156. Shitara Y, Hirano M, Sato H, Sugiyama Y (2004) Gemfibrozil and its glucuronide inhibit the organic anion transporting polypeptide 2 (OATP2/OATP1B1:SLC21A6)-mediated hepatic uptake and CYP2C8-mediated metabolism of cerivastatin: analysis of the mechanism of the clinically relevant drug-drug interaction between cerivastatin and gemfibrozil. J Pharmacol Exp Ther 311:228–236
157. Niemi M, Backman JT, Neuvonen M, Neuvonen PJ (2003) Effects of gemfibrozil, itraconazole, and their combination on the pharmacokinetics and pharmacodynamics of repaglinide: potentially hazardous interaction between gemfibrozil and repaglinide. Diabetologia 46:347–351
158. Meyer zu Schwabedissen HE, Kim RB (2009) Hepatic OATP1B transporters and nuclear receptors PXR and CAR: interplay, regulation of drug disposition genes, and single nucleotide polymorphisms. Mol Pharm 6:1644–1661
159. Zhang L, Zhang Y, Huang SM (2009) Scientific and regulatory perspectives on metabolizing enzyme-transporter interplay and its role in drug interactions: challenges in predicting drug interactions. Mol Pharm 6:1766–1774
160. Hagenbuch B, Gui C (2008) Xenobiotic transporters of the human organic anion transporting polypeptides (OATP) family. Xenobiotica 38:778–801
161. Hsiang B, Zhu Y, Wang Z et al (1999) A novel human hepatic organic anion transporting polypeptide (OATP2). Identification of a liver-specific human organic anion transporting polypeptide and identification of rat and human hydroxymethylglutaryl-CoA reductase inhibitor transporters. J Biol Chem 274:37161–37168
162. Kobayashi D, Nozawa T, Imai K et al (2003) Involvement of human organic anion transporting polypeptide OATP-B (SLC21A9) in pH-dependent transport across intestinal apical membrane. J Pharmacol Exp Ther 306:703–708
163. Konig J, Cui Y, Nies AT, Keppler DA (2000) Novel human organic anion transporting polypeptide localized to the basolateral hepatocyte membrane. Am J Physiol Gastrointest Liver Physiol 278:G156–G164
164. Kullak-Ublick GA, Hagenbuch B, Stieger B et al (1995) Molecular and functional characterization of an organic anion transporting polypeptide cloned from human liver. Gastroenterology 109:1274–1282
165. Tamai I, Nezu J, Uchino H et al (2000) Molecular identification and characterization of novel members of the human organic anion transporter (OATP) family. Biochem Biophys Res Commun 273:251–260
166. Martin PD, Warwick MJ, Dane AL et al (2003) Metabolism, excretion, and pharmacokinetics of rosuvastatin in healthy adult male volunteers. Clin Ther 25:2822–2835
167. Kitamura S, Maeda K, Wang Y, Sugiyama Y (2008) Involvement of multiple transporters in the hepatobiliary transport of rosuvastatin. Drug Metab Dispos 36:2014–2023
168. Huang L, Wang Y, Grimm S (2006) ATP-dependent transport of rosuvastatin in membrane vesicles expressing breast cancer resistance protein. Drug Metab Dispos 34:738–742
169. Dingemanse J, van Giersbergen PL (2004) Clinical pharmacology of bosentan, a dual endothelin receptor antagonist. Clin Pharmacokinet 43:1089–1115
170. Treiber A, Schneiter R, Hausler S, Stieger B (2007) Bosentan is a substrate of human OATP1B1 and OATP1B3: inhibition of hepatic uptake as the common mechanism of its interactions with cyclosporin a, rifampicin, and sildenafil. Drug Metab Dispos 35:1400–1407
171. El-Sheikh AA, Masereeuw R, Russel FG (2008) Mechanisms of renal anionic drug transport. Eur J Pharmacol 585:245–255

172. Jariyawat S, Sekine T, Takeda M et al (1999) The interaction and transport of beta-lactam antibiotics with the cloned rat renal organic anion transporter 1. J Pharmacol Exp Ther 290:672–677

173. Cha SH, Sekine T, Fukushima JI et al (2001) Identification and characterization of human organic anion transporter 3 expressing predominantly in the kidney. Mol Pharmacol 59:1277–1286

174. Race JE, Grassl SM, Williams WJ, Holtzman EJ (1999) Molecular cloning and characterization of two novel human renal organic anion transporters (hOAT1 and hOAT3). Biochem Biophys Res Commun 255:508–514

175. Sekine T, Cha SH, Tsuda M et al (1998) Identification of multispecific organic anion transporter 2 expressed predominantly in the liver. FEBS Lett 429:179–182

176. Sekine T, Watanabe N, Hosoyamada M, Kanai Y, Endou H (1997) Expression cloning and characterization of a novel multispecific organic anion transporter. J Biol Chem 272:18526–18529

177. Burnell JM, Kirby WM (1951) Effectiveness of a new compound, benemid, in elevating serum penicillin concentrations. J Clin Invest 30:697–700

178. Takeda M, Narikawa S, Hosoyamada M et al (2001) Characterization of organic anion transport inhibitors using cells stably expressing human organic anion transporters. Eur J Pharmacol 419:113–120

179. Brown GR (1993) Cephalosporin-probenecid drug interactions. Clin Pharmacokinet 24:289–300

180. Cunningham RF, Israili ZH, Dayton PG (1981) Clinical pharmacokinetics of probenecid. Clin Pharmacokinet 6:135–151

181. Cundy KC, Petty BG, Flaherty J et al (1995) Clinical pharmacokinetics of cidofovir in human immunodeficiency virus-infected patients. Antimicrob Agents Chemother 39:1247–1252

182. Mulato AS, Ho ES, Cihlar T (2000) Nonsteroidal anti-inflammatory drugs efficiently reduce the transport and cytotoxicity of adefovir mediated by the human renal organic anion transporter 1. J Pharmacol Exp Ther 295:10–15

183. Jung N, Taubert D (2009) Organic cation transporters and their roles in antiretroviral drug disposition. Expert Opin Drug Metab Toxicol 5:773–787

184. Koepsell H, Lips K, Volk C (2007) Polyspecific organic cation transporters: structure, function, physiological roles, and biopharmaceutical implications. Pharm Res 24:1227–1251

185. Gorboulev V, Ulzheimer JC, Akhoundova A et al (1997) Cloning and characterization of two human polyspecific organic cation transporters. DNA Cell Biol 16:871–881

186. Grundemann D, Babin-Ebell J, Martel F et al (1997) Primary structure and functional expression of the apical organic cation transporter from kidney epithelial LLC-PK1 cells. J Biol Chem 272:10408–10413

187. Busch AE, Karbach U, Miska D et al (1998) Human neurons express the polyspecific cation transporter hOCT2, which translocates monoamine neurotransmitters, amantadine, and memantine. Mol Pharmacol 54:342–352

188. van Crugten J, Bochner F, Keal J, Somogyi A (1986) Selectivity of the cimetidine-induced alterations in the renal handling of organic substrates in humans. Studies with anionic, cationic and zwitterionic drugs. J Pharmacol Exp Ther 236:481–487

189. Tanihara Y, Masuda S, Sato T et al (2007) Substrate specificity of MATE1 and MATE2-K, human multidrug and toxin extrusions/H(+)-organic cation antiporters. Biochem Pharmacol 74:359–371

190. Ito S, Kusuhara H, Kuroiwa Y et al (2010) Potent and specific inhibition of mMate1-mediated efflux of type I organic cations in the liver and kidney by pyrimethamine. J Pharmacol Exp Ther 333:341–350

191. German P, Liu HC, Szwarcberg J et al (2012) Effect of cobicistat on glomerular filtration rate in subjects with normal and impaired renal function. J Acquir Immune Defic Syndr 61:32–40

192. Koteff J, Borland J, Chen S et al (2013) A phase 1 study to evaluate the effect of dolutegravir on renal function via measurement of iohexol and para-aminohippurate clearance in healthy subjects. Br J Clin Pharmacol 75:990–996

193. Yao SY, Ng AM, Sundaram M et al (2001) Transport of antiviral 3'-deoxy-nucleoside drugs by recombinant human and rat equilibrative, nitrobenzylthioinosine (NBMPR)-insensitive (ENT2) nucleoside transporter proteins produced in Xenopus oocytes. Mol Membr Biol 18:161–167

194. Young JD, Yao SY, Baldwin JM, Cass CE, Baldwin SA (2013) The human concentrative and equilibrative nucleoside transporter families, SLC28 and SLC29. Mol Asp Med 34:529–547

195. Endres CJ, Moss AM, Govindarajan R, Choi DS, Unadkat JD (2009) The role of nucleoside transporters in the erythrocyte disposition and oral absorption of ribavirin in the wild-type and equilibrative nucleoside transporter 1–/– mice. J Pharmacol Exp Ther 331:287–296

196. Anand BS, Patel J, Mitra AK (2003) Interactions of the dipeptide ester prodrugs of acyclovir with the intestinal oligopeptide transporter: competitive inhibition of glycylsarcosine transport in human intestinal cell line-Caco-2. J Pharmacol Exp Ther 304:781–791

197. Terada T, Inui K (2004) Peptide transporters: structure, function, regulation and application for drug delivery. Curr Drug Metab 5:85–94

198. Brandsch M (2009) Transport of drugs by proton-coupled peptide transporters: pearls and pitfalls. Expert Opin Drug Metab Toxicol 5:887–905

199. Shen H, Smith DE, Yang T et al (1999) Localization of PEPT1 and PEPT2 proton-coupled oligopeptide transporter mRNA and protein in rat kidney. Am J Phys 276:F658–F665

200. Bretschneider B, Brandsch M, Neubert R (1999) Intestinal transport of beta-lactam antibiotics: analysis of the affinity at the H+/peptide symporter (PEPT1), the uptake into Caco-2 cell monolayers and the transepithelial flux. Pharm Res 16:55–61

201. Juliano RL, Ling VA (1976) Surface glycoprotein modulating drug permeability in Chinese hamster ovary cell mutants. Biochim Biophys Acta 455:152–162

202. Cordon-Cardo C, O'Brien JP, Casals D et al (1989) Multidrug-resistance gene (P-glycoprotein) is expressed by endothelial cells at blood-brain barrier sites. Proc Natl Acad Sci U S A 86:695–698

203. Klimecki WT, Futscher BW, Grogan TM, Dalton WS (1994) P-glycoprotein expression and function in circulating blood cells from normal volunteers. Blood 83:2451–2458

204. Sugawara I, Kataoka I, Morishita Y et al (1988) Tissue distribution of P-glycoprotein encoded by a multidrug-resistant gene as revealed by a monoclonal antibody, MRK 16. Cancer Res 48:1926–1929

205. Thiebaut F, Tsuruo T, Hamada H et al (1987) Cellular localization of the multidrug-resistance gene product P-glycoprotein in normal human tissues. Proc Natl Acad Sci U S A 84:7735–7738

206. Schinkel AH, Smit JJ, van Tellingen O et al (1994) Disruption of the mouse mdr1a P-glycoprotein gene leads to a deficiency in the blood-brain barrier and to increased sensitivity to drugs. Cell 77:491–502

207. Schinkel AH, Wagenaar E, van Deemter L, Mol CA, Borst P (1995) Absence of the mdr1a P-glycoprotein in mice affects tissue distribution and pharmacokinetics of dexamethasone, digoxin, and cyclosporin a. J Clin Invest 96:1698–1705

208. Sparreboom A, van Asperen J, Mayer U et al (1997) Limited oral bioavailability and active epithelial excretion of paclitaxel (Taxol) caused by P-glycoprotein in the intestine. Proc Natl Acad Sci U S A 94:2031–2035

209. Choo EF, Leake B, Wandel C et al (2000) Pharmacological inhibition of P-glycoprotein transport enhances the distribution of HIV-1 protease inhibitors into brain and testes. Drug Metab Dispos 28:655–660

210. Lin JH (2007) Transporter-mediated drug interactions: clinical implications and in vitro assessment. Expert Opin Drug Metab Toxicol 3:81–92

211. de Lannoy IA, Silverman M (1992) The MDR1 gene product, P-glycoprotein, mediates the transport of the cardiac glycoside, digoxin. Biochem Biophys Res Commun 189:551–557

212. Kawahara M, Sakata A, Miyashita T, Tamai I, Tsuji A (1999) Physiologically based pharmacokinetics of digoxin in mdr1a knockout mice. J Pharm Sci 88:1281–1287

213. Gutmann H, Fricker G, Drewe J, Toeroek M, Miller DS (1999) Interactions of HIV protease inhibitors with ATP-dependent drug export proteins. Mol Pharmacol 56:383–389

214. Maxwell DL, Gilmour-White SK, Hall MR (1989) Digoxin toxicity due to interaction of digoxin with erythromycin. BMJ 298:572

215. Kiso S, Cai SH, Kitaichi K et al (2000) Inhibitory effect of erythromycin on P-glycoprotein-mediated biliary excretion of doxorubicin in rats. Anticancer Res 20:2827–2834

216. Takara K, Tanigawara Y, Komada F et al (1999) Cellular pharmacokinetic aspects of reversal effect of itraconazole on P-glycoprotein-mediated resistance of anticancer drugs. Biol Pharm Bull 22:1355–1359

217. Hamman MA, Bruce MA, Haehner-Daniels BD, Hall SD (2001) The effect of rifampin administration on the disposition of fexofenadine. Clin Pharmacol Ther 69:114–121

218. Hebert MF, Roberts JP, Prueksaritanont T, Benet LZ (1992) Bioavailability of cyclosporine with concomitant rifampin administration is markedly less than predicted by hepatic enzyme induction. Clin Pharmacol Ther 52:453–457

219. Westphal K, Weinbrenner A, Zschiesche M et al (2000) Induction of P-glycoprotein by rifampin increases intestinal secretion of talinolol in human beings: a new type of drug/drug interaction. Clin Pharmacol Ther 68:345–355

220. Hebert MF, Fisher RM, Marsh CL, Dressler D, Bekersky I (1999) Effects of rifampin on tacrolimus pharmacokinetics in healthy volunteers. J Clin Pharmacol 39:91–96

221. Tong L, Phan TK, Robinson KL et al (2007) Effects of human immunodeficiency virus protease inhibitors on the intestinal absorption of tenofovir disoproxil fumarate in vitro. Antimicrob Agents Chemother 51:3498–3504

222. Doyle LA, Yang W, Abruzzo LV et al (1998) A multidrug resistance transporter from human MCF-7 breast cancer cells. Proc Natl Acad Sci U S A 95:15665–15670

223. Mao Q, Unadkat JD (2005) Role of the breast cancer resistance protein (ABCG2) in drug transport. AAPS J 7:E118–E133

224. Eisenblatter T, Galla HJA (2002) New multidrug resistance protein at the blood-brain barrier. Biochem Biophys Res Commun 293:1273–1278

225. Maliepaard M, Scheffer GL, Faneyte IF et al (2001) Subcellular localization and distribution of the breast cancer resistance protein transporter in normal human tissues. Cancer Res 61:3458–3464

226. van Herwaarden AE, Schinkel AH (2006) The function of breast cancer resistance protein in epithelial barriers, stem cells and milk secretion of drugs and xenotoxins. Trends Pharmacol Sci 27:10–16

227. Cole SP, Bhardwaj G, Gerlach JH et al (1992) Overexpression of a transporter gene in a multidrug-resistant human lung cancer cell line. Science 258:1650–1654

228. Zhou SF, Wang LL, Di YM et al (2008) Substrates and inhibitors of human multidrug resistance associated proteins and the implications in drug development. Curr Med Chem 15:1981–2039

229. Dazert P, Meissner K, Vogelgesang S et al (2003) Expression and localization of the multidrug resistance protein 5 (MRP5/ABCC5), a cellular export pump for cyclic nucleotides, in human heart. Am J Pathol 163:1567–1577

230. Flens MJ, Zaman GJ, van der Valk P et al (1996) Tissue distribution of the multidrug resistance protein. Am J Pathol 148:1237–1247

231. Keppler D, Konig J, Buchler M (1997) The canalicular multidrug resistance protein, cMRP/MRP2, a novel conjugate export pump expressed in the apical membrane of hepatocytes. Adv Enzym Regul 37:321–333

232. Schaub TP, Kartenbeck J, Konig J et al (1997) Expression of the conjugate export pump encoded by the mrp2 gene in the apical membrane of kidney proximal tubules. J Am Soc Nephrol 8:1213–1221

233. van Aubel RA, Smeets PH, Peters JG, Bindels RJ, Russel FG (2002) The MRP4/ABCC4 gene encodes a novel apical organic anion transporter in human kidney proximal tubules: putative efflux pump for urinary cAMP and cGMP. J Am Soc Nephrol 13:595–603

234. Zhang Y, Han H, Elmquist WF, Miller DW (2000) Expression of various multidrug resistance-associated protein (MRP) homologues in brain microvessel endothelial cells. Brain Res 876:148–153

235. Westley IS, Brogan LR, Morris RG, Evans AM, Sallustio BC (2006) Role of Mrp2 in the hepatic disposition of mycophenolic acid and its glucuronide metabolites: effect of cyclosporine. Drug Metab Dispos 34:261–266
236. Staatz CE, Tett SE (2007) Clinical pharmacokinetics and pharmacodynamics of mycophenolate in solid organ transplant recipients. Clin Pharmacokinet 46:13–58
237. Wolff NA, Burckhardt BC, Burckhardt G, Oellerich M, Armstrong VW (2007) Mycophenolic acid (MPA) and its glucuronide metabolites interact with transport systems responsible for excretion of organic anions in the basolateral membrane of the human kidney. Nephrol Dial Transplant 22:2497–2503
238. Kuypers DR, Verleden G, Naesens M, Vanrenterghem Y (2005) Drug interaction between mycophenolate mofetil and rifampin: possible induction of uridine diphosphate-glucuronosyltransferase. Clin Pharmacol Ther 78:81–88
239. Cui Y, Konig J, Keppler D (2001) Vectorial transport by double-transfected cells expressing the human uptake transporter SLC21A8 and the apical export pump ABCC2. Mol Pharmacol 60:934–943
240. Stieger B, Beuers U (2011) The canalicular bile salt export pump BSEP (ABCB11) as a potential therapeutic target. Curr Drug Targets 12:661–670
241. Yanni SB, Augustijns PF, Benjamin DK Jr et al (2010) In vitro investigation of the hepatobiliary disposition mechanisms of the antifungal agent micafungin in humans and rats. Drug Metab Dispos 38:1848–1856
242. Byrne JA, Strautnieks SS, Mieli-Vergani G et al (2002) The human bile salt export pump: characterization of substrate specificity and identification of inhibitors. Gastroenterology 123:1649–1658
243. Moss DM, Marzolini C, Rajoli RK, Siccardi M (2015) Applications of physiologically based pharmacokinetic modeling for the optimization of anti-infective therapies. Expert Opin Drug Metab Toxicol 11:1203–1217
244. Siccardi M, Marzolini C, Seden K et al (2013) Prediction of drug-drug interactions between various antidepressants and efavirenz or boosted protease inhibitors using a physiologically based pharmacokinetic modelling approach. Clin Pharmacokinet 52:583–592
245. Jones HM, Barton HA, Lai Y et al (2012) Mechanistic pharmacokinetic modeling for the prediction of transporter-mediated disposition in humans from sandwich culture human hepatocyte data. Drug Metab Dispos 40:1007–1017
246. Burt HJ, Neuhoff S, Almond L et al (2016) Metformin and cimetidine: physiologically based pharmacokinetic modelling to investigate transporter mediated drug-drug interactions. Eur J Pharm Sci 88:70–82
247. Colbers A, Greupink R, Litjens C, Burger D, Russel FG (2016) Physiologically based modelling of Darunavir/ritonavir pharmacokinetics during pregnancy. Clin Pharmacokinet 55:381–396
248. De Sousa Mendes M, Hirt D, Urien S et al (2015) Physiologically-based pharmacokinetic modeling of renally excreted antiretroviral drugs in pregnant women. Br J Clin Pharmacol 80:1031–1041

Chapter 4
Drug-Food Interactions

Eric Wenzler, Kelly Sprandel-Harris, and Keith A. Rodvold

4.1 Introduction

Drug–food interactions are an often unrecognized source of pharmacokinetic variability and can have detrimental outcomes on patient care if ignored. This chapter is a comprehensive summary of the literature regarding interactions between antimicrobials and food. The magnitude of the interaction is discussed along with the clinical significance and subsequent dosing recommendations.

It is important to be cognizant of the specific dosage formulations being discussed throughout this chapter as the effect of food on the pharmacokinetics of an agent can vary significantly between capsules, tablets, and suspensions. This is particularly true for antiretroviral medications used to treat human immunodeficiency virus (HIV). Additionally, the composition and size of the meal in terms of caloric content are extremely important when assessing and predicting the magnitude and variability of food–drug interactions. Administering medications with food results in a more uniformly acidic gastric environment, typically reducing the pharmacokinetic variability relative to the fasting state. Furthermore, light meals have a reduced capacity to buffer gastric acid secretion, while large, protein-rich meals tend to increase gastric pH. Drugs with high aqueous solubility under acidic conditions may therefore be affected by a light meal to a much greater extent than a high-calorie,

E. Wenzler
University of Illinois at Chicago, College of Pharmacy, Chicago, IL, USA

K. Sprandel-Harris
Achaogen, Inc, San Fransisco, CA, USA

K.A. Rodvold (✉)
Colleges of Pharmacy and Medicine, University of Illinois at Chicago,
Chicago, IL, USA
e-mail: kar@uic.edu

© Springer International Publishing AG 2018
M.P. Pai et al. (eds.), *Drug Interactions in Infectious Diseases: Mechanisms and Models of Drug Interactions*, Infectious Disease,
https://doi.org/10.1007/978-3-319-72422-5_4

high-fat, or protein-rich meal. If food-effect studies are not available for a given agent, preclinical solubility and dissolution data should be considered when attempting to evaluate the potential effect of food on absorption and exposure.

Finally, the term bioavailability will be avoided to the extent possible throughout this chapter. Absolute bioavailability refers to the oral formulation relative to the intravenous formulation, and relative bioavailability is between two dosage forms, neither of which is evaluated in food-effect studies. The pharmacokinetic parameters will be reported and compared between the fasted and fed state, stratified by the caloric content of the meal administered in the study.

4.2 Mechanisms of Drug–Food Interactions

4.2.1 Physiologic Effects of Food

The vast majority of medications are absorbed in the duodenum, with very little absorption occurring directly via the stomach during digestion. However, changes in gastrointestinal secretions and gastric pH can have an effect on the subsequent absorption of medications [1]. Gastrointestinal secretions increase in response to food ingestion, which increases hydrochloric acid in the stomach, thus lowering stomach pH. This acidic environment can accelerate the dissolution and absorption of alkaline drugs while increasing the degradation of acid labile drugs.

The volume of a meal may also affect the absorption of a drug. Large fluid volumes tend to increase gastric emptying rates, whereas large solid-food consumption tends to slow gastric emptying [1]. Delayed emptying can increase the degradation of drugs that are unstable at low pH. Conversely, longer transit time may increase absorption for drugs that take more time to dissolve by increasing the percentage of the drug in solution.

Finally, the components of food may interact directly with medications in a number of ways. Foods may chelate a drug if they contain polyvalent metal ions or act as a mechanical barrier to inhibit the absorption of food across the mucosal surface of the intestines. Thus, the physiologic effect of food may have variable effects on drug absorption, depending on the characteristics of each individual drug and the type of food consumed.

4.2.2 Effects of Food on Drug Absorption

Drug–food interactions can be divided into three possible outcomes. Drug absorption may be increased, decreased, or unaffected. Decreased absorption can be further subclassified into reduced versus delayed absorption. Reduced absorption is reflected by a decrease in the area under the concentration–time curve (AUC) of the drug in plasma. Delayed absorption is reflected by an increase in the time to reach maximum concentration (t_{max}) of the drug. Alterations in the rate of drug absorption caused by the ingestion of food are generally not considered as clinically significant as changes in the extent of drug absorption [2].

4.2.3   Effects of Food on Drug Metabolism

A number of dietary factors are known to have potential for altering the metabolism of drugs [3], such as protein, cruciferous vegetables, and charcoal-broiled beef. Contrarily, malnutrition has been shown to alter the metabolism of certain drugs [4]. Grapefruit juice has been demonstrated to increase the bioavailability of drugs that are known to be metabolized by cytochrome P450 (CYP) 3A4 enzymes [5–7]. It appears that grapefruit juice interactions are mediated by inhibition of gut-wall metabolism, which results in reduced pre-systemic drug metabolism resulting in an increase in drug bioavailability. Studies have demonstrated the effect of grapefruit juice on HIV protease inhibitors and macrolides, among others. The effect of grapefruit juice on P-glycoprotein (P-gp)-mediated drug transport is controversial [8, 9]. A study reported that grapefruit juice, Seville orange juice, and apple juice were more potent inhibitors of the organic anion transporting polypeptides (OATPs) than of P-gp [7]. Although it appears that drug-metabolizing enzymes and transporters determine drug disposition, further research in this field is necessary. A more complete review of transport proteins is provided in Chap. 3.

4.3   Drug–Food Interaction Studies

The Food-Effect Working Group of the Biopharmaceutics Coordinating Committee in the Center for Drug Evaluation and Research (CDER) at the FDA published draft guidelines for food-effect bioavailability and bioequivalence studies for oral immediate-release or modified-release dosage forms in 2002. The guidance paper provides recommendations for study design, data analysis, and labeling, as well as specifying areas in which food-effect studies may not be important. These guidelines can be accessed at http://www.fda.gov/downloads/RegulatoryInformation/Guidances/UCM126833.pdf.

4.3.1   Test Meal

The FDA guidance paper recommends that food-effect studies should be conducted under conditions expected to maximally affect systemic drug availability. For this effect, they recommend a high-fat (50% of total caloric value from the meal), high-calorie (approximately 800–1000 calories) meal deriving approximately 150, 250, and 500–600 calories from protein, carbohydrates, and fat, respectively. The specific caloric breakdown of the meal used should be provided in the study report.

4.3.2 Study Design

The recommended study design for assessing the effect of food on bioavailability is a randomized, single-dose, crossover study in which the test formulation is administered under fasting conditions in one study regimen and following a test meal in the other regimen with an adequate washout period in between. These studies should be conducted in healthy volunteers with at least 12 subjects completing the study for appropriate statistical comparison.

4.3.3 Treatment Arms

Following an overnight fast of at least 10 h, subjects in the fasted arm should take the drug formulation with a full glass of water (180 mL or 6 fl oz.). No food should be allowed for the following 4 h, after which normally scheduled meals should be permitted. For fed subjects, following an overnight fast of at least 10 h, subjects should be fed the test meal over not more than 30 min. The drug formulation should be given with a full glass of water no later than 5 min after finishing the test meal. As before, no other meals should be allowed for the following 4 h, after which scheduled meals are permitted.

4.3.4 Pharmacokinetic Analysis

During fasted and fed study regimens, serial plasma samples should be collected post-dose in order to adequately characterize the concentration–time profile and determine pharmacokinetic parameters. The appropriate sampling scheme will be dependent on the specific agent being tested and may need to be altered between fasted and fed regimens as coadministration with food may alter the timing and extent of plasma exposure.

4.3.5 Data and Statistical Analysis

For statistical analysis of pharmacokinetic parameters, the fasted state should serve as the reference, and the geometric means (with 90% confidence intervals) of AUC_{0-inf}, AUC_{0-t}, and C_{max} should be compared between groups. A significant food effect will be concluded when the 90% confidence interval falls outside 80–125% for AUC and C_{max}. Clinical relevance of the observed magnitude should be indicated by the sponsor of the study.

4.4 Anti-infectives and Drug–Food Interactions Studies

The following sections detail drug–food interaction studies of anti-infective agents by drug class. It is important to recognize that many of the earlier studies were completed prior to the abovementioned FDA guidance paper. In addition, data have frequently been obtained in only one or two clinical studies, and observations made under these particular situations may not be relevant to the current clinical care of patients. A summary of selected studies reporting the effect of food on the C_{max}, t_{max}, and AUC of oral antimicrobial agents is shown in Table 4.1.

4.4.1 Penicillins

4.4.1.1 Penicillin

The absorption of penicillin and penicillin V potassium is decreased with the coadministration of food [108]. In a study performed in the late 1950s, six groups of ten volunteers were given a standard meal served simultaneously with 15, 30, or 60 min before dosing or 1 or 2 h after the dose of antibiotic. Blood concentrations of penicillin or penicillin V potassium were obtained at 0.5, 1, and 2 h after dosing. Lower concentrations were observed with both drugs when given with food, although the effect was greater for penicillin V potassium. In another study, healthy nurses were given 150 mg doses of penicillin V (K), potassium V (Ca), and potassium V (acid), with or without a standard meal [1]. Observed C_{max} was markedly reduced with all formulations of penicillin when given with a meal. Thus, penicillin V potassium, the only oral formulation currently available, should be taken on an empty stomach 1 h before or 2 h after meals to increase absorption.

4.4.1.2 Ampicillin

The AUC of ampicillin is decreased by approximately 50% when given with food [109]. This effect was evident when volunteers were given ampicillin with a high-carbohydrate, high-protein, or high-fat meal, a standard breakfast, or a Sudanese diet (rich in wheat, flour, and corn) [10, 110].

4.4.1.3 Amoxicillin

Early research with amoxicillin demonstrated no effect on the absorption when given with food [111]. In two follow-up studies, one demonstrated decreased absorption when amoxicillin was given with food in 6 healthy volunteers, and another showed no effect in a crossover study of 16 healthy volunteers [10, 109].

Table 4.1 Effect of food on the select pharmacokinetic parameters of antimicrobials from representative food-effect studies

Antimicrobial	References	Dosage form	Test meal[a] (composition)	Oral dose (mg)	C_{max} (µg/ml)		AUC (µg h/mL)		t_{max} (h)	
					Fasting	Fed	Fasting	Fed	Fasting	Fed
Penicillins										
Ampicillin	[10]	Capsule	SB	500	5.9	4.6	19.8	13.7	1.49	2.5
Amoxicillin	[10]	Capsule	SB	500	8.9	8.8	30.9	29.2	1.86	2.4
Amoxicillin	[11]	Tablet	SB	750	8.5	8.6	21.9	21.0	1	1
Clavulanic acid					6.6	6.3	14.7	13.8	1	1
Cephalosporins										
Cephalexin	[12]	Capsule	SB	1000	38.8	23.1	93	70	0.93	1.9
Cefadroxil	[12]	Capsule	SB	1000	33.0	32.7	108.5	NR	1.71	2
Cefaclor	[13]	Capsule	SB	250	8.7	4.3	8.6	7.6	0.6	1.3
Cefprozil	[13]	Capsule	SB	250	6.1	5.3	15	14.9	1.2	2
Cefprozil	[14]	Capsule	SB	1000	15.5	16.8	52.2	55.6	1.5	2.3
Cefuroxime	[15]	Tablet	SB	500	4.9	7.0	18.9	27.4	2.3	3.0
Cefuroxime	[16]	Tablet	SB	1000	1.5	1.5	23.5	39.8	7.3	13.6
Cefpodoxime	[17]	Tablet	SM	200	2.6	3.1	13.5	17.6	2.8	3.3
			HPM			3.2		16.9		3.8
			LPM			3.1		17.0		3.4
			HFM			3.0		16.3		3.2
			LFM			3.3		18.0		3.5
Cefixime	[18]	Tablet	SB	100	1.0	1.4	8.7	10.5	4.2	4.2
Cefixime	[19]	Capsule	SB	400	4.4	4.2	33.1	30.2	3.7	4.8
Cefditoren	[20]	Tablet	NR	200	2.5	2.7	7.9	10.8	1.5	1.8
Ceftibuten	[21]	Capsule	HFB	200	9.9	6.6	42.1	33.7	1.8	~4

Drug	Ref	Formulation	Meal	Dose						
Macrolides										
Erythromycin base	[22]	Enteric-coated pellets in capsule	Not standardized	250	2.3	1.6	8.5	6.3	NR	NR
		Enteric-coated tablet			1.0	1.4	4.7	6.2	NR	NR
Erythromycin base	[23]	Unprotected tablet	SM	500	1.6	0.7	7.3	3.3	3	3
		Film-coated tablet			1.0	1.2	5.0	5.1	3	4.5
		Enteric-coated tablet			1.0	0.80	4.9	5.0	4.5	4.5
Erythromycin base	[24]	Enteric-coated tablet	SB	500	2.0	1.7	7.1	5.3	3.5	4.9
Erythromycin base	[25]	Enteric-coated pellets in capsule	SB	500	1.8	1.9	4.9	5.0	4.4	4.3
Erythromycin stearate	[24]	Tablet	SB	500	3.6	1.7	9.5	4.1	1.2	2.4
Erythromycin stearate	[25]	Tablet	SB	500	2.1	0.4	5.0	1.0	1.3	2.3
Erythromycin stearate	[26]	Tablet	HCM	500	1.4	1.3	9.3	4.8	2.0	3.3
			HFM			1.4		5.2		2.3
			HPM			1.2		4.0		2.2
Erythromycin ethylsuccinate	[27]	Tablet	SB	800	2.7	1.5	7.5	4.9	1.3	2.4
Clarithromycin	[28]	Tablet	SB	500	2.5	1.7	15.7	12.6	2	2.8
Clarithromycin extended release	[29]	Tablet	HFB (1000 kcal, 56 g fat)	1000	2.3	3.9	35.9	49.2	5.5	5.6
Azithromycin	[30]	Tablet	HFB (≥50 g fat)	500	0.3	0.4	2.5	2.40	NR	NR
		Suspension			0.3	0.5	3.2	3.60	NR	NR
		Sachet		1000	0.7	1.1	6.5	7.37	NR	NR
Telithromycin	[31]	Tablet	HFB (850 kcal, 55 g fat)	800	1.4	1.5	7.0	7.4	2.5	2.3
Tetracyclines										
Doxycycline	[32]	Capsule	HCM (80% carbohydrates)	200	4.1	2.6	74.7	27	4.2	3.8
			HFM (50% fat)			2.7		31.7		4.7
			HPM (53% protein)			2.0		25.5		4.5

(continued)

Table 4.1 (continued)

Antimicrobial	References	Dosage form	Test meal[a] (composition)	Oral dose (mg)	C_{max} (µg/ml) Fasting	Fed	AUC (µg h/mL) Fasting	Fed	t_{max} (h) Fasting	Fed
Minocycline	[33]	Capsule	SB	100	1.8	1.4	20.5	19.9	1.8	3.1
Fluoroquinolones										
Ciprofloxacin	[34]	Tablet	SB (12.5 g fat)	750	2.2	2.5	12.7	13.7	1.4	1.6
			High fat, high Ca^{2+} (37 g fat, 729 mg Ca^{2+})			2.4		12.6			1.2
Gemifloxacin	[35]	Tablet	HFB	320	1.2	1.1	7.6	7.4	1.5	2.0
				640	2.3	1.9	15.9	13.5	1.5	2.0
Levofloxacin	[36]	Tablet	HFB	500	5.9	5.1	50.5	45.6	1.0	2.0
Moxifloxacin	[37]	Tablet	HFB	400	2.8	2.5	38.5	37.7	1.0	2.5
Ofloxacin	[38]	Tablet	HFB	200	2.24	1.6	13.2	11.3	0.8	1.9
Anthelmintics										
Albendazole	[39]	Capsule	HFM (1399 kcal, 57 g fat)	10 mg/kg	0.2	1.6	2.1	19.6	2.5	5.3
Ivermectin	[40]	Tablet	HFB (784 kcal, 48.6 g fat)	30	0.1	0.3	1.7	4.6	4.3	4.6
Praziquantel	[41]	Tablet	HFM (656 kcal, 23.6 g fat)	1800	0.3	1.1	0.9	2.5	1.4	1.9
			HCM (674.5 kcal, 125 g carbohydrates)			2.0		3.3			1.5
Antimalarials										
Primaquine	[42]	Tablet	SM (28 g fat)	30	0.1	0.2	1.2	1.4	2.0	1.5
Chloroquine	[43]	Tablet	SB (490 kcal, 10 g fat)	600	0.5	0.7	4.5	6.4	4.0	4.0
Mefloquine	[44]	Tablet	SB (800 kcal, 37 g fat)	750	0.9	1.5	461	645	36	17
Atovaquone	[45]	Suspension	SB (23 g fat)	500	8.4	14.8	161	270	3.9	7.0
				750	12.4	15	238	301	6.5	8.9
				1000	13	17	234	325	3.8	6.1

Miscellaneous antimicrobials										
Clindamycin	[46]	Capsule	SB	500	5.6	5.2	NR	55.5	1.0	2.0
Fosfomycin	[47]	Sachet	SM	1000	12.1	7.8	77	168.5	2.7	3.2
	[48]	Sachet	Normal meal	1000	22.6	12.7	227.9	76.7	2.5	4.0
Nitazoxanide	[49]	Tablet	SB	1000	12.3	15.9	50.6	110	2.0	4.0
				2000	9.1	15.8	59.2	95.3	2.0	3.0
				3000	7.4	10	52.9	192	3.0	5.5
				4000	10.5	17.5	88.5	57.7	3.5	8
Metronidazole	[50]	Tablet	SB (440 kcal)	400	9.1	8.0	55.9	50	1.2	2.3
Linezolid	[51]	Tablet	HFB (850 kcal, 55 g fat)	375	7.6	6.2	51.7	81.8	1.5	2.2
Tedizolid	[52]	Disodium salt capsule	HCHFB	600	6.4	4.7	79.9	28.9	2.0	8.0
Trimethoprim	[53]	Suspension	Standard diabetic breakfast	3 mg/kg	2.35	1.84	37.1		2.7	2.8
Antimycobacterials										
Isoniazid	[54]	Tablet	HFM (792 kcal, 51 g fat)	300	5.5	2.7	20.2	17.7	1.0	1.9
Rifampin	[55]	Capsule	HFM (792 kcal, 51 g fat)	600	10.9	7.3	57.2	55.2	2.3	4.4
Rifabutin	[56]	Capsule	HFB	150	0.2	0.2	2.5	2.6	3.0	5.4
Ethambutol	[57]	Tablet	HFM (792 kcal, 51 g fat)	25 mg/kg	4.6	3.8	29.8	27.5	2.5	3.2
Pyrazinamide	[58]	Tablet	HFM (792 kcal, 51 g fat)	30 mg/kg	53.4	45.6	673	687	1.4	3.1
Para-aminosalicylic acid	[59]	Granules	HFB (792 kcal, 51 g fat)	6000	21.4	32.5	140	240	4.4	6.6
Cycloserine	[60]	Capsule	HFB (792 kcal, 51 g fat)	500	14.8	12.4	214	217	0.8	3.5
Ethionamide	[61]	Tablet	HFB (792 kcal, 51 g fat)	500	2.3	2.3	10.0	10.0	1.7	2.6
Thalidomide	[62]	Capsule	HFB (860 kcal, 57.3 g fat)	200	2.0	2.2	24.7	23.5	4.0	6.1
Clofazimine	[63]	Tablet	HFB (792 kcal, 51 g fat)	200	0.1	0.4	1.5	3.7	6.2	6.6
Antifungals										
Terbinafine	[64]	Tablet	NR	250	1.6	1.9	10.1	11.7	1.9	2.8

(continued)

Table 4.1 (continued)

Antimicrobial	References	Dosage form	Test meal[a] (composition)	Oral dose (mg)	C_{max} (µg/ml)		AUC (µg h/mL)		t_{max} (h)	
					Fasting	Fed	Fasting	Fed	Fasting	Fed
Ketoconazole	[65]	Tablet	SB (603 kcal, 24.5 g fat)	200	4.4	3.3	12.9	13.6	1.6	2.6
				400	9.1	10.6	37.2	59.2	16.2	3.0
				600	15.1	15.4	74.5	107.9	1.9	2.8
				800	21.0	19.2	148.8	151.2	2.8	2.9
Fluconazole	[66]	Capsule	LM (239 kcal)	100	2.3	2.3	113	101	3.1	3.1
			FM (860 kcal)			2.2		106		3.5
Itraconazole	[66]	Capsule	LM (239 kcal)	100	0.1	0.2	1.6	2.3	3.3	3.7
			FM (860 kcal)			0.2		2.8		4.2
Voriconazole	[67]	Tablet	HFB (45 g fat)	200	2.0	1.3	19.3	13.1	1.5	2.6
Posaconazole	[68]	Tablet	HFB (841 kcal, 48.6 fat)	200	NR	0.4	NR	10.3	NR	5.5
		Suspension	HFB (841 kcal, 48.6 g fat)	200	0.1	0.5	3.6	13.9	5.0	4.8
			NFB (461 kcal, 0 g fat)			0.4		9.5		4.1
Posaconazole	[69]	Suspension	HFB (54 g fat)	100	0.1	0.2	3.4	8.8	4.0	6.0
		Tablet A			0.4	0.3	11.7	11.9	5.0	8.0
		Tablet B			0.4	0.3	11.3	12.4	5.0	6.0
		Capsule			0.3	0.3	11.0	12.3	5.0	8.0
Posaconazole	[70]	Suspension	HFM (50 g fat)	400	0.2	0.5	4.3	21.0	5.0	6.0
Posaconazole	[71]	Delayed-release tablet	HFB (66.7 g fat)	300	0.9	1.04	25.6	38.7	5.0	6.0
Isavuconazole	[72]	Capsule	HFB (936 kcal, 60.4 g fat)	400	3.8	3.5	183.8	201.5	5.0	3.0
Griseofulvin	[73]	Ultramicrosize tablet	SB	125	0.4	0.7	10.8	12.7	9.4	3.0
		Microsize tablet			0.5	0.7	11.9	13.1	4.3	3.0
HIV nucleoside reverse transcriptase inhibitors										
Didanosine	[74]	Tablet	SB	375	2.8	1.3	3.9	2.1	0.5	0.5

Didanosine enteric coated	[75]	Capsule	HFM (757 kcal, 49% fat)	400	1.2	0.7	3.2	2.6	2.0	5.3
			LM (373 kcal, 41.2 g fat)		1.2	1.0	3.2	2.9	2.0	2.3
Zidovudine	[76]	Capsule	HFB (800 kcal, 47 g fat)	100	0.8	0.3	0.9	0.8	0.7	1.7
Lamivudine	[77]	Tablet	HFB (1000 kcal, 67 g fat)	150	1.6	1.4	6.1	6.0	0.9	1.9
Stavudine	[78]	Capsule	HFB (773 kcal, 45.5 g fat)	70	1.4	0.8	2.5	2.4	0.7	1.7
Abacavir	[79]	Tablet	HFB	300	2.6	1.9	5.5	5.3	0.6	1.4
Emtricitabine	[80]	Tablet	SM (540 kcal, 21 g fat)	200	2.1	2.0	11	10.3	2.0	2.0
			LM (390 kcal, 12 g fat)			2.1		10.6		2.0
Tenofovir disoproxil fumarate	[80]	Tablet	SM (540 kcal, 21 g fat)	300	0.4	0.5	2.6	3.6	1.0	1.5
			LM (390 kcal, 12 g fat)			0.4		3.4		1.5

HIV non-nucleoside reverse transcriptase inhibitors

Efavirenz	[81]	Tablet	SM (650 kcal, 19 g fat)	600	3.1	4.6	46.3	52.2	3.0	3.0
Efavirenz	[82]	Capsule	Applesauce	600	3.1	2.5	170.1	138.9	3.8	4.0
			Grape jelly			2.7		144.4		3.0
			Yogurt			4.1		219		4.0
			Infant formula			3.8		203		4.0
Etravirine	[83]	Tablet	SB (561 kcal, 15.3 g fat)	100	0.1	0.1	0.9	1.4	2.0	4.0
			LB (345 kcal, 17.4 g fat)			0.1		1.2		4.0
			High-fiber breakfast (685 kcal, 3.1 g fat)			0.1		0.9		3.0
			HFB (1160 kcal, 70.3 g fat)			0.1		1.2		4.0
Rilpivirine	[84]	Tablet	SB (533 kcal, 21 g fat)	75	0.2	0.3	7.2	11.5	4.0	5.0
			HFB (928 kcal, 56 g fat)			0.3		10.7		5.0
			PRD (300 kcal, 18.8 g protein)			0.2		6.1		5.0
Rilpivirine	[85]	Tablet	LFB (353 kcal, 11 g fat)	25	0.2	0.2	2.0	2.6	4.0	3.0
			MFB (589 kcal, 19 g fat)			0.2		2.4		4.0

(continued)

Table 4.1 (continued)

Antimicrobial	References	Dosage form	Test meal[a] (composition)	Oral dose (mg)	C_{max} (μg/ml) Fasting	Fed	AUC (μg h/mL) Fasting	Fed	t_{max} (h) Fasting	Fed
Rilpivirine	[80]	Tablet	SB (540 kcal, 21 g fat)	25	0.1	0.1	2.8	3.0	3.8	4.8
			LB (390 kcal, 12 g fat)			1.0		2.9		3.5
HIV protease inhibitors										
Indinavir	[86]	Sulfate salt in dry-filled capsule	HCHFM (784 kcal, 48.6 g fat)	200	0.8	0.3	1.0	0.7	0.9	2.8
			HCHFM (784 kcal, 48.6 g fat)	400	4.5	0.6	6.9	1.5	0.7	2.0
			LFB (292 kcal, 2.13 g fat)	800	11.7	9.4	23.2	22.7	0.8	1.4
			LCLFB (141 kcal, 1 g fat)	800		8.9		21.4		1.4
Saquinavir	[87]	Capsule	SB	600	0.01	0.05	0.03	1.6	2.9	4.8
Saquinavir	[88]	Capsule	SB (600 kcal, 22 g fat)	1200	0.2	0.5	7.3	1.3	NR	1.5
			HFB (1040 kcal, 62 g fat)			0.9		2.4		1.5
Nelfinavir	[89]	Tablet	SB (820 kcal, 44.4 g fat)	1250	0.8	3.9	3.5	29.4	2.0	5.0
Ritonavir	[90]	Capsule	LFB	100	NR	11.7	NR	142.5	NR	2.4–3.3
				200		12.9		187.5		
				400		10.9		163.6		
			HFB	100		9.6		119.7		3.4–3.6
				200		11.6		173.8		
				400		10.3		161.6		
Ritonavir	[91]	Capsule	SB (20 g fat)	100	Nr	1.0	NR	6.5	NR	NR
			HFB (45 g fat)			1.0		7.0		NR
			SB (20 g fat)	400		7.1		43.4		NR

Drug	Ref	Formulation	Meal	Dose						
Fosamprenavir	[92]	Suspension	HFB (1000 kcal, 67 g fat)	1728	5.0	3.0	22.5	18	4.0	2.0
		Tablet	HFB (1000 kcal, 67 g fat)		4.6	3.9	20.2	17.6	1.3	2.5
		Tablet	LFB (400 kcal, 11 g fat)			4.4		18.9		2.5
Lopinavir	[93]	Tablet	MFM (665 kcal, 20 g fat)	800	10.2	9.9	90.3	89.0	3.0	4.0
			HFB (840 kcal, 36 g fat)			8.7		77.2		4.0
Lopinavir	[94]	Softgel capsule	SM (530 kcal, 20 g fat)	400	6.9	7.6	86	102.1	3.2	5.6
Atazanavir	[95]	ILL	ILL							
Darunavir	[96]	Tablet	SB (533 kcal 21 g fat)	400	3.6	5.3	46.8	71.9	1.5	3.0
			HFB (928 kcal, 56 g fat)			5.9		68.7		3.0
			LFB (240 kcal, 12 g fat)			5.4		76.7		3.0
			PRD (250 kcal, 10.5 g protein)			5.5		80.3		3.0
Cobicistat	259		ILL							
HIV integrase strand transfer inhibitors										
Raltegravir	[97]	Tablet	LFM (300 kcal, 2.5 g fat)	400	2.7	1.3	10.0	5.4	3.0	3.0
			MFM (600 kcal, 21 g fat)			2.9		11.3		4.0
			HFM (825 kcal, 52 g fat)			5.3		21.2		4.0
Dolutegravir	[98]	Tablet	LFM (300 kcal, 2.3 g fat)	50	2.7	3.9	50.3	66.7	2.1	3.0
			MFM (600 kcal, 20 g fat)			4.0		71.0		4.0
			HFM (870 kcal, 51.2 g fat)			4.4		83.6		5.0
Elvitegravir	[99]	Tablet	SB (413 kcal, 9.6 g fat)	150	1.1	2.3	14.9	28.9	4.0	3.5
			PRD (250 kcal, 8.8 g protein)			2.6		32.2		4.0
Hepatitis B antivirals										
Adefovir	[100]	Tablet	LFM	10	24.8	22.9	225	200	1.0	2.8
Telbivudine	[101]	Tablet	HCHFB (1000 kcal, 66.7 g fat)	600	2.7	2.8	21.8	23	3.0	3.0

(continued)

Table 4.1 (continued)

Antimicrobial	References	Dosage form	Test meal[a] (composition)	Oral dose (mg)	C_{max} (μg/ml) Fasting	Fed	AUC (μg h/mL) Fasting	Fed	t_{max} (h) Fasting	Fed
Hepatitis C antivirals										
Telaprevir	[102]	Tablet	SB (533 kcal, 21 g fat)	750	0.5	2.2	4.7	14.9	4.0	4.0
			HCHFB (928 kcal, 56 g fat)			2.3		19.4		5.0
			LCHPB (260 kcal, 9 g fat, 30 g protein)			1.7		11.2		4.5
			LCLFB (249 kcal, 3.6 g fat)			1.5		9.6		3.5
Ledipasvir	[103]	Tablet	NR	90	0.348	0.4	8.5	8.4	1.0	2–2.5
Sofosbuvir	[103]	Tablet	NR	400	0.767	0.6	1.3	1.4	1.0	2–2.5
Velpatasvir	[104]	Tablet	LM (400 kcal, 13.3 g fat)	100	0.6	0.8	6.0	7.1	2.5	3.3
			HCHFM (800 kcal, 44.4 g fat)		0.5	0.4	4.1	3.6	2.5	3.5
Other antivirals										
Rimantadine	[105]	Tablet	SB	100	0.1	0.1	4.1	4.1	4.3	3.4
Valganciclovir	[106]	Tablet	SB (569 kcal, 18.9 g fat)	450	0.2	0.2	0.2	0.2	0.5	1.0
				875	0.3	0.3	0.3	0.4	0.5	1.0
				1750	0.5	0.5	0.7	0.8	0.8	1.0
				2625	0.7	0.6	1.1	1.3	1.3	1.5
Famciclovir	[107]	Tablet	NR	250	1.9	1.0	4.6	4.5	0.6	2.3

[a]The type of meal used in the food-effect study and its relevant caloric composition is reported, if available. Abbreviations: *SB* standard breakfast, *HPM* high-protein meal, *LPM* low-protein meal, *HFM* high-fat meal, *LFM* low-fat meal, *HFB* high-fat breakfast, *SM* standard meal, *HCM* high-calorie meal, *HCHFB* high-calorie, high-fat breakfast, *HCHFM*, high-calorie, high-fat meal, *LCLFB* low-calorie, low-fat breakfast, *LCHPB* low-calorie, high-protein breakfast, *LM* light meal, *FM* full meal, *NFB* nonfat breakfast, *LB* light breakfast, *PRD* protein-rich drink, *MFB* moderate-fat breakfast, *MFM* moderate-fat meal

NR not reported

In both studies the authors concluded that the effect was not clinically significant, and it was suggested that amoxicillin could be administered without regard to meals. Interestingly, the absorption of amoxicillin was decreased when given with 25 mL of water as compared to 250 mL. Thus, it is recommended that amoxicillin be taken with a full glass (250 mL) of water or other suitable liquid. Moxatag™ extended-release tablets are intended to provide once-daily dosing of amoxicillin in the treatment of tonsillitis and/or pharyngitis secondary to *Streptococcus pyogenes* [112]. Administration of Moxatag™ with food decreases the rate, but not the extent, of amoxicillin absorption. The manufacturer recommends that Moxatag™ be taken within 1 h of finishing a meal.

4.4.1.4 Amoxicillin–Clavulanate

Gastrointestinal side effects appear to be reduced when the combination of amoxicillin and clavulanate potassium (Augmentin®) is administered with food [11]. In one study, after the administration of two 500 mg Augmentin® tablets, no significant difference was seen in the AUC, C_{max}, or t_{max} for either amoxicillin or clavulanate between the fed and fasted state [1, 11, 113, 114]. According to the manufacturer, Augmentin® tablets, powder, and chewable tablets may be administered without regard to meals. The effect of food on the oral absorption of Augmentin-ES has not been evaluated.

4.4.1.5 Dicloxacillin

The prescribing information for dicloxacillin states that "food in the gastrointestinal tract decreases the absorption of dicloxacillin," but no specific data are given [115]. It recommends that dicloxacillin be taken on an empty stomach, at least 1 h prior to or 2 h after a meal.

The manufacturers' dosing recommendations for penicillin antibiotics with regard to food are shown in Table 4.2.

4.4.2 Cephalosporins

4.4.2.1 First-Generation Oral Cephalosporins

The concomitant administration of cephalexin and food delayed the t_{max}, although the delay was minor and not considered clinically significant [116, 117]. The rate and extent of absorption of cefadroxil was not affected by the administration of a standard breakfast [12]. Thus, cephalexin and cefadroxil can be administered without regard to meals.

Table 4.2 Dosing recommendations for the penicillins with regard to food

Antimicrobial	Formulation	Manufacturer recommendations
Penicillin V	Penicillin VK tablets, powder for oral solution	May be given with meals; however, blood levels are slightly higher when given on an empty stomach
Ampicillin	Capsules	Administer 1/2 h before or 2 h after meals for maximal absorption
	Powder for oral suspension	Administer 1/2 h before or 2 h after meals for maximal absorption
Amoxicillin	Capsules, tablets, and chewable tablets	Can be given without regard to meals
	Extended-release tablets	Should be taken within 1 h of finishing a meal
Amoxicillin/ Clavulanate	Tablets and chewable tablets	May be given without regard to meals. Should be taken at the start of meals to minimize GI upset.
	Extended-release tablets	Should be taken at the start of a meal to enhance absorption of amoxicillin and to minimize GI upset (should not be taken with high-fat meals because clavulanate absorption is decreased)
	Powder for oral suspension	Can be given without regard to meals
Dicloxacillin	Capsules	Food in the gastrointestinal tract decreases absorption; therefore should be taken on an empty stomach 1 h prior to or 2 h after a meal

4.4.2.2   Second-Generation Oral Cephalosporins

A number of studies have examined the effect of food on the absorption of cefaclor [13, 118, 119]. When given with food, the C_{max} of cefaclor capsules is reduced by approximately 50%, the t_{max} is prolonged, and the AUC is decreased slightly by 10–20% [13, 120]. Contrarily, the AUC of the controlled-release formulation is increased with food [13, 121]. The administration of a standard breakfast did not affect the C_{max} or the AUC for cefprozil capsules but delayed the t_{max} by approximately 50 min [13, 14]. This delay in absorption was not found to be significant.

The absorption of cefuroxime axetil tablets is increased with food or milk [15, 16, 122]. Administration with a standard breakfast caused an almost 100% increase in the C_{max} and the AUC; however, trough concentrations were similar in both groups [16]. Likewise, administration of cefuroxime tablets with milk causes a 25–88% increase in the AUC and C_{max} [122]. Despite these changes in the pharmacokinetic profile, the manufacturer recommends that cefuroxime axetil tablets may be given with or without food.

4.4.2.3   Third-Generation Oral Cephalosporins

The effect of food on the pharmacokinetics of third-generation cephalosporins can be summarized by dividing this generation into ester and non-ester formulations. The systemic availability of the ester cephalosporins is increased by the presence of food [123]. This effect is likely secondary to increased contact time between the drug and esterases of the intestinal mucosa secondary to delayed gastric emptying resulting from food consumption. The non-ester cephalosporins, on the other hand, display a decrease in the AUC and C_{max} when given with food.

The systemic availability of cefpodoxime proxetil, an ester cephalosporin, is increased when given with food [17, 124]. A four-way crossover study assessed the absorption of cefpodoxime after a high- or low-fat and high- or low-protein meal compared to fasting conditions. In all cases, administering cefpodoxime with any meal increased the C_{max} and the AUC by approximately 22% and 34%, respectively [17]. Absorption of cefixime, a non-ester cephalosporin, is unaffected by food other than a slight delay in t_{max} [18, 19].

When cefdinir capsules were administered with a high-fat meal, the C_{max} and AUC were reduced by only 16% and 10%, respectively, therefore they may be administered without regard to meals [125, 126]. In contrast, the administration of cefdinir with 60 mg of ferrous sulfate or a vitamin with 10 mg of elemental iron reduced the systemic availability by 80% and 31%, respectively. The manufacturer recommends administering cefdinir at least 2 h before or after iron supplements [126].

When administered with a low-fat meal, the systemic availability of cefditoren, a prodrug ester cephalosporin, increased from approximately 14% to 16% [20]. A moderate or high-fat meal resulted in a 70% increase in AUC and a 50% increase in C_{max} compared with the fasted state. As a result, cefditoren should be taken with food to enhance absorption [127].

The administration of a standard meal (530 kcal) had no effect on the pharmacokinetics of ceftibuten, besides a slight increase in t_{max} [128]. However, the administration of a high-fat breakfast resulted in an approximate 20% and 33% decrease in the AUC and C_{max}, respectively [21]. The official labeling for ceftibuten suspension recommends that the drug be taken on an empty stomach 1 h before or 2 h after a meal.

The manufacturers' dosing recommendations for cephalosporin antibiotics with regard to food are shown in Table 4.3.

4.4.3   Macrolides

4.4.3.1   Erythromycin

A variety of dosage forms of erythromycin have been developed to improve the stability and absorption when given with food [1], including an enteric-coated formulation and relatively acid-fast esters designed to resist acid degradation in the stomach.

Erythromycin-base coated tablets improved the overall absorption of erythromycin compared to non-coated tablets, and food tended to simply delay the occurrence of t_{max} [129].

Food did not significantly affect the absorption of erythromycin base given in the form of enteric-coated pellets in a capsule or an enteric-coated tablet in healthy volunteers allowed to eat a non-standardized meal of their choosing [22]. Another study demonstrated that film-coated erythromycin-base tablets produce a more optimal pharmacokinetic profile in the fed state compared to unprotected base tablets or

Table 4.3 Dosing recommendations for the cephalosporins with regard to food

Antimicrobial	Formulation	Manufacturer recommendations
First generation		
Cephalexin	Oral suspension, capsules	Can be given without regard to meals
	Tablets	Absorption may be delayed by food but the amount absorbed is not affected
Cefadroxil	Capsules, powder for oral suspension, tablets	Can be given without regard to meals
Second generation		
Cefaclor	Chewable tablets	Can be given without regard to meals (total absorption is same); well absorbed in fasting subjects (see PI)
	Capsules, powder for oral suspension	Can be given without regard to meals
Cefprozil	Oral suspension	Can be given without regard to meals
	Tablets	Absorption may be delayed by food but the amount absorbed is not affected
Cefuroxime axetil	Oral suspension	Must be administered with food
	Tablets	Can be given without regard to meals
Third generation		
Cefpodoxime	Tablets	Should be administered with food to enhance absorption
Cefixime	Oral suspension, tablets	Can be given without regard to meals
Cefdinir	Capsules, oral suspension	Can be given without regard to meals
Cefditoren	Tablets	Should be administered with meals to enhance absorption
Ceftibuten	Capsules	Absorption may be delayed by food but the amount absorbed is not affected
	Oral suspension	Suspension must be administered at least 2 h before or 1 h after a meal

enteric-coated base tablets after a single dose, but the enteric-coated capsules were more favorable after multiple doses [23]. When compared to the fed stated, administering base film-coated tablets improved the systemic availability and as such should be taken on an empty stomach. The enteric-coated capsules also produced less intersubject variability in pharmacokinetic parameters.

The absorption of erythromycin stearate was reduced when given after meals in single- and multiple-dose studies [26, 130, 131]. In the fasting state, erythromycin stearate tablets have demonstrated improved absorption over erythromycin-base enteric-coated pellets in a capsule [24]. Food effectively negates the difference in bioavailability between these formulations by decreasing the bioavailability of the stearate formulation while not affecting the base formulation [25]. Erythromycin ethylsuccinate is an ester of the erythromycin base and was developed to improve bioavailability when coadministered with food. This ester is less water soluble and more resistant to acid degradation. Studies have demonstrated either no effect or slightly increased absorption when erythromycin ethylsuccinate is given with food

[27, 132]. The C_{max} and AUC were significantly increased when erythromycin was administered with grapefruit juice compared with water due to the inhibition of first-pass metabolism of CYP3A in the small intestine [133]. The t_{max} and half-life values were not significantly different.

4.4.3.2 Clarithromycin

In a study of healthy volunteers given a single dose of 500 mg of clarithromycin, food increased the absorption of clarithromycin by 25% [28]. The effect of grapefruit juice on the pharmacokinetics of clarithromycin and its active metabolite, 14-OH clarithromycin, has been evaluated in 12 healthy subjects [134]. After an overnight fast of at least 8 h, subjects received a single 500 mg dose of clarithromycin with 240 ml of either water or freshly squeezed white grapefruit juice at time 0 and 2 h after administration in a randomized, crossover fashion. Although administration of grapefruit juice significantly delayed the t_{max} of the parent and active metabolite, it did not affect the extent of absorption of clarithromycin [134]. In contrast to the immediate-release formulation, the manufacturer recommends that clarithromycin extended-release tablets be taken with food [135]. Results from a study of thirty-six healthy subjects administered two 500 mg clarithromycin extended-release tablets once daily for 5 days in the fasting state and 30 min after starting a high-fat breakfast (1000 kcal) showed that the AUC was 30% lower under fasting condition compared to fed state [29].

4.4.3.3 Azithromycin

Confusion has existed as to the absorption of azithromycin with food. Early studies with azithromycin capsules demonstrated a 50% decrease in the overall absorption of azithromycin with food [136]. However, research with the currently marketed tablet and suspension has shown little effect on the bioavailability when coadministered with a high-fat meal [30].

4.4.3.4 Telithromycin

Telithromycin is a semisynthetic ketolide analogue of the macrolide class of antibacterials that maintains activity against macrolide-resistant bacterial pathogens due to alterations in its chemical structure [137]. A food-effect study examined the impact of a high-fat (850 kcal, 55 g fat) breakfast on the pharmacokinetics of a single 800 mg dose of telithromycin compared to the fasted state in 18 healthy male subjects and found no appreciable impact, with geometric mean ratios of 98 for C_{max} and 111 for $AUC_{0-\infty}$ [31]. These results indicate that telithromycin may be taken without regard to meals, as recommended by the manufacturer [138]. On March 11, 2016, the FDA announced that both the 300 and 400 mg tablets had been permanently discontinued by Sanofi as the result of a business decision [139].

Table 4.4 Dosing recommendations for the macrolides with regard to food

Antimicrobial	Formulation	Manufacturer recommendations
Erythromycin base	Delayed-release tablets	Well absorbed and may be given without regard to meals
	Dispertab tablets	Optimal blood levels are obtained when taken on an empty stomach (at least 30 min and preferably 2 h before meals)
	Filmtab tablets	Optimum blood levels are obtained when doses are given on an empty stomach (2 h before or after a meal)
	Delayed-release capsules	Optimum blood levels are obtained on a fasting stomach (administer at least 1/2 h and preferably 2 h before or after a meal)
Erythromycin stearate	Erythrocin stearate filmtab tablets	Optimal serum levels of erythromycin are reached when taken in the fasting state or immediately before meals
Erythromycin ethylsuccinate	Liquid suspension	Can be given without regard to meals
	Filmtab tablets	Can be given without regard to meals
	Granules	Can be given without regard to meals
	Powder for suspension	Can be given without regard to meals
	Drops	Can be given without regard to meals
Clarithromycin	Filmtab tablets, granules for oral suspension	Can be given without regard to meals
	Biaxin XL filmtab	Should be taken with food
Azithromycin	Oral suspension	Can be given without regard to meals
	Tablets	Can be given without regard to meals
	Powder for suspension extended release	Taken on an empty stomach (at least 1 h before or 2 h following a meal)
Telithromycin	Tablets	May be taken without regard to meals

The manufacturers' dosing recommendations for macrolide antibiotics with regard to food are shown in Table 4.4.

4.4.4 Tetracyclines

In general, the tetracyclines are affected to various degrees by food, milk, and iron products. Tetracycline, the prototype antibiotic for this class, has amassed a substantial body of literature concerning its food and supplement interactions. Studies involving doxycycline and minocycline are plentiful as well, comparing their food, milk, and iron interactions with that of tetracycline. The reduced bioavailability of the tetracyclines is most likely due to chelation of the antibiotic with heavy metals such as iron and calcium and binding to macromolecules found in food [1]. Iron preparations and antacids containing calcium, magnesium, and aluminum cations form poorly soluble complexes that decrease absorption, to varying degrees, all of the tetracyclines [140, 141]. It has been hypothesized that the tetracyclines with higher degrees of lipophilicity may display the least interaction with food or milk due to increased absorption and a lesser tendency to form complexes [142]. Of the three main tetracyclines, minocycline is the most lipophilic, followed by doxycycline and then tetracycline. Given the lack of clinical use of tetracycline, only doxycycline and minocycline are discussed in this section. The drug–food interactions of tetracycline have been discussed in detail in previous editions of the book.

4.4.4.1 Doxycycline

Doxycycline is less affected than tetracycline by coadministration with food or milk [32]. The coadministration of doxycycline with meals high in fat, carbohydrates, and protein produced an approximately 20% decrease in bioavailability. Another study reported a 30% decrease in AUC and a 24% decrease in the C_{max} of doxycycline after it was administered with 300 ml of milk compared to water [143]. The authors concluded that, similar to tetracycline, doxycycline should not be administered with milk. The coadministration of doxycycline with ferrous sulfate not only causes decreased absorption of doxycycline but also reduces the half-life of the drug from 17 to 11 h due to decreased enterohepatic recirculation secondary to chelation with iron salts in the gastrointestinal tract [144].

4.4.4.2 Minocycline

Minocycline also is minimally affected when given with food or milk, but coadministration with antacids or other divalent cations caused significantly decreased absorption and is contraindicated [33, 140, 145]. Although not as well documented, the enterohepatic recirculation interaction probably occurs to the same extent with minocycline as doxycycline. Thus, it is recommended that minocycline and doxycycline be given with food to decrease incidence of gastrointestinal upset but that the administration of all tetracyclines be spaced by at least 2 h with antacids [146]. Due to the significant gastrointestinal transit time of iron preparations, concomitant prescribing is contraindicated with the tetracyclines.

4.4.4.3 Demeclocycline

The prescribing information for demeclocycline indicates that oral forms of tetracyclines should be administered 1 h before or 2 h after meals, with no specific information regarding demeclocycline [147].

Table 4.5 Dosing recommendations for the tetracyclines with regard to food

Antimicrobial	Formulation	Manufacturer recommendations
Doxycycline	Tablet (as monohydrate)	Administration with adequate amounts of fluid is recommended. The absorption of doxycycline is not markedly influenced by simultaneous ingestion of food or milk
	Delayed-release tablets	Administration with adequate amounts of fluid is recommended. The absorption of doxycycline is not markedly influenced by simultaneous ingestion of food or milk
	Capsules	May be given with food if GI upset occurs. Administration with adequate amounts of fluid is recommended
	Capsules (sugar spheres)	Should be taken at least 1 h prior to or 2 h after meals
	Tablet (as hyclate)	Administration with adequate amounts of fluid is recommended Should be taken at least 1 h prior to or 2 h after meals
	Oral syrup	May be given with food if GI upset occurs administration with adequate amounts of fluid is recommended
	Oral suspension	May be given with food if GI upset occurs administration with adequate amounts of fluid is recommended
	Film-coated tablets(as hyclate)	May be given with food if GI upset occurs administration with adequate amounts of fluid is recommended
Minocycline	Capsules, film-coated tablets	Can be given without regards to meals
	Oral suspension	Can be given without regards to meals
	Pellet-filled capsules	Can be given without regards to meals
	Extended-release tablets	Taking with food may lower the chances of getting irritation or ulcers in the esophagus
Demeclocycline	Tablet	Should be taken 1 h before or 2 h after meals

The manufacturers' dosing recommendations for the tetracyclines with regard to food are shown in Table 4.5.

4.4.5 Fluoroquinolones

In general, the fed state itself has little clinical effect on the pharmacokinetics of the fluoroquinolones [34, 148, 149]. However, they are affected by chelation with divalent and trivalent cations, so food, enteral feeds, and supplements containing heavy metal ions have been shown to have a significant effect on the bioavailability of the fluoroquinolones.

For example, the $AUC_{0-\infty}$ and C_{max} of moxifloxacin were reduced by 61% and 41% when coadministered with two consecutive doses of iron sulfate [150]. It is recommended that moxifloxacin be administered at least 4 h before or 8 h after products containing multivalent cations [151], while only 2 h before or after is necessary for levofloxacin [152] and 2 h before or 6 h after for ciprofloxacin [149]. The C_{max} and AUC of ofloxacin are reduced by 70% and 61% when administered with sucralfate, although the addition of food to this coadministration can decrease these reductions to 39% and 31% [153]. Fluoroquinolones also inhibit the liver enzymes responsible for caffeine metabolism, creating another potential interaction. Given the abundance of literature published with regard to fluoroquinolone–food interactions, this section will be subdivided to detail specific interactions.

4.4.5.1 Food

A 1987 study examined the pharmacokinetics of a 200 mg single dose of ofloxacin in 12 healthy adult male volunteers in the fasted state and after ingestion of a "fat-rich" breakfast (250 mL milk, coffee, two pieces of bread with butter and jam) [38]. The C_{max} in fasting subjects was significantly lower than that after a fat-rich breakfast, and the t_{max} was approximately 1 h sooner, although the AUCs were not significantly different. In another study evaluating a standard breakfast, the C_{max} and AUC_{0-28} of a 300 mg dose of ofloxacin were not clinically significantly different from the fasted state in 12 healthy male volunteers [154]. This lack of clinically significant food effect was also observed in a study of serum and skin blister fluid pharmacokinetics of ofloxacin [155] and in another study evaluating 21 healthy male volunteers given a standard breakfast [156]. Given these data the manufacturer recommends that ofloxacin may be administered without regard to meals [157].

In 12 healthy male volunteers administered a single oral 750 mg dose of ciprofloxacin immediately after or 2 h after a standard (12.5 g fat) breakfast or halfway through a high-fat, high-calcium (37 g fat, 729 mg Ca^{2+}) breakfast, there were no significant differences observed in $AUC_{0-\infty}$, C_{max}, or t_{max} between any of the study periods compared to the fasted state [34]. In ten healthy volunteers, ingestion of a standard breakfast (120 g white bread, 10 g butter, 10 g jam, and 150 mL rose hip tea) did not significantly affect the pharmacokinetics of a 1750 mg single oral dose of ciprofloxacin, although the t_{max} was slightly delayed by 1 h at steady state [158]. Similar results were seen with a 500 mg/10 mL dose of the oral suspension in which food did not affect the bioavailability in 68 healthy volunteers [159]. The overall pharmacokinetic profile of either dosage form of ciprofloxacin is therefore not significantly affected by coadministration with food and can be given without regard for meals [149].

Levofloxacin tablets are well absorbed after oral administration, with a bioavailability of greater than 90% [160]. Administration with food decreases the C_{max} by approximately 14% and lengthens the t_{max} by approximately 1 h [36]. This is not considered clinically significant, and levofloxacin tablets can be administered without regard to meals. The manufacturer recommends that levofloxacin oral solution be given 1 h before or 2 h after eating [152].

Similar to levofloxacin, moxifloxacin has excellent oral absorption with an absolute bioavailability of approximately 90% [161]. After the administration of a high-fat breakfast, the absorption of moxifloxacin is slightly delayed. The median t_{max} values were 1 h under fasting conditions and 2.5 h in the fed state. The C_{max} and AUC were decreased by approximately 12% and 3%, respectively, after the administration of a high-fat meal. The magnitude of these effects is not considered clinically significant [37]. The absolute bioavailability of gemifloxacin is approximately 71% and does not appear to be significantly altered by the administration of a high-fat meal [35, 162].

4.4.5.2 Milk or Yogurt

Coadministration with milk or yogurt significantly decreased the C_{max} and AUC of ciprofloxacin in two healthy volunteer studies [163, 164]. The effect of milk and yogurt on the absorption of norfloxacin was investigated in two other healthy volunteer trials [165, 166]. The administration of milk caused a greater-than-50% decrease in the C_{max} and AUC of norfloxacin.

Dairy products did not significantly affect the pharmacokinetics of moxifloxacin or ofloxacin in healthy volunteer studies [156, 167, 168]. In general, patients should be counseled to avoid coadministration of milk with all the fluoroquinolones.

4.4.5.3 Vitamin- or Mineral-Fortified Foods

Most calcium-fortified food products have more calcium per serving than the milk or yogurt used in dietary calcium interaction studies [169]. Several studies have evaluated the effect of calcium-fortified orange juice on the bioavailability of the fluoroquinolones [170, 171]. In a randomized, three-way crossover study, 15 healthy subjects received a single dose of ciprofloxacin with 12 ounces of water, orange juice, and calcium-fortified orange juice. The C_{max} and AUC decreased significantly by 41% and 38%, respectively, when ciprofloxacin was administered with calcium-fortified orange juice compared to water [169].

After administering a single 500 mg dose of levofloxacin with either water or calcium-fortified orange juice to 16 healthy subjects, the C_{max} was reduced by 18% and t_{max} increased by 58%, with no significant change in AUC [172, 173]. Interestingly, there were no significant differences in pharmacokinetic changes after plain and calcium-fortified orange juice consumption, suggesting that inhibition of P-gp or OATP in the gastrointestinal tract by the orange juice may play a role along with chelation. These studies highlight the need for cognizance when administering fluoroquinolones with any types of juice or food, as the interactions with multivalent cations are significant and the number of fortified foods being produced is increasing.

4.4.5.4   Caffeine

In vitro experiments utilizing human liver microsomes have assessed the inhibitory potency of various fluoroquinolones against CYP1A2. Ciprofloxacin and norfloxacin were the strongest inhibitors of CYP1A2, followed by ofloxacin [174]. Inhibition of CYP1A2 activity results in decreased clearance and results in an increase in the AUC of caffeine [175]. In human studies, norfloxacin significantly altered the pharmacokinetics of caffeine, causing similar changes in the clearance and AUC [176]. Concomitantly administered ciprofloxacin caused approximately a 50% increase in the AUC and 50% decrease in the clearance of caffeine. Thus, caffeine should be avoided in patients with liver disorders, cardiac arrhythmias, latent epilepsy, or critical illness while undergoing treatment with fluoroquinolones known to interact with caffeine [177].

4.4.5.5   Enteral Feeds

Studies on the effect of enteral feeds on the absorption of fluoroquinolones have produced conflicting results. Importantly, enteral feeds contain various amounts of multivalent cations that may affect the quinolones to different extents, so it is important to review the product labeling to estimate the degree of interaction and determine whether tube feeds should be held around the administration of fluoroquinolones [178]. One enteral feeding product, Ensure®, reduced the relative oral bioavailability of ciprofloxacin by 28% in 13 healthy volunteers [179]. Another study showed either no effect or increased ciprofloxacin absorption with concomitant enteral feeds Pulmocare® or Osmolite® in six healthy volunteers [180]. There was no clinically relevant change in the rate or extent of absorption of moxifloxacin in 12 healthy volunteers when Isosource® Energy enteral feed was administered for 30 min prior to moxifloxacin and immediately resumed after administration for another 2 h [181]. Feeding into the jejunum has been shown to produce a larger reduction in bioavailability compared to feeding into the stomach [182].

While not contraindicated, it is prudent to avoid the simultaneous administration of enteral feeds and quinolones to assure adequate absorption. It is recommended to hold enteral feeding for 2 h before and after administration of quinolones [183].

The manufacturers' dosing recommendations for fluoroquinolone antibiotics with regard to food are shown in Table 4.6.

4.4.6   Miscellaneous Antibiotics

4.4.6.1   Anthelmintics

The mean C_{max} of albendazole was increased 6.5-fold, and the AUC increased 9.5-fold in six healthy male subjects after the administration of a fatty meal [39, 184]. Therefore, albendazole tablets are recommended to be administered with meals to

Table 4.6 Dosing recommendations for the fluoroquinolones with regard to food

Antimicrobial	Formulation	Manufacturer recommendations
Ofloxacin	Tablets	Can be given without regard to meals
Ciprofloxacin	Tablets, oral suspension	Can be given without regards to meals
	Extended-release tablets	Can be given without regard to meals
Levofloxacin	Tablets, oral solution	Can be given without regard to meals
Moxifloxacin	Tablets	Can be given without regard to meals
Gemifloxacin	Tablets	Can be given without regard to meals
Norfloxacin	Tablets	Administer 1 h before or 2 h after meals. Patients should be well hydrated

enhance absorption. Compared to the fasting state, the administration of grapefruit juice enhanced the C_{max} and AUC of albendazole by 3.2-fold and 3.1-fold, respectively. Literature reports suggest that coadministration of mebendazole with a fatty meal increases the peak concentrations and overall absorption [185, 186]. The manufacturer of thiabendazole recommends that it be administered after meals [187].

Administration of ivermectin to healthy volunteers following a high-fat meal increased the systemic exposure 2.5 times compared to the fasted state [40]. This increased exposure is an undesirable effect with ivermectin; therefore, the manufacturer recommends that it be administered on an empty stomach with water [188].

Food has also been reported to increase the bioavailability of praziquantel [41]; therefore, it should be administered with meals [189]. Praziquantel mean C_{max} and AUC were also increased by 1.62-fold and 1.9-fold after administration with grapefruit juice [190].

4.4.6.2 Antimalarials

Unpublished data indicate that the C_{max} and AUC_{0-24} were not significantly different among 26 healthy volunteers given a single 324 mg oral dose of quinine sulfate with and without a high-fat breakfast, and thus quinine may be given without regard to meals but is recommended to be taken with food to minimize gastric irritation [191].

Food significantly increased the C_{max} and AUC of primaquine by 26% and 14%, respectively, when a 30 mg dose was administered to healthy volunteers [42]. The administration of half-concentrated grapefruit juice also increased the C_{max} and AUC of primaquine; however, large intersubject variability was observed. Neither food nor grapefruit juice changed the C_{max} or AUC of the primary metabolite, carboxyprimaquine.

The systemic availability of chloroquine was also increased in healthy volunteers by the administration of a meal [43]. Therefore, primaquine and chloroquine should be taken with food to minimize gastrointestinal upset and increase systemic availability.

Hydroxychloroquine has efficacy as prophylaxis for malaria in some instances, but there are no data regarding the effect of food. The manufacturer's prescribing information recommends that each dose be taken with a meal or glass of milk [192].

Pyrimethamine and sulfadoxine–pyrimethamine should be administered after a meal [193, 194].

Food increases the absorption of artemether and lumefantrine. In healthy volunteers, the relative bioavailability of artemether was increased between two- to three-fold and that of lumefantrine sixteen-fold when artemether–lumefantrine tablets were given after a high-fat meal compared to fasted conditions [195].

The systemic availability of mefloquine is improved ~40% when administered with food [44], and the manufacturer recommends that it be taken immediately after a meal [196].

Atovaquone has very poor oral bioavailability, and therapeutic concentrations may not be achieved when it is taken while fasting. Food, especially fatty food, increases the systemic availability of atovaquone by two- to threefold. Thus, it is recommended that atovaquone always be taken with a meal or nutritional supplement with at least a moderate amount of fat [45, 197, 198].

Similarly, the combination of atovaquone and proguanil should also be administered with food [199].

4.4.6.3 Clindamycin

Food does not affect the absorption of clindamycin granules or capsules [46], although there are no specific recommendations for dosing with regard to food in the package insert [200].

4.4.6.4 Fidaxomicin

Fidaxomicin is a macrocyclic antibacterial indicated for the treatment of *Clostridium difficile*-associated diarrhea in adults. In a food-effect study of 28 healthy volunteer subjects, the C_{max} of fidaxomicin was decreased 21.5% when given with a high-fat meal relative to the fasted state, while AUC_{0-t} remained unchanged [201]. This change in C_{max} is not considered clinically significant, and therefore fidaxomicin may be given without regard to food.

4.4.6.5 Fosfomycin

A standard meal has been shown to decrease the mean C_{max} of fosfomycin after a 1 g dose of fosfomycin tromethamine by approximately 36%, although overall absorption in terms of AUC was not affected [47]. Another study demonstrated a decrease in the rate of absorption of a 50 mg/kg dose when given with food [48]. The prescribing information indicates that the rate of urinary excretion, but not the cumulative amount excreted, was reduced after administration with a high-fat meal and designates that fosfomycin tromethamine may be taken without regard to food [202].

4.4.6.6 Nitazoxanide

Nitazoxanide is a broad-spectrum antiparasitic agent primarily used for the treatment of diarrhea due to *Cryptosporidium parvum* or *Giardia lamblia*. In 32 healthy male subjects given ascending doses of 1 g, 2 g, 3 g, and 4 g of nitazoxanide tablets with and without a standardized breakfast, the coadministration with food effectively doubled the plasma concentrations regardless of the dose given [49]. When the oral suspension was administered with food, the AUC_{0-t} and C_{max} increased by 45–50% and ≤10%, respectively [203]. It is recommended that both formulations of nitazoxanide be administered with food.

4.4.6.7 Nitrofurantoin

Food tends to enhance the absorption of nitrofurantoin [204–206]. The increased dissolution time resulting from coadministration of food with nitrofurantoin has been hypothesized as the mechanism behind this increased absorption. Interestingly, food tends to have more of an effect on the urinary levels of nitrofurantoin than on the corresponding serum levels. An explanation for this phenomenon is that food increases the fraction of drug excreted by potentially saturating metabolic pathways in kidney.

4.4.6.8 Nitroimidazoles

The absorption of metronidazole is delayed but not reduced by the presence of food [50].

Administration of tinidazole tablets with food resulted in a delay in t_{max} of ~2 h and a decrease in C_{max} of ~10%, with no effect on AUC. The manufacturer recommends taking tinidazole with food to minimize gastrointestinal side effects [207].

4.4.6.9 Oxazolidinones

When administered with a high-fat meal (850 calories), linezolid required a slightly longer t_{max} than when given under fasting conditions and C_{max} was significantly lower. No difference was observed in mean AUC values under fasted and fed conditions; therefore, linezolid may be taken without regard to meals [51]. Linezolid is a weak, competitive, reversible inhibitor of human monoamine oxidase-A (MOA-A) [208]. When linezolid is administered at the approved dose, dietary restriction of tyramine containing foods is generally not necessary. However, patients should be advised to avoid consuming large amounts of foods high in tyramine (i.e., aged cheeses, fermented meats, sauerkraut, soy sauce, draught beers, and red wines).

Tedizolid is a novel oxazolidinone antibiotic with improved in vitro activity against *Staphylococcus aureus* and vancomycin-resistant enterococci compared to linezolid. A phase I, randomized, open-label, crossover study conducted in 12 healthy volunteer subjects evaluated the pharmacokinetics of a single oral dose of tedizolid after an overnight 10 h fast or with a high-calorie, high-fat breakfast [52]. In the fasted state, the mean tedizolid C_{max} was approximately 26% higher than after a high-fat breakfast, and the t_{max} was decreased by 6 h (2 h vs. 8 h). The overall plasma exposure in terms of $AUC_{0-\infty}$ was not changed (geometric mean ratio % 102.3), suggesting that tedizolid may be administered without regard to meals. Similar to linezolid, tedizolid is a weak, reversible inhibitor of MOA-A and MOA-B. The inhibition of MOA-A and MOA-B was evaluated via in vitro and in vivo methods [209]. In the in vitro study, the mean IC50 of tedizolid for MOA-A was over fivefold lower, while the IC50 for MOA-B was approximately twofold higher. Importantly, the free-drug concentration at C_{max} of tedizolid is severalfold lower than this IC50, while the linezolid-free C_{max} is close to and greater than this threshold. Animal studies evaluated serotonergic activity via head twitch response in adult male mice. In this murine head twitch serotonergic model, linezolid did significantly elevate the number of head twitches while tedizolid did not. Human studies evaluated the interaction with oral tyramine or pseudoephedrine in randomized trials. Thirty subjects were enrolled in the tyramine challenge study, and seven of them exceeded a predefined systolic blood pressure increase threshold of ≥ 30 mmHg following tyramine administration, with no difference between the tedizolid and placebo groups. Mean maximum increases in blood pressure and heart rate were not significantly different between tedizolid and placebo in the pseudoephedrine challenge study.

A tyramine-rich meal is expected to contain no more than 40 mg of tyramine, which is well below the lowest dose used in these human studies (275 mg). Therefore, meals containing tyramine with tedizolid should not cause any adverse reactions.

4.4.6.10 Rifaximin

Rifaximin is a structural analog of rifampin indicated for the treatment of traveler's diarrhea caused by noninvasive strains of *Escherichia coli* and has garnered some use for *Clostridium difficile*-associated diarrhea. Compared to the fasting state, a high-fat meal ingested with a single oral dose of rifaximin increased the AUC by twofold and doubled t_{max} [210]. This increase was not considered clinically significant as rifaximin is minimally absorbed systemically so the drug may be given with or without food.

4.4.6.11 Sulfadiazine

No information about the effect of food on sulfadiazine is available, and dosing recommendations with regard to food are not included in the prescribing information [211].

4.4.6.12   Trimethoprim–Sulfamethoxazole

The effect of food (orange juice, bread, butter, cheese, tomato, and sour milk) and guar (5 g guar gum) on the absorption of a single 3 mg/kg dose of trimethoprim suspension was studied in 12 healthy volunteers and demonstrated that food decreased the C_{max} and AUC by 22%. The addition of guar alone or with the food did not have any significant effect [53]. No food-effect studies in humans are available for sulfamethoxazole. The manufacturer of the combination product sulfamethoxazole–trimethoprim does not provide specific recommendations about dosing with food or meals [212].

4.4.6.13   Vancomycin

No food-effect studies in humans are available for oral vancomycin. The manufacturer of the oral vancomycin capsules does not provide specific recommendations about dosing with food or meals [213].

The manufacturers' dosing recommendations for miscellaneous antibiotics with regard to food are shown in Table 4.7.

4.4.7   Antimycobacterials

4.4.7.1   Isoniazid

Peak concentrations and the relative bioavailability of isoniazid decreased by 70% and 40% with the addition of food, respectively, which suggests that isoniazid always be given on an empty stomach [214]. Another study in 14 healthy volunteers investigated the effect of a high-fat breakfast on the absorption of isoniazid [54]. Relative to fasting, the high-fat meal reduced C_{max} by 51%, doubled t_{max}, and reduced AUC by 12%. The manufacturer advises that isoniazid can be given with food if stomach upset occurs but should preferentially be given on an empty stomach if tolerable. Because isoniazid is a weak MAO inhibitor, several case reports have described adverse reactions in patients taking isoniazid who have ingested foods high in monoamines (e.g., tyramine) [215]. Flushing of the arms, face, and upper body was observed in patients after ingestion of cheese or red wine during isoniazid therapy [216–218]. Other possible symptoms include palpitations, headache, and mild increases in systolic blood pressure. Isoniazid also inhibits histaminase. At least 30 cases of adverse reactions after ingestion of fish with high histamine contents (e.g., tuna, mackerel, salmon, and skipjack) have been reported in patients taking isoniazid. Patients should be cautioned about the potential for adverse reactions with certain cheeses, red wine, and fish with high tyramine and/or histamine content while taking isoniazid.

Table 4.7 Dosing recommendations for miscellaneous antimicrobials with regard to food

Antimicrobial	Formulation	Manufacturer recommendations
Anthelmintics		
Albendazole	Tablets	Should be taken with food
Mebendazole	Tablets	Not required to be taken with food
Thiabendazole	Chewable tablets	Give after meals if possible
Ivermectin	Tablets	Should be taken with water
Praziquantel	Tablets	Should be taken with water during meals
Antimalarials		
Quinine	Capsules	May be taken without regard to meals, although food may reduce gastric irritation
Primaquine	Tablets	Can be given without regard to meals, although food may reduce gastric irritation
Chloroquine	Tablet	Administer with food or milk to reduce gastric irritation
Hydroxychloroquine	Tablets	Administer with food or milk to reduce gastric irritation
Pyrimethamine	Tablets	Should be taken after a meal
Sulfadoxine	Tablets	Should be taken after a meal
Artemether–lumefantrine	Tablets	Should be taken with food
Mefloquine	Tablets	Should be taken immediately after a meal
Atovaquone	Suspension	Administer with a meal or nutritional supplement with at least a moderate amount of fat
Atovaquone–proguanil	Tablet	Take with food
Other		
Fidaxomicin	Tablets	May be given without regard to food
Fosfomycin	Sachet	May be given without regard to food
Nitazoxanide	Tablets, suspension	Administer with food
Nitrofurantoin	Capsules	Should be taken with food to improve absorption and tolerance
Tinidazole	Tablets	Should be taken with food to minimize the incidence of epigastric discomfort and other gastrointestinal side effects
Linezolid	Tablets	May be taken with or without food
Tedizolid	Tablets	May be taken with or without food
Rifaximin	Tablets	May be taken with or without food

4.4.7.2   Rifampin

In a healthy volunteer study performed in the 1970s, the coadministration with food caused a 25% reduction in the C_{max} and urinary excretion of rifampicin [219]. In a subsequent analysis with 14 healthy volunteers, the addition of a high-fat meal reduced the C_{max} by 36% and the overall AUC by 6% [55]. An aluminum–magnesium antacid had no effect on the bioavailability of rifampin. Thus, rifampin should be taken on an empty stomach whenever possible but may be taken with food if stomach upset occurs.

4.4.7.3 Rifabutin

The effect of a high-fat meal on the pharmacokinetics of rifabutin was studied in 12 healthy male volunteers [56]. Although a delay was seen in the t_{max} (5.4 versus 3.0 h), little effect on overall systemic exposure was seen with the addition of food.

4.4.7.4 Ethambutol

A standardized breakfast produced little to no effect on the mean AUC of ethambutol in 11 normal healthy volunteers [220]. A subsequent study in 12 male and female volunteers showed similar results with the coadministration of a high-fat meal [57]. However, the coadministration of an aluminum–magnesium antacid caused a 29% decrease in the C_{max} and a 10% decrease in AUC. The authors of this paper suggested that antacids should be avoided near the time of ethambutol dosing.

4.4.7.5 Pyrazinamide

A study of 12 healthy volunteers demonstrated that a high-fat meal or aluminum–magnesium antacid had minimal effects on the pharmacokinetics of pyrazinamide [58].

4.4.7.6 Bedaquiline

Bedaquiline is novel diarylquinoline antimycobacterial agent with a unique mechanism of action involving specific inhibition of mycobacterial ATP synthase [221]. In healthy adult subjects, the mean AUC of bedaquiline increased ~2–2.4-fold after administration with a standard (533 kcal, 21 g fat) meal relative to fasted conditions. The manufacturer recommends that bedaquiline be taken with food, with no specific requirements to the type or content of food [222].

4.4.7.7 Aminosalicylic Acid

Para-aminosalicylic acid (PAS) is a second-line antitubercular agent marketed in a granule formulation to improve the gastrointestinal tolerability [223]. A pharmacokinetic study evaluated the effect of a high-fat (792 kcal, 51 g fat) breakfast, orange juice (240 mL), and antacids (Mylanta® maximum strength) on the plasma exposure of PAS in a randomized, four-period crossover design [59]. Twelve subjects received a single dose of 6 g PAS after a 12 h overnight fast or with the high-fat breakfast or orange juice. The antacids were given 9 h before dosing, immediately after each meal, and at bedtime on the dosing day. The t_{max} of PAS was delayed 1.5-fold when administered with food but was unaffected by orange juice or antacids. Compared to the fasting state, the high-fat breakfast significantly increased C_{max} and $AUC_{0-\infty}$ by 1.5- and 1.7-fold, respectively. Neither orange juice nor antacids had clinically significant effects on plasma exposure of PAS, although the large intersubject variability in this study caused the 90% confidence intervals to fall outside of the FDA proposed boundaries of 80–125%. The increase in C_{max} and AUC is a desirable effect in the case of PAS, so the drug should be administered with food, particularly a high-fat meal. The manufacturer's prescribing information states that the protective acid-resistant outer coating of PAS granules, designed to protect against degradation in the stomach, is rapidly dissolved within 1 min in neutral media [224]. Therefore, a mildly acidic food (pH < 5) like orange juice, apple juice, yogurt, or apple sauce should be used to maintain the granules in an acidic food during dosage administration. The granules should be sprinkled on one of these foods or suspended in a fruit drink and will last at least 2 h. If patients have taken antacids or proton-pump inhibitors, this step is not necessary as the granules will not be degraded in the absence of stomach acid.

4.4.7.8 Cycloserine

Cycloserine is a bacteriostatic cell wall synthesis inhibitor used primarily for the treatment of multidrug-resistant *Mycobacterium tuberculosis* infections. Twelve healthy subjects received a single 500 mg dose of cycloserine under fasted conditions and with a high-fat breakfast, orange juice, or antacids in a pharmacokinetic study identical to the PAS study above [60]. In the case of cycloserine, the C_{max} was statistically, but not clinically, significantly decreased by the high-fat meal (14.8 vs 12.4 mg/L) but not by either the orange juice or antacids. None of the three fed states affected the AUC, and therefore the drug can be given without regard to food or antacids, although a particularly high-fat meal should potentially be avoided. The prescribing information for cycloserine does not address administration with food [225].

4.4.7.9 Ethionamide

Ethionamide is one of only four drugs recommended by the World Health Organization for the combined treatment of leprosy [226] and also maintains activity against tuberculosis. A single 500 mg dose of ethionamide was given to the same 12 healthy volunteers in the same fashion as discussed for PAS and cycloserine [61]. Food, orange juice, or antacids showed no significant effect on the C_{max}, AUC, or t_{max} of ethionamide. The manufacturer recommends that the drug be taken with meals to reduce gastrointestinal intolerance [227].

4.4.7.10 Rifapentine

Rifapentine is a cyclopentyl rifamycin indicated for the treatment of pulmonary tuberculosis. A study designed to evaluate the effect of food on the pharmacokinetics of rifapentine and its major metabolite 25-desacetyl rifapentine included meals comprised largely of maize as this is a dietary staple in many parts of Africa and South Africa [228]. A single 900 mg dose of rifapentine was administered to 34 healthy adult male volunteers with one of four meals: a high-fat (English breakfast; 469.9 kcal, 27 g fat), low-fat bulky (maize meal porridge; 307.1 kcal, 3 g fat), high-fat bulky (maize meal porridge with lard; 532.7 kcal, 28 g fat), and a low-fat, high-fluid (reconstituted chicken noodle soup; 184.9 kcal, 4 g fat). The four meals increased the bioavailability of rifapentine 85.7%, 32.7%, 45.7%, and 48.9%, respectively. These findings are consistent with a previous study demonstrating that a high-fat English breakfast, particularly one including eggs, significantly increased the bioavailability [229]. The prescribing information reports that a high-fat (850 kcal, 55 g fat) increased the C_{max} and AUC of rifapentine by approximately 51% and 53%, respectively, in asymptomatic HIV-infected volunteers [230]. The manufacturer recommends that rifapentine be taken with food, but does not specify the fat content.

4.4.7.11 Dapsone

The manufacturer's prescribing information for dapsone does not include guidance for administration with food, and no published data are available [231].

4.4.7.12 Thalidomide

Thalidomide is a glutamic acid derivative approved for the treatment of erythema nodosum leprosum [232]. A high-fat meal resulted in a 62% delay in t_{max}, an 8.54% increase in C_{max}, and a 5.5% decrease in the AUC of thalidomide [62]. The manufacturer recommends that thalidomide be taken at bedtime at least 1 h after the evening meal [233].

4.4.7.13 Clofazimine

Clofazimine has demonstrated efficacy in the treatment of several mycobacterial diseases, including leprosy and *Mycobacterium avium* complex [234]. It was studied in the same pharmacokinetic study as PAS, cycloserine, and ethionamide discussed previously. The high-fat breakfast significantly increased C_{max} and AUC_{0-t} by approximately 2- and 2.5-fold, respectively [63]. Orange juice and antacids decreased the AUC, with geometric mean ratios of 93.0 and 65.2 compared to the fasted state. In a small pharmacokinetic study of only three subjects, food increased the AUC 60% and the C_{max} 30% [235]. The manufacturer recommends that clofazimine be taken with meals, although the composition of the meal is not specified [236].

The manufacturers' dosing recommendations for antimycobacterial antibiotics with regard to food are shown in Table 4.8.

Table 4.8 Dosing recommendations for the antimycobacterials with regard to food

Antimicrobial	Formulation	Manufacturer recommendations
Isoniazid	Tablet	Should not be administered with food
Rifampin	Capsules	Take on empty stomach, either 1 h before or 2 h after a meal, with a full glass of water
Rifabutin	Capsules	May be taken with meals if GI upset occurs
Ethambutol	Tablets	Can be given without regard to meals
Bedaquiline	Tablets	Take with food
Para-aminosalicylic acid	Granules	A mildly acidic food (pH < 5) like orange juice, apple juice, yogurt, or apple sauce should be used to maintain the granules in an acidic food during dosage administration. The granules should be sprinkled on one of these foods or suspended in a fruit drink and will last at least 2 h. If patients have taken antacids or proton-pump inhibitors, this step is not necessary as the granules will not be degraded by stomach acid
Ethionamide	Film-coated tablet	Take with food to reduce gastrointestinal intolerance
Rifapentine	Tablets	Take with food
Thalidomide	Capsules	Take at bedtime at least 1 h after the evening meal
Clofazimine	Capsules	Take with meals

4.4.8 Antifungals

4.4.8.1 Flucytosine

The manufacturer's information for flucytosine does not include recommendations for dosing with regard to meals [237], and no published data are available.

4.4.8.2 Terbinafine

A food-effect study evaluated the impact of fed condition on the pharmacokinetics of terbinafine in 15 elderly and 15 young healthy subjects after a single oral dose of 250 mg [64]. Exposures were generally higher under the fed condition but not to a statistically or clinically significant extent (AUC increased <20%). Therefore, terbinafine tablets may be taken with or without food [238].

4.4.8.3 Ketoconazole

A number of healthy volunteer studies have investigated the influence of food on the pharmacokinetics of ketoconazole, with conflicting results.

A crossover study of 12 volunteers showed a 55–60% decrease in C_{max} and AUC as well as a delayed t_{max} when 200 mg of ketoconazole was given immediately after a low-fat breakfast [239]. Another study in 18 volunteers investigated the influence of a high-fat breakfast on the pharmacokinetics of ketoconazole over a wider dosing range (200–800 mg) [65] and determined that food did not reduce AUC or C_{max} but did tend to lengthen t_{max}. Finally, a third study of 12 volunteers showed that a high-fat meal significantly prolonged t_{max} and a high-carbohydrate meal significantly decreased C_{max} [240]. There was a non-statistically significant trend toward increased AUC values with the high-fat meal and decreased AUC values with the high-carbohydrate meals. The manufacturer recommends that ketoconazole be given with food, which appears reasonable given the conflicting results from pharmacokinetic studies.

4.4.8.4 Fluconazole

The influence of a low-fat (1000-kcal) and a high-fat (3600-kcal) meal on the pharmacokinetics of 100 mg of fluconazole tablets and 100 mg of itraconazole capsules was investigated in 24 healthy volunteers [66]. The C_{max}, AUC, and t_{max} of fluconazole were not significantly affected after meals compared to fasting. In contrast, the AUC of itraconazole when given on an empty stomach was approximately 40% lower than when given with a high-fat meal.

4.4.8.5 Itraconazole

Similar results were seen when itraconazole capsules were given to patients with superficial fungal infections [241]. Contrarily, the C_{max} and AUC decreased by 44 and 30%, respectively, when 200 mg of itraconazole oral solution was given with a high-fat meal to 30 healthy volunteers [242]. Thus, itraconazole capsules and tablets should be given with food, while the oral solution should be given on an empty stomach.

The effect of acidic cola beverages on the absorption of 100 and 200 mg doses of itraconazole has been assessed in two separate healthy volunteer studies [243, 244]. Results from these studies showed that the addition of a cola product increased the AUC and C_{max} of itraconazole by up to 100%. Thus, the addition of an acidic beverage should be recommended to improve the absorption of itraconazole. Regardless of dosage form, itraconazole should not be taken with antacids. Importantly, the capsule and oral solution formulation are not bioequivalent, so they should not be used interchangeably.

The effect of grapefruit juice on the pharmacokinetics of itraconazole capsules has also been evaluated in healthy volunteers. In one study, single-strength grapefruit juice had no effect on the pharmacokinetics of a 100 mg dose of itraconazole [245]. In the second study, administration of double-strength grapefruit juice (concentrated with half the recommended amount of water) resulted in a decrease in the mean AUC_{0-48} of itraconazole by 43% and a decrease in the mean AUC_{0-72} of the hydroxy-metabolite by 47% after administration of a 200 mg dose [246]. The mechanism by which double concentrated grapefruit juice reduces the absorption of itraconazole capsules is unknown, but the authors suggest a number of possibilities including a reduction in duodenal pH causing an increase in the amount of ionized itraconazole, increased intestinal P-gp mediated efflux of itraconazole, decreased intestinal CYP3A4 expression, a delay in gastric emptying, and interindividual differences in intestinal CYP3A4 and P-gp content between study subjects. Repeated administration of single-strength grapefruit juice with itraconazole oral solution in healthy volunteers increased the AUC_{0-48} and AUC_{0-inf} of itraconazole by 15.8% and 19.5%, respectively, with no change in the exposure to the hydroxy-metabolite [247]. These findings suggest inhibition of intestinal CYP3A4.

4.4.8.6 Voriconazole

The effect of a high-fat breakfast on the pharmacokinetics of voriconazole was evaluated in 12 healthy male subjects [67]. At steady state, the bioavailability of voriconazole was reduced by approximately 22% when taken with food compared to fasting. The rate of absorption was also significantly delayed after administering voriconazole with food. Therefore, voriconazole tablets should be taken at least 1 h before or 2 h after a meal.

4.4.8.7 Posaconazole

Posaconazole is a lipophilic second-generation antifungal triazole with a similar molecular structure to that of itraconazole. Of all antimicrobials discussed in this chapter, posaconazole, particularly its oral suspension formulation, has the most clinically important and variable drug–food and gastric interactions. Patient counseling is essential when prescribing the oral suspension form of this agent. Following the administration of a single 200 mg dose of the oral suspension in 20 healthy male volunteers, the AUC and C_{max} of posaconazole were approximately 4 times higher when administered with a high-fat meal (841 calories, 52% fat) and approximately 3 times higher when administered with a nonfat meal (461 calories, 0% fat) when compared to the fasted state [68]. Additionally, the effect of administration of a nutritional supplement (Boost® Plus) on posaconazole pharmacokinetics was evaluated in 24 healthy volunteers [248]. Each subject received a single 400 mg dose of posaconazole oral suspension in combination with 8 fluid ounces of the supplement (360 calories, 16% protein, 34% fat, and 50% carbohydrates) and a single 400 mg dose after an overnight fast. Administration with the nutritional supplement increased the C_{max} and AUC_{0-72} approximately 3.4- and 2.6-fold, respectively. Another study evaluated the effect of varying amounts of a nutritional supplement on posaconazole bioavailability in 30 healthy volunteers, to determine if an amount less than 8 ounces would also be effective in enhancing absorption [249]. Following administration of a single 400 mg dose of the oral suspension, posaconazole bioavailability increased roughly linearly with increasing amounts of supplement. The AUC of posaconazole was 35% (fasting), 48% (1 ounce), 60% (2 ounces), and 77% (4 ounces) compared to the AUC achieved with 8 ounces. A comprehensive four-part, randomized, crossover study in healthy volunteers evaluated the effect of gastric pH, dosing frequency, prandial state, food consumption timing, and gastric motility on the absorption of posaconazole oral suspension [70]. Compared to a fasting state, the administration of posaconazole with an acidic carbonated beverage increased the mean C_{max} and AUC by 92% and 70%, respectively. Administration under increased gastric pH conditions, induced by coadministration with esomeprazole, decreased the C_{max} and AUC by 46% and 32%. This study also confirmed previous findings that posaconazole administration during or immediately after a meal or nutritional supplement provides larger increases in AUC than that observed after administration before a meal, likely due to improved solubility rather than a delay in gastric emptying.

Posaconazole has recently been formulated into a delayed-release tablet formulation that demonstrates improved bioavailability and allows once-daily dosing. This tablet formulation consists of the active moiety, posaconazole, combined with a pH-sensitive polymer hypromellose acetate succinate via a hot-melt extrusion technique [250]. This strategy allows for enhanced solubility and bioavailability of poorly soluble drugs by creating a molecularly dispersed drug–polymer combination that releases only in the elevated pH environment of the intestine where absorption is maximized. A 100 mg dose of this tablet formulation has been shown to achieve substantially higher mean plasma posaconazole exposure compared to the oral suspension in the fasted state [69], and C_{max} and AUC were not significantly affected by food. The food-effect of this novel formulation has been evaluated in a phase I study of healthy volunteer subjects given a 300 mg dose [71]. This randomized, open-label, single-dose, crossover study included 16 subjects given a single oral dose of posaconazole after a 10 h overnight fast or with a high-fat (70 g fat) meal. Serial blood samples were collected up to 72 h post-dose for pharmacokinetic analysis. When administered with the high-fat meal, the C_{max} and AUC_{0-72} of posaconazole increased 16% and 51% compared to the fasted state. This modest 1.5-fold increase in AUC is in contrast to the fourfold difference observed when the suspension was given with a high-fat meal. These results suggest that posaconazole tablets may be taken without regard to food, although the manufacturer recommends that they be taken with food citing the same study [251]. Further improving on the shortcomings of the oral suspension, the pharmacokinetics of the novel tablet formulation are not affected by coadministration with antacids, H_2 receptor antagonists, or proton-pump inhibitors in healthy subjects [252].

4.4.8.8 Isavuconazole

Isavuconazonium sulfate, the prodrug of the active agent isavuconazole, is a newly approved triazole antifungal indicated for the treatment of invasive aspergillosis or mucormycosis in adults. The manufacturer recommends that the capsules may be taken with or without food. The package insert states coadministration of 400 mg isavuconazonium sulfate with a high-fat meal reduced the C_{max} by 9% and increased AUC by 9%. Isavuconazonium sulfate also demonstrated no effect on the CYP1A2 substrate caffeine after a 200 mg dose.

Three open-label studies were conducted in healthy adult subjects to assess the absolute bioavailability, effects of food, and of elevated gastric pH on the absorption of isavuconazole [72]. Fourteen subjects completed the absolute bioavailability study, which indicated an absolute F of 0.98 ± 0.07 based on $AUC_{0-\infty}$. After a single oral dose of 400 mg of isavuconazole, 24 subjects in the food-effect study received a high-fat breakfast (936 kcal, 37.1 g protein, 60.4 g fat, and 60.5 g fiber). Mean plasma pharmacokinetic parameters were similar over the 36 day sampling period between fasted and fed conditions, with the exception of a longer median t_{max} under fed conditions (5 vs. 3 h). The geometric mean ratios of AUC and C_{max} between fed and fasted were 110% and 92%, respectively. In order to study the effect of basic pH, 24 subjects received esomeprazole 40 mg once per day on days 1–10 along with 200 mg isavuconazole oral 3 times daily on days 6 and 7 and then once per day on days 8–10. Concomitant esomeprazole administration did not significantly affect the systemic exposure of isavuconazole, as the geometric mean ratios between fed and fasted for AUC and C_{max} were 108% and 105%, respectively.

After 200 mg of caffeine given concomitantly with 200 mg isavuconazole to healthy adult volunteers, the geometric mean ratio in AUC and C_{max} was 104% and 99%, respectively, indicating no significant effect [253].

4.4.8.9 Griseofulvin

The effect of food on the pharmacokinetics of microsized and ultramicrosized griseofulvin was studied in nonfasting volunteers [254]. The results showed similar systemic exposures between the two products when given with food. A study from the early 1960s showed that serum griseofulvin concentrations were higher when given with a high-fat meal, and thus it is recommended that griseofulvin be given with a meal high in fat [255]. A subsequent study confirmed these findings with the microsized and ultramicrosized tablets after a standard breakfast [73].

The manufacturers' dosing recommendations for antifungal agents with regard to food are shown in Table 4.9.

4.4.9 HIV Nucleoside Reverse Transcriptase Inhibitors

4.4.9.1 Didanosine

Didanosine is variably absorbed after oral administration due to its poor solubility at acidic pH, with bioavailability ranging from 25% to 43% [256, 257]. Food alters the absolute bioavailability of didanosine by approximately 50%, most likely due to increased hydrolysis at lower pH and delayed gastric emptying [74]. Acid-catalyzed hydrolysis results in significant degradation of the drug, which was slightly overcome by the introduction of the buffered didanosine formulation [258]. In healthy volunteers, and in subjects infected with HIV, the AUC was equivalent for didanosine administered as the enteric-coated formulation (Videx® EC) relative to a

Table 4.9 Dosing recommendations for the antifungals with regard to food

Antimicrobial	Formulation	Manufacturer recommendations
Terbinafine	Tablets	Take with or without food
Ketoconazole	Tablets	Administration with a meal may decrease absorption
Itraconazole	Capsules	Should be taken with a full meal to ensure maximal absorption
	Oral solution	If possible, do not take with food
Voriconazole	Tablets	Should be taken at least 1 h before or 1 h after a meal
	Powder for oral suspension	Should be taken at least 1 h before or 1 h after a meal
Posaconazole	Powder for oral suspension	Administer with a full meal or liquid nutritional supplement
	Delayed-release tablets	Take with food
Isavuconazole	Capsules	Take with or without food

buffered tablet formulation [259]. The effect of food and timing of meals on the bioavailability of didanosine from encapsulated enteric-coated beads was evaluated in healthy subjects [75]. Concomitant administration with either a high-fat (757 calories) or low-calorie meal (373 calories) decreased the rate of absorption. The overall reduction in AUC was approximately 20–25% when didanosine was administered with food, regardless of the timing with meals. Although this reduction is moderate, it is recommended to administer this formulation on an empty stomach [260].

4.4.9.2 Zidovudine

Zidovudine is fairly well absorbed after oral administration, with an average bioavailability of 60–70% [261]. However, considerable interpatient variability does exist, and the bioavailability can range from 40% to 100% [262]. Several studies have examined the effect of certain types of food on zidovudine absorption. Overall, food consumption (especially high-fat meals) tends to decrease the rate, but not to the extent, of absorption of zidovudine [263]. In a study of 13 patients with acquired immunodeficiency syndrome (AIDS) who were either fasting or taking a standard breakfast, the mean AUC in the fed state was 24% lower than the fasted and demonstrated greater interpatient variability [264]. In a study by Shelton et al. [76], a high-fat breakfast significantly reduced the C_{max} of zidovudine, but did not significantly affect total systemic exposure (AUC).

4.4.9.3 Lamivudine

The administration of lamivudine with a standard breakfast (55% fat, 20% carbohydrates, 13% proteins) significantly increased t_{max} and decreased C_{max}, but had no significant effect on the overall AUC. Administration of a high-fat breakfast

(1000 kcal) did not affect the extent of absorption of lamivudine or zidovudine from the combined tablet, Combivir® [77]. Food slowed the rate of absorption, delaying the t_{max} and decreasing the C_{max} of lamivudine and zidovudine, but these changes were not considered clinically significant. Thus, lamivudine can be taken without regard to meals. Administration with meals, however, may decrease the likelihood of gastrointestinal upset.

4.4.9.4 Stavudine

In patients with HIV given a high-fat (773 kcal, 53% fat) breakfast, the C_{max} was significantly lower and the t_{max} was significantly longer, although the overall $AUC_{0-\infty}$ was not significantly different compared to the fasted state [78]. Given that the overall absorption of stavudine is not affected by food, it can be taken without regard to meals [265].

4.4.9.5 Abacavir

After single doses of abacavir taken with food, the C_{max} was reduced by 26–35% and the AUC by up to 5% [79, 266]. This was not considered clinically significant, and abacavir can be taken without regard to meals. The extent of absorption of Trizivir® tablets (abacavir, lamivudine, and zidovudine) is not affected by the administration of a meal, and this formulation can be given with or without food [267]. Ethanol decreases the elimination of abacavir. Coadministration of ethanol and abacavir resulted in a 41% increase in abacavir AUC and a 26% increase in abacavir $t_{1/2}$.

Abacavir is now co-formulated with dolutegravir and lamivudine in a one-tablet once-daily regimen called Triumeq®. In a study of the effect of food on the pharmacokinetics of this fixed-dose combination regimen, 12 healthy adult subjects received the combination product after fasting and with a high-fat (869 kcal, 53% fat) meal. Coadministration with this meal decreased the abacavir C_{max} by 23% compared to fasting but had no effect on any other pharmacokinetic parameter [268]. These differences were not statistically or clinically significant, so the combination product may also be administered with or without food.

4.4.9.6 Emtricitabine

The systemic exposure of emtricitabine was not affected by the administration of a high-fat meal (1000 kcal), although the C_{max} was reduced by 29% compared to the fasting state [269]. Truvada® tablets, the combination of emtricitabine and tenofovir, may be taken without regard to meals [270].

4.4.9.7 Tenofovir

Following a high-fat meal (700–1000 kcal), the AUC of tenofovir disoproxil fumarate increased by approximately 40%, and the C_{max} was increased approximately 14%. Administration with a light meal does not appear to significantly affect the pharmacokinetics of tenofovir [271, 272]. Thus, tenofovir can be administered with or without food.

A new salt formulation of tenofovir, tenofovir alafenamide, is approved as part of several combination HIV antiretroviral medications discussed above and is also marketed alone for the treatment of chronic hepatitis B virus in adults with compensated liver disease [273]. The alafenamide salt form produces much higher intracellular concentrations compared to the disoproxil fumarate formulation, which in turn reduces systemic exposure and toxicities, particularly nephrotoxicity, and decreases in bone mineral density. The manufacturer recommends that tenofovir alafenamide be taken with food, but there are no data provided in the package insert to support this recommendation.

4.4.10 HIV Non-nucleoside Reverse Transcriptase Inhibitors

Delavirdine is no longer recommended for use in the United States by the Department of Health and Human Services medical practice guidelines for the treatment of HIV/AIDS (https://aidsinfo.nih.gov/contentfiles/lvguidelines/adultandadolescentgl.pdf) and thus has been removed from this chapter. Please refer to the 3rd edition for information regarding drug–food interactions of delavirdine.

4.4.10.1 Efavirenz

When efavirenz capsules were administered with a high-fat meal (894 kcal) or a standard meal (440 kcal), respectively, the AUC was increased by 22% and 17%, and the C_{max} was increased by 39% and 51% compared to fasting. This increase was more pronounced with the administration of a 600 mg efavirenz tablet with a high-fat meal (1000 kcal), leading to a 28% increase in AUC and a 79% increase in C_{max} relative to fasting conditions. In Ugandan adults administered a traditional Ugandan meal containing a moderate amount of fat (19 g), the C_{max} of efavirenz was significantly increased approximately 1.5-fold [81]. There was no difference in the AUC_{0-24} and t_{max} was identical at 3 h. Opening efavirenz capsules and mixing the powder with either applesauce, grape jelly, yogurt, or infant formula did not affect the bioavailability compared to ingesting an intact capsule in a fasted state [82]. To avoid an increase in the frequency of adverse events, it is recommended that efavirenz be administered on an empty stomach, preferably at bedtime [274]. The effect of food on the combination product Atripla® (efavirenz, tenofovir, and emtricitabine) has not been studied; however this formulation should also be administered on an empty stomach, preferably at bedtime to minimize adverse effects of efavirenz [275].

4.4.10.2   Nevirapine

The absorption of nevirapine is not affected by food, and thus the drug can be taken without regard to meals [276].

4.4.10.3   Etravirine

The effect of various food compositions on the pharmacokinetics of etravirine was evaluated in 12 healthy male volunteers [83]. Administration of etravirine in a fasted state reduced the AUC by ~50% when compared to dosing after a standard breakfast or other types of food intake (high-fat, enhanced-fiber, or light breakfast). Therefore, etravirine should be administered after a meal to improve absorption [277].

4.4.10.4   Rilpivirine

Rilpivirine is a potent diarylpyrimidine non-nucleoside reverse transcriptase inhibitor (NNRTI) with a long elimination half-life allowing for once-daily dosing. Rilpivirine is poorly soluble and bioavailability is highest at pH 2 and decreased as pH increases. A phase IIb study demonstrated that a standard breakfast increased the bioavailability by 50% by increasing $AUC_{0\text{-last}}$ and C_{max} by 33% and 41%, respectively, compared to fasting [278]. In the clinical trial phases, patients were instructed to take rilpivirine with a meal. A further phase I, open-label, randomized, four-period crossover study investigated the bioavailability of a 75 mg single dose of rilpivirine administered under fasting conditions, after a normal- (533 kcal, 21 g fat) and high-fat (928 kcal, 56 g fat) breakfast and with a protein-rich (300 kcal, 7.9 g fat) drink to healthy volunteer subjects [84]. Blood samples were collected serially up to 168 h after dosing. Eighteen male and two female volunteers completed the study. Administration after fasting or with a protein-rich drink led to lower mean plasma concentration–time profiles of rilpivirine compared to a normal-fat meal. A normal-fat breakfast increased the $AUC_{0\text{-last}}$ by 43% and 50% compared to fasting and the protein-rich drink. The high-fat breakfast did not significantly change these values from the normal-fat. The protein-rich drink produced the longest lag time, approximately 1–2 h compared to 0.5 for the meals. The impact of food on rilpivirine is analogous to other poorly soluble, lipophilic NNRTIs such as etravirine. The impact of fat-rich versus protein-rich food in this study also argues to the dissolution of rilpivirine under acidic conditions. A subsequent study in HIV-infected Ugandan adults investigated the effect of a low-fat (353 kcal, 11 g fat) and moderate-fat (589 kcal, 19 g fat) meal on the steady-state pharmacokinetics of rilpivirine after administration as a multidrug combination treatment with tenofovir disoproxil fumarate and efavirenz [85]. In this study, the $AUC_{0\text{-}24}$ was significantly decreased by 16% in the fasted state compared to after a moderate-fat meal. Trough concentrations were also significantly lower in the fasted state, while C_{max} was similar between all three groups. These data together indicate that patients should be advised to take rilpivirine with a meal and not solely with a nutritional supplement high in protein.

The effect of food on rilpivirine as part of a three-drug combination regimen of emtricitabine, rilpivirine, and tenofovir disoproxil fumarate (trade name Complera®) has also been evaluated in 24 healthy subjects [80]. The phase I, randomized, open-label, three-period crossover study randomized subjects to treatment regimens consisting of a single dose of emtricitabine–rilpivirine–tenofovir given in the fasting state, with a standard-fat (540 kcal, 21 g fat) meal or a low-fat (390 kcal, 12 g fat) meal. Serial blood samples were obtained up to 192 h post-dose for pharmacokinetic evaluation. As observed in previous studies, C_{max} and $AUC_{0-\infty}$ of rilpivirine were significantly higher following a standard- or low-fat meal when administered as part of the combination drug regimen. In this study, there was a significant food effect on rilpivirine, but the magnitude of that effect was less than observed in studies when rilpivirine was administered alone. The fat content of the meal did not significantly affect the plasma exposure of rilpivirine, so a full or high-fat meal is not required, and this single-tablet regimen may be taken with only a light meal or snack.

4.4.11   HIV Protease Inhibitors

Amprenavir is no longer recommended for use in the United States by the Department of Health and Human Services medical practice guidelines for the treatment of HIV/AIDS (https://aidsinfo.nih.gov/contentfiles/lvguidelines/adultandadolescentgl.pdf) and thus has been removed from this chapter. Please refer to the 3rd edition for information regarding drug–food interactions of amprenavir.

4.4.11.1   Indinavir

The absolute oral bioavailability of indinavir is approximately 65% [279]. Eight healthy volunteers received indinavir with or without a high-fat meal consisting of eggs, toast, butter, bacon, whole milk, and hash browns [86]. The high-fat meal caused a significant reduction in the C_{max} and AUC by 84% and 77%, respectively. A similar study in 12 healthy volunteers investigated the influence of various low-fat meals on the pharmacokinetics of indinavir. In this study the meal consisted of toast, jelly, apple juice, coffee, skim milk, and sugar or cornflakes, sugar, and skim milk. These low-fat meals caused no significant reduction in the C_{max} or AUC. When indinavir is administered every 8 h, it should be taken on an empty stomach (1 h before or 2 h after meals). Alternatively, administration with liquids such as skim milk, juice, coffee, tea, or a low-fat meal should not affect absorption. Indinavir should not be taken with or immediately after a heavy, high-fat meal (>2 g of fat) [276]. The addition of ritonavir, a known inhibitor of CYP3A4, at doses of 100–200 mg twice daily increases the AUC of indinavir by two- to threefold, respectively, and is not affected by the administration of food [90]. This pharmacokinetic interaction is advantageous because it eliminates the indinavir food restrictions and allows for twice daily dosing. The manufacturer reports a decrease in the indinavir AUC by 26% ± 18% after a single 400 mg dose was administered to healthy volunteers with 8 ounces of single-strength grapefruit juice [280]. This is in contrast to two other studies where the administration of grapefruit juice and Seville orange juice had no effect on the bioavailability of indinavir 800 mg doses in HIV-infected patients and healthy volunteers [281, 282]. These results are consistent with findings that although indinavir undergoes extensive first-pass metabolism, intestinal metabolism accounts for less than 10%.

4.4.11.2 Saquinavir

Saquinavir hard-gel capsule is poorly absorbed due to high first-pass metabolism and incomplete absorption, with an oral bioavailability of only 4% following a high-fat breakfast [283]. The mean AUC of saquinavir after a 600 mg dose in healthy volunteers was increased from 24 ng*h/mL in the fasting state to 161 ng*h/mL when administered after a high-fat breakfast (48 g protein, 60 g carbohydrate, 57 g fat; 1006 kcal). Following administration of a higher calorie meal (943 kcal, 54 g fat), the C_{max} and AUC were roughly twice of that observed after administration of a lower fat meal (355 kcal, 8 g fat). Grapefruit juice also increased the bioavailability by approximately twofold in eight healthy volunteers, which the authors' attributed to inhibition of intestinal CYP3A4 [284]. In December 2004, a new tablet formulation of saquinavir was approved by the FDA. A similar bioavailability was achieved when the tablets (2 × 500 mg) and capsules (5 × 200 mg) were administered with ritonavir under fed conditions. Thus, it is recommended that the hard-gel capsule and tablet formulations of saquinavir be administered with ritonavir and taken within 2 h after a meal [285]. The mechanism for the profound increase in bioavailability of saquinavir when administered with food is not due to changes in gastric pH [87] but likely due to more rapid disintegration of the capsules and prolonged gastric emptying time in the fed state [286]. The softgel capsule formulation of saquinavir is also significantly affected by high-fat meals [88].

4.4.11.3 Nelfinavir

The absolute oral bioavailability of nelfinavir has not been studied in humans, but increased systemic concentrations were noted when the drug was taken concurrently with food [89]. Nelfinavir AUC values in six fasted volunteers were 27–59% of those achieved in fed volunteers after administration of single doses of 400 and 800 mg [287]. Thus, it is recommended that nelfinavir be administered with food.

4.4.11.4 Ritonavir

The administration of ritonavir with food appears to increase the absorption of the capsule while decreasing the absorption of the liquid formulation [91, 276]. However, neither change is considered significant nor therefore it is recommended that ritonavir be given without regard to meals. However, it is most commonly administered with meals to improve gastrointestinal tolerability.

4.4.11.5 Fosamprenavir

To reduce the pill burden associated with amprenavir, a phosphate ester prodrug, fosamprenavir, was FDA approved in 2003. The administration of a high-fat meal had no influence on the AUC of fosamprenavir tablets compared to the fasting state [92]. Contrarily, administration of fosamprenavir oral suspension with a standard high-fat meal reduced the amprenavir C_{max} by 46% and the AUC by 28% compared to the fasted state. The manufacturer recommends that the suspension be administered without food in adults and with food in pediatric patients [288].

4.4.11.6 Lopinavir

The systemic availability of the combination product lopinavir/ritonavir capsules or liquid was increased with the administration of a moderate-to-high-fat meal [94, 289]. A tablet formulation of lopinavir/ritonavir was approved in December 2005. A high-fat (840 kcal, 36 g fat) breakfast reduced the C_{max} and AUC_{0-12} of lopinavir tablets by 14% in HIV-infected adults in Uganda [93]. The tablet formulation can be administered without regard to meals, does not have to be refrigerated, and reduces the pill burden from three capsules twice daily to two tablets twice daily [290].

4.4.11.7 Atazanavir

There is a clinically significant increase in the absorption of atazanavir capsules when administered with food. After a single 400 mg dose, the AUC of atazanavir was increased by 35% with a light meal and by 70% with a high-fat meal [291]. The manufacturer recommends that atazanavir capsules be administered with food.

Atazanavir is also co-formulated with the pharmacokinetic enhancer cobicistat and marketed as a fixed-dose combination regimen under the trade name Evotaz®. A randomized, open-label, single-dose, five-period crossover study of 62 healthy subjects examined the effect of food on the pharmacokinetics of atazanavir and cobicistat when given as a fixed-dose combination with a light (336 kcal, 5.1 g fat) or high-fat (1038 kcal, 59 g fat) meal or in the fasted state [95]. The light meal increased the C_{max}, AUC, and C_{24} of atazanavir by approximately 42%, 28%, and 35%, respectively, and the C_{max} and AUC of cobicistat by 30% and 24%, respectively.

The high-fat meal did not significantly affect plasma exposure of either agent compared to the fasting state. As recommended for atazanavir capsules alone, the fixed-dose combination regimen of atazanavir–cobicistat should be taken with food.

4.4.11.8 Tipranavir

Tipranavir is a P-gp substrate, a weak P-gp inhibitor, and a potent P-gp inducer. As a result, tipranavir must be administered with ritonavir to achieve effective tipranavir plasma concentrations. When tipranavir capsules or solution is coadministered with ritonavir capsules, food has no clinically significant effect on the C_{max} or AUC compared to the fasted state. Tipranavir capsules or solution taken with ritonavir capsules or solution can be taken without regard to meals, while tipranavir coadministered with ritonavir tablets must be taken with meals [292].

4.4.11.9 Darunavir

The effect of various meal types on the pharmacokinetics of darunavir in combination with ritonavir was evaluated in 24 healthy volunteers [96]. Compared to the fasted state, the C_{max} and AUC of darunavir were ~30% higher when administered with food. Darunavir exposure was comparable regardless of the type of meal administered (standard breakfast, high-fat breakfast, nutritional protein drink, or croissant with coffee).

4.4.11.10 Cobicistat

Cobicistat is a structural analogue of ritonavir and a potent CYP3A4 inhibitor and also inhibits CYP2D6, P-gp, and several other transporters. It is co-formulated into fixed-dosed combinations with several HIV antiretrovirals as a pharmacokinetic booster to reduce dosing frequency. A phase I study evaluated the effect of food on the pharmacokinetics of the fixed-dose combination of darunavir and cobicistat [293]. This randomized, open-label, three-period crossover study in which a single dose of the single agents darunavir and cobicistat were given with or without a standard (533 kcal, 21 g fat) breakfast and a single dose of the fixed-dose combination was given with and without a standard or high-fat (928 kcal, 56 g fat) breakfast. When darunavir and cobicistat were given alone, the C_{max} and $AUC_{0-\infty}$ were within the 80–125% allowable error range between fasted and fed states. When the combination was administered with food, the high-fat meal significantly increased darunavir C_{max} and $AUC_{0-\infty}$ by 2.27 and 1.7-fold, respectively. Neither meal significantly affected the pharmacokinetics of cobicistat (geometric least-squares mean ratio of C_{max} and $AUC_{0-\infty}$ 1.06 and 1.04). These results demonstrate that the fixed-dose combination regimen of darunavir–cobicistat (marketed under the trade name Prezcobix®) should be taken with food.

4.4.12 HIV Integrase Strand Transfer Inhibitors

4.4.12.1 Raltegravir

The effect of a low-, moderate-, and high-fat meal on steady-state raltegravir pharmacokinetics was assessed in 20 healthy volunteers [97]. When administered with a low-fat meal (~300 kcal, 7% fat), the AUC and C_{max} were reduced by ~50%. A moderate-fat meal (~600 kcal, 31% fat) had a minimal effect on raltegravir absorption (AUC and C_{max} increased by 13% and 5%, respectively), while a high-fat meal (~825 kcal, 57% fat) increased AUC and C_{max} by almost twofold. Because of the considerable intersubject variability observed with all meal types, the modest magnitude of the varying effects of food on absorption, and the fact that raltegravir was administered without regard to meals in the pivotal safety and efficacy studies in HIV-1-infected patients, the authors conclude that the pharmacokinetic differences observed with various meals are not of clinical importance and support the current recommendation that raltegravir can be administered with or without food.

The effect of antacids on the pharmacokinetics of raltegravir has also been evaluated in 12 healthy volunteer subjects [294]. This randomized, crossover study utilized a single 400 mg dose of raltegravir alone or with 30 mL of Maalox® Plus Extra Strength Liquid antacid after an 8 h overnight fast. Administration with the antacid did not significantly affect AUC_{0-12} (measured as raltegravir is dosed twice daily) or C_{max} but did significantly decrease t_{max} from 3 to 1 h and decreased the 12 h plasma concentration from 37 ng/mL to 13 ng/mL. Similar to previous studies, the intersubject variability was high especially in the fasted state which prohibited calculation of all pharmacokinetic parameters (half-life, volume, clearance, and $AUC_{0-\infty}$). The decreased t_{max} in this study purported to be due to the fact that raltegravir is preferentially absorbed at higher gastric pH values. Given the decrease in C12 concentrations, patients should be cautioned on the use of antacids concomitantly with raltegravir.

4.4.12.2 Dolutegravir

Dolutegravir is a second-generation HIV integrase strand transfer inhibitor that maintains activity against raltegravir-resistant strains [295]. A randomized, open-label, crossover study of healthy adult subjects evaluated the effect of food on the pharmacokinetics of dolutegravir [98]. In the first phase of the study, 24 subjects received a single 50 mg dose of dolutegravir after an overnight fast of at least 6 h. Eighteen of these 24 subjects were then enrolled in phase two of the study in which they received a single 50 mg dose on three separate occasions with either a low-fat (300 kcal, 7% fat), moderate-fat (600 kcal, 30% fat), or high-fat (870 kcal 53% fat) meal with a 7 day washout period in between. Serial blood samples were collected up to 48 h post-dose. Coadministration of dolutegravir with food increased plasma exposures and slowed the rate of absorption. These increases were modest and were proportional to the fat content of the meal. The $AUC_{0-\infty}$ increased by 33%, 41%, and 66% when dolutegravir was given with a low-, moderate-, or high-fat meal, respectively, compared to the fasting state, while C_{max} increased 46%, 52%, and 67%. The t_{max} was prolonged from 2.1 h to 3, 4, and 5 h following a low-, moderate-, and high-fat meal, respectively. The ratios of geometric least-squares mean values for these parameters ranged from 1.33 to 1.67 and were considered statistically and clinically insignificant. Dolutegravir can be dosed without regard to meals or their fat content.

The mechanism of action of the integrase inhibitors involves binding to magnesium in the active site of the integrase enzyme, subsequently making these agents susceptible to chelation interactions with metal cations. The pharmacokinetics of dolutegravir when coadministered with mineral supplements has also been evaluated in healthy adult subjects [296]. This study evaluated both the effect of coadministration with iron and calcium supplements and strategies for mitigating the effect of these interactions but timing of administration and administration with a meal to offset the interaction. The study was an open-label, randomized, four-period crossover study evaluating the effect of calcium carbonate and ferrous fumarate on the pharmacokinetics of a single dose of 50 mg of dolutegravir. The four study periods consisted of dolutegravir administered under fasting conditions, followed by coadministration with either calcium carbonate or ferrous fumarate, followed by coadministration with calcium carbonate or ferrous fumarate and a moderate-fat (30%) meal, followed lastly by dolutegravir administered under fasting conditions 2 h prior to either calcium carbonate or ferrous fumarate. A washout period of 7 days separated each dosing regimen. Coadministration with either calcium carbonate or ferrous fumarate in the fasting state resulted in significant reductions in $AUC_{0-\infty}$, C_{max}, and C_{24} concentrations. The ratios of geometric least-squares means for these three parameters with calcium carbonate and ferrous fumarate were 0.61, 0.63, and 0.61 and 0.46, 0.43, and 0.44, respectively. Adding a moderate-fat meal to this regimen counteracted the interaction and normalized plasma exposures compared to the fasting state alone. Administering dolutegravir 2 h prior to either calcium carbonate or ferrous fumarate also negated this interaction. These results indicate that coadministration of dolutegravir with calcium and/or iron supplements under fasted conditions is not recommended, although separation by at least 2 h before or 6 h after or addition of a moderate-fat meal with the supplement is advisable.

Another study evaluated the effect of a multivitamin (One A Day®), an antacid (Maalox® Advanced Maximum Strength Liquid), and a proton-pump inhibitor (omeprazole 40 mg) on the pharmacokinetics of dolutegravir in 28 healthy adult subjects in and open-label, randomized, four-period crossover study [297]. Subjects received dolutegravir alone, with a single multivitamin tablet, with a single 20 mL dose of antacid, 2 h prior to a 20 mL dose of antacid, and 2 h the dose on day 5 after 5 days of daily dosing of omeprazole. Coadministration with a multivitamin had a modest effect on the pharmacokinetic parameters of dolutegravir, while simultaneous administration with an antacid significantly reduced $AUC_{0-\infty}$, C_{max}, and C_{24}. The geometric least-squares mean ratios compared to the fasted state were 0.26, 0.28, and 0.26 for $AUC_{0-\infty}$, C_{max}, and C_{24}, respectively. These effects were nullified when dolutegravir was given 2 h prior to antacid. Administration with omeprazole had no effect on the plasma exposure of dolutegravir. The findings from this study indicate that dolutegravir may be taken with a multivitamin, a proton-pump inhibitor, and at least 2 h before or 6 h after, but not with, an antacid with high metal cation content.

Dolutegravir is also co-formulated with abacavir and lamivudine in a one tablet once-daily regimen called Triumeq®. In a study of the effect of food on the pharmacokinetics of this fixed-dose combination regimen, 12 healthy adult subjects received the combination product after fasting and with a high-fat (869 kcal, 53% fat) meal. Coadministration with this meal increased the dolutegravir AUC 48% and C_{max} by 37% compared to fasting [268]. These differences were not statistically or clinically significant, so the combination product may also be administered with or without food.

4.4.12.3 Elvitegravir

Elvitegravir is a novel low molecular weight integrase strand transfer inhibitor available as part of a four-drug, fixed-dose combination regimen of elvitegravir–cobicistat–emtricitabine–tenofovir disoproxil fumarate (Stribild®) and elvitegravir–cobicistat–emtricitabine–tenofovir alafenamide (Genvoya®).

In a single-dose study of elvitegravir alone without the pharmacokinetic enhancer cobicistat, administration with food increases C_{max} and $AUC_{0-\infty}$ 3.3- and 2.7-fold, respectively, compared to the fasted state [298].

The C_{max} and $AUC_{0-\infty}$ of a single dose of the elvitegravir-containing combination product Stribild® and Genvoya® was increased by 22% and 34%, respectively, after a low-fat (373 kcal, 20% fat) and 56% and 87%, respectively, after a high-fat (800 kcal, 50% fat) meal and therefore should be administered with food [299, 300]. Absorption of elvitegravir is decreased when coadministered with antacids but not with histamine H_2 receptor blockers or proton-pump inhibitors, similar to raltegravir and dolutegravir.

A phase I randomized, open-label, single-dose, three-way crossover study evaluated the effects of a protein-rich drink or a standard meal on the pharmacokinetics of elvitegravir when administered as part of a four-drug combination regimen of elvitegravir–cobicistat–emtricitabine–tenofovir disoproxil famurate (Stribild®) in 12 healthy Japanese male subjects [99]. Subjects received a single dose of the combination regimen after a 10 h overnight fast or with a standard breakfast (413 kcal, 11.4 g protein, 9.6 g fat) or protein-rich nutritional drink (Ensure® 250 kcal, 8.8 g protein, 8.8 g fat), and serial blood samples were collected up to 48 h post-dose. Administration of elvitegravir in the fasted state resulted in significantly lower plasma exposure relative to either fed state, which did not significantly differ from each other. The C_{max} and $AUC_{0-\infty}$ in the fasted state were 55% and 50% lower, respectively, than following a standard breakfast. These results confirm that elvitegravir given as a part of the combination product Stribild® should be administered with food without specific restrictions to the type of food ingested.

In a real-world evaluation of the fixed-dose combination product elvitegravir–cobicistat–emtricitabine–tenofovir disoproxil fumarate (Stribild®), plasma trough concentrations of elvitegravir were measured in 75 HIV-infected adult patients taking Stribild® alone or as part of their antiretroviral therapy regimen for at least 30 days prior to the study [301]. Twelve of the 75 patients had elvitegravir concentrations below the lower limit of detection (25 ng/mL), and all 12 of these patients reported taking Stribild® under fasted conditions. Importantly, these concentrations are also below the free IC90 pharmacodynamic values for elvitegravir of 50 ng/mL. Trough concentrations in the remaining 63 patients ranged from 50 to 2311 ng/mL, although the exact timing of medication in relation to meals and the composition of the meals were not reported.

4.4.13   HIV Entry Inhibitors

In a phase I food-effect study, a high-fat meal food reduced the exposure of maraviroc by ~33%, primarily by reduction of C_{max}. The effect of food was also assessed in a 10-day phase IIa study to determine if these effects translated into an effect on antiviral activity. The results of this study demonstrated that when administered at 150 mg twice daily, food reduced the C_{max} and AUC of maraviroc by ~60% and 50%, respectively. However, there was little effect of food on the short-term antiviral activity (change from baseline in viral load) of maraviroc. Therefore, there were no food restrictions in the phase III safety and efficacy studies [302].

The manufacturers' dosing recommendations for HIV antiretrovirals with regard to food are shown in Table 4.10.

4.4.14   Hepatitis B Antivirals

4.4.14.1   Adefovir

The effect of food on the pharmacokinetics of adefovir was evaluated in ten healthy male Chinese subjects in a randomized, open-label, single-dose, two-period crossover study [100]. Subjects received 10 mg of adefovir after a 10 h overnight fast and with a 1000 kcal high-fat meal. Serial blood samples were collected up to 24 h post-dose for pharmacokinetic assessment. Food had no significant effect on the pharmacokinetic parameters of interest in this study but delayed t_{max} from 1 h to 2.75 h. The manufacturer recommends that adefovir may be taken without regard to food [303].

4.4.14.2   Entecavir

According to the manufacturer, the administration of 0.5 mg of entecavir with a light (379 kcal, 8.2 g fat) or high-fat (945 kcal, 54.6 g fat) meal decreased the C_{max} and AUC approximately 44–46% and 18–20%, respectively, and doubled t_{max} from 0.75 h to 1–1.5 h [304]. Given these data the recommendation is that entecavir be taken on an empty stomach, at least 2 h before or 2 h after a meal.

Table 4.10 Dosing recommendations for the HIV antivirals with regard to food

Antimicrobial	Formulation	Manufacturer recommendations
Nucleoside reverse transcriptase inhibitors		
Didanosine	Delayed-release capsule, enteric-coated beadlet	Take on an empty stomach
Zidovudine	Tablets, capsules, syrup	May be taken with or without food
Lamivudine	Tablets	May be taken without regard to meals
Stavudine	Capsules, oral solution	May be taken without regard to meals
Abacavir	Tablets	Can be taken with or without food
Emtricitabine	Capsules, oral solution	May be taken without regard to meals
Tenofovir disoproxil fumarate	Tablets, powder	May be administered with or without food
Tenofovir alafenamide	Tablets	Take with food
Non-nucleoside reverse transcriptase inhibitors		
Efavirenz	Capsules, tablets	Should be taken on an empty stomach, preferably at bedtime
Nevirapine	Tablets and oral suspension	Can be given without regard to meals
Etravirine	Tablets	Administer after a meal to improve absorption
Rilpivirine	Tablets	Take with a light meal or snack not high in protein
Protease inhibitors		
Indinavir	Capsules	For optimal absorption, should be administered without food but with water 1 h before or 2 h after a meal
Saquinavir	Capsules	Take within 2 h after a full meal
Nelfinavir	Tablets, powder	Should be taken with a meal
Ritonavir	Capsules, oral solution	Take with meals if possible
Fosamprenavir	Tablets	Can be taken without regard to meals
	Oral suspension	Adults should take the suspension without food children should take the suspension with food
Lopinavir	Capsules, oral solution	Should be taken with food
Lopinavir	Tablets	Can be taken without regard to food
Atazanavir	Capsules	Should be taken with food
Tipranavir	Capsules, solution	May be taken without regard to meals if coadministered with ritonavir capsules or solution. If given with ritonavir tablets, then take with food
Darunavir	Tablets	Should be taken with food

(continued)

Table 4.10 (continued)

Antimicrobial	Formulation	Manufacturer recommendations
Cobicistat	Tablets	Should be taken with food
Integrase strand transfer inhibitors		
Raltegravir	Tablets	Can be given without regard to meals
Dolutegravir	Tablets	Can be given without regard to meals
Elvitegravir	Tablets	Take with food
Entry inhibitors		
Maraviroc	Tablets	Take with or without food

4.4.14.3 Telbivudine

A phase I, randomized, open-label, single-dose, crossover study was completed to assess the impact of food on the pharmacokinetics of telbivudine in 24 healthy adult subjects [101]. Subjects were given a single 600 mg dose of telbivudine after a 10 h overnight fast or with a high-calorie, high-fat (1000 kcal, 600 kcal fat) meal. The plasma concentration–time profile of telbivudine in the fasted and fed state was virtually superimposable, with no significant differences in any pharmacokinetic parameter. Geometric least-squares mean ratios for C_{max} and $AUC_{0-\infty}$ between fasted and fed were 103.8% and 106.4%, indicating that a high-calorie, high-fat meal does not alter the pharmacokinetics of telbivudine. Although only a high-calorie, high-fat meal has been studied, the manufacturer recommends that telbivudine may be taken with or without food [305].

4.4.15 Hepatitis C (HCV) Antivirals

4.4.15.1 Ribavirin

The bioavailability of a single dose of oral ribavirin was increased when administered with a high-fat meal. Three studies have evaluated the effect of a high-fat meal on the pharmacokinetics of a single oral dose of ribavirin. These studies demonstrated that C_{max} increased between 16 and 70% and AUC between 4% and 70% [306]. It is recommended that ribavirin be taken with food.

Ribavirin is known to be actively transported across the intestinal mucosa by the human concentrative nucleoside transporter 2 (hCNT2) and is likely responsible for the saturable uptake of ribavirin and subsequent lack of pharmacokinetic linearity above oral doses of 800 mg. Purine nucleosides strongly inhibit the uptake of ribavirin by the hCNT2 in vitro, and therefore dietary purines may reduce the absorption of ribavirin in vivo. A phase I, randomized, crossover study was completed in order to evaluate this interaction in 20 healthy adult subjects administered either a high (192.1 mg) or low (7.56 mg) purine meal followed by a single dose of ribavirin [306]. The caloric value and fat content of the two meals were roughly identical, with tuna, ham, and soy milk added to increase purine content of the high-purine meal. In this analysis, administration of ribavirin with a low-purine meal resulted in significantly higher C_{max} and $AUC_{0-\infty}$ values than a higher purine meal, with geometric mean ratios of 1.36 and 1.39, respectively. These results suggest that high dietary purine content competes with absorption and transportation of ribavirin across the intestinal mucosa, leading to decreased plasma exposure. This is important to consider especially in Western countries as the purine content for a typical meal exceeds 300 mg.

4.4.15.2 Boceprevir

Boceprevir was one of the first two novel direct-acting antivirals approved for the treatment of chronic HCV in 2011 and the first treatment option since the pegylation of interferon in 2001 [307]. It is a NS3 serine protease inhibitor that when given after a meal, the AUC increased by up to 65% compared to the fasting state [308]. The type of meal, caloric content, and timing did not make a significant difference. It is advised that boceprevir be taken with a meal or light snack.

4.4.15.3 Telaprevir

Telaprevir is an NS3/4a protease inhibitor of HCV similar to boceprevir [309]. The manufacturer reports that absorption of telaprevir is reduced on fasting and with a low-fat meal. It is also a substrate of P-gp. In healthy volunteers administered a single dose of 750 mg of telaprevir, a significant proportional relationship was observed between systemic exposure of telaprevir and fat content of co-ingested food [102]. A high-fat meal increased exposure by 20% compared to a standard meal while fasting, a low-calorie high-protein meal, and a low-calorie low-protein meal which each decreased exposure by 73%, 26%, and 39%, respectively. Therefore, telaprevir should be given with food with adequate fat content.

In a phase I, open-label, randomized, five-way crossover study in 28 healthy subjects, the five study phases consisted of a single oral dose of 750 mg of telaprevir given under fasting conditions or with a standard breakfast (533 kcal, 21 g fat); high-calorie, high-fat breakfast (928 kcal, 56 g fat); low-calorie, high-protein breakfast (260 kcal, 9 g fat); and a low-calorie, low-fat breakfast (249 kcal, 3.6 g fat). Blood samples for pharmacokinetic analysis were obtained serially up to 24 h post-dose. The high-calorie, high-fat breakfast resulted in the highest C_{max} and AUC values and the longest t_{max}. The geometric mean values of C_{max} and $AUC_{0-\infty}$ were significantly decreased by 83% and 73%, respectively, when telaprevir was given under fasted conditions compared to a standard breakfast. The low-calorie, high-protein and low-calorie, low-fat breakfasts also decreased exposure by 25–26% and 38–39%, respectively, compared to a standard breakfast indicating that telaprevir should be administered with food, particularly food not low in fat [310].

4.4.15.4 Simeprevir

Ingestion of a high-fat, high-calorie (928 kcal) or standard (533 kcal) breakfast increased the AUC of simeprevir by 61% and 69%, respectively, and thus the drug should be taken with food [311].

4.4.15.5 Ledipasvir and Sofosbuvir

Ledipasvir is an HCV NS5A inhibitor, while sofosbuvir inhibits HCV NS5B polymerase. These two agents are combined in a fixed-dose combination tablet marked under the trade name Harvoni® for the treatment of patients with chronic HCV. A phase I study in healthy subjects demonstrated that a moderate-fat (600 kcal, 25–30% fat) or high-fat, high-calorie (1000 kcal, 50% fat) meal did not significantly alter the C_{max}, $AUC_{0-\infty}$, or t_{max} of ledipasvir–sofosbuvir [103]. A post hoc analysis of the phase III clinical trial data was performed to evaluate the effect of food on the pharmacokinetics and clinical outcomes of ledipasvir–sofosbuvir and revealed no significant effects [103].

Ledipasvir demonstrates pH-dependent solubility in vitro and therefore was evaluated in two phase I studies examining the effects of coadministration with a histamine H_2-receptor antagonist (famotidine 40 mg) and a proton-pump inhibitor (omeprazole 20 mg) [103]. Administration of a single dose of the combination product ledipasvir–sofosbuvir with famotidine or omeprazole and food did not significantly alter the AUC or C_{max} of either agent. Ledipasvir–sofosbuvir may be administered without regard to meals or timing of acid-reducing agents.

4.4.15.6 Paritaprevir, Ombitasvir, Dasabuvir, and Ritonavir

Paritaprevir is a HCV NS3/4A protease inhibitor. Ombitasvir is an HCV NS5A inhibitor. Dasabuvir is an HCV non-nucleoside NS5B palm polymerase inhibitor. Ritonavir is a CYP3A4 inhibitor and pharmacokinetic booster. Paritaprevir–ombitasvir–dasabuvir–ritonavir is a four-drug, fixed-dose combination regimen indicated for the treatment of chronic HCV and marketed under the trade name Viekira Pak®. The manufacturer's prescribing information indicates that the regimen should be taken with a meal without regard to its specific fat or caloric content [312]. The geometric least-squares mean ratio of the AUC after a moderate-fat (600 kcal, 20–30% fat) or a high-fat (900 kcal, 60% fat) meal compared to the fasting state ranged from 1.22 to 1.82. A moderate-fat meal had a more pronounced effect of the pharmacokinetics of each individual agent than did a high-fat meal, and the effect was most significant on paritaprevir (ratio 3.11). Omeprazole did not significantly affect C_{max} or AUC of any agent, suggesting a lack of pH effect on absorption.

The effect of food on dasabuvir has been studied alone and demonstrated that the AUC and C_{max} were 22–42% higher after a high-fat (850 kcal, 59% fat) meal and 30–53% higher after a moderate-fat (612 kcal, 21% fat) meal compared to fasting [313]. Therefore, dasabuvir should be administered with a meal without regard to the fat content.

4.4.15.7 Velpatasvir

Velpatasvir is the first HCV NS5a inhibitor with pangenotypic activity. Preclinical studies indicate that the drug demonstrates high aqueous solubility under acidic conditions and in the presence of intestinal bile salts [104]. The food-effect portion of a phase I study evaluated a single dose of 100 mg in 72 healthy adult subjects of velpatasvir under fasted conditions, after a light meal (400 kcal, 30% fat) and after a high-calorie, high-fat meal (800 kcal, 50% fat). Blood samples were collected for pharmacokinetic analysis up to 96 h post-dose. Administration of velpatasvir with food slightly prolonged median t_{max} from 2.5 h to 3.25 and 3.5 h following the light and high-calorie meals, respectively. The $AUC_{0-\infty}$ and C_{max} were increased 25% and 35%, respectively, after a light meal but were decreased 14% and 25% after the high-calorie meal. These changes were consistent with the physiochemical properties of velpatasvir and are not considered to be of clinical relevance.

4.4.15.8 Daclatasvir

A high-fat meal reduced the C_{max} and AUC of daclatasvir by 28% and 23%, respectively. These reductions are not considered clinically significant, and the drug can be administered without regard to meals.

4.4.15.9 Elbasvir and Grazoprevir

Elbasvir is a hepatitis C virus (HCV) nonstructural (NS) 5a inhibitor, and grazoprevir is an NS3/4A HCV protease inhibitor. The fixed-dose combination product marketed as Zepatier® is indicated for the treatment of chronic HCV genotype 1 or 4 infection in adults [314]. It is the first HCV antiviral FDA approved that may be administered to patients with renal dysfunction. The absolute F of elbasvir and grazoprevir is low at 32% and 27%, respectively. According to the manufacturer, the administration of a single dose of elbasvir 50 mg and grazoprevir 100 mg with a high-fat (900 kcal, 500 from fat) meal to healthy subjects reduced the $AUC_{0-\infty}$ and C_{max} approximately 11% and 15%, respectively. Their decreases were not considered clinically relevant, and therefore the drug may be taken without regard to meals. If elbasvir–grazoprevir is administered with ribavirin, then the recommendations for giving ribavirin with food apply.

The manufacturers' dosing recommendations for hepatitis antivirals with regard to food are shown in Table 4.11.

4.4.16   Other Antivirals

The manufacturing of oral ganciclovir capsules was discontinued in 2013. Please refer to previous editions of the book for information regarding drug–food interactions of oral ganciclovir.

Table 4.11 Dosing recommendations for the hepatitis antivirals with regard to food

Antimicrobial	Formulation	Manufacturer recommendations
Hepatitis B antivirals		
Adefovir	Tablets	May be taken without regard to food
Entecavir	Tablets, oral solution	Take on an empty stomach, at least 2 h before or 2 h after a meal
Telbivudine	Film-coated tablet	May be taken with or without food
Hepatitis C antivirals		
Ribavirin	Tablets	Take with food
Boceprevir	Capsules	Take with a meal or light snack
Telaprevir	Film-coated tablets	Administer with food not low in fat
Simeprevir	Capsules	Take with food
Ledipasvir and sofosbuvir	Tablets	May be administered without regard to meals
Paritaprevir, ombitasvir, dasabuvir, and ritonavir	Tablets	Take with a meal
Velpatasvir and sofosbuvir	Tablets	Take with or without food
Daclatasvir	Tablets	Can be administered without regard to meals
Elbasvir and grazoprevir	Tablets	May be taken without regard to meals

4.4.16.1   Amantadine

Amantadine can be taken without regard to meals [315].

4.4.16.2   Rimantadine

The manufacturer of rimantadine has no specific recommendations for administration with food in the prescribing information [316]. In a study of 12 healthy volunteer subjects, the pharmacokinetics of a single 100 mg dose of rimantadine were not significantly affected by coadministration with a standard breakfast [105].

4.4.16.3   Acyclovir and Valacyclovir

Valacyclovir and the prototype, acyclovir, can be given without regard to meals [317, 318].

4.4.16.4   Valganciclovir

Because of the low bioavailability of oral ganciclovir, a prodrug has been developed, valganciclovir. The absolute bioavailability of oral valganciclovir is approximately tenfold higher than with oral ganciclovir [319, 320]. Compared to the fasted state, the administration of valganciclovir with a standard breakfast increased the AUC by 23–57% depending on the dose administered [106]. Valganciclovir should be taken with food.

Table 4.12 Dosing recommendations for the other antivirals with regard to food

Antimicrobial	Formulation	Manufacturer recommendations
Amantadine	Tablets, syrup	May be taken without regard to meals
Acyclovir	Capsules, tablets, and suspension	Can be given without regard to meals
Valacyclovir	Caplets	Can be given without regard to meals
Valganciclovir	Tablets, oral solution	Should be administered with food
Famciclovir	Tablets	May be taken with or without food

4.4.16.5 Famciclovir

The effect of food was evaluated in two separate studies involving healthy volunteers given 250 or 500 mg of famciclovir [107]. Administration with food decreased the C_{max} by approximately 53% and prolonged the t_{max} by approximately 2 h. However, the AUC was unchanged in the fed-versus-fasting group, and the authors hypothesized that famciclovir could be given without regard to meals.

The manufacturers' dosing recommendations for other antivirals with regard to food are shown in Table 4.12.

4.5   Antimicrobials and Disulfiram-Like Reactions

The drug disulfiram is a therapeutic option in the treatment of alcoholism that acts to deter further ingestion of alcohol [321]. Disulfiram is a remarkably effective agent for inhibiting aldehyde dehydrogenase. By the same mechanism, other compounds have been linked with causing a disulfiram-like reaction, including antimicrobials. Cephalosporins, chloramphenicol, metronidazole, and other antibiotics have been associated with causing a disulfiram-like reaction. In general, these reactions are rare and spontaneously occurring [322]. Although all patients should be counseled and warned of this potential interaction, it appears that patients who chronically consume large amounts of alcohol may be at higher risk of developing these reactions, due to greater accumulation of acetaldehyde. The likelihood of a reaction exists while the drug is still present in the body, and reactions have occurred with minimal amounts of alcohol up to a day after the last dose of an antibiotic [323]. Thus, in general it is recommended that patients abstain from alcohol during and for 2–3 days after therapy with any agents implicated in causing a disulfiram-like reaction.

4.5.1   Nitroimidazoles

Disulfiram-like reactions and a decreased desire to consume alcoholic beverages have been described with metronidazole [324, 325]. Although rare, patients should still be informed about the possible disulfiram-like reaction when metronidazole is combined with alcohol. The effect of alcohol and disulfiram was not specifically studied with tinidazole; however, since adverse reactions have been reported with metronidazole, the manufacturer recommends that patients avoid alcoholic beverages and preparations containing alcohol during therapy and for 3 days afterward. Likewise, tinidazole should not be administered to patients who have taken disulfiram within the last 2 weeks [207].

4.5.2 Cephalosporins

The majority of case reports and research involving disulfiram-like reactions and antimicrobials have focused on the cephalosporins and other beta-lactams. Anecdotal reports have described a disulfiram reaction with cefmenoxime, cefotetan, cefoperazone, cefamandole, and moxalactam after the ingestion of an alcoholic beverage. In general, cephalosporins that have been implicated in causing a disulfiram-like reaction have in common a methyl-tetrazolethiol (MTT) side chain [326–328]. A hypothesis for the mechanism of this effect is that the MTT side chain becomes liberated from the rest of the cephalosporin molecule in vivo and is oxidized to a molecule that is structurally similar to disulfiram [329]. Thus, it appears that cephalosporins that contain the MTT side chain are at higher risk of precipitating a disulfiram-like reaction. Most case reports have involved patients receiving moxalactam, cefoperazone, and cefamandole; however, all cephalosporins with this side chain are likely to provide an increased risk [330]. All patients receiving these medications should be advised of the possibility of a disulfiram-like reaction. Chronic abusers of alcohol appear to be at the most risk of displaying a disulfiram-like reaction to these antibiotics, and an alternative agent may be prudent unless the patient can abstain from alcohol during therapy.

4.5.3 Other Antibiotics

Isolated case reports have described disulfiram-like reactions with trimethoprim–sulfamethoxazole, chloramphenicol, griseofulvin, or furazolidone when combined with alcohol [331, 332]. Although most of these reports hypothesized that the reaction was secondary to an accumulation of acetaldehyde, the exact mechanism is unknown.

4.5.4 Ritonavir Oral Solution

Ritonavir oral solution contains alcohol, and thus a potential interaction is possible when the solution is combined with disulfiram or anti-infectives associated with a disulfiram-like reaction [333]. It is advisable to avoid coadministration of disulfiram with ritonavir solution and to be aware of the potential interaction when ritonavir oral solution is co-prescribed with metronidazole or cephalosporins containing the MTT side chain.

References

1. Welling PG, Tse FLS (1982) The influence of food on the absorption of antimicrobial agents. J Antimicrob Chemother 9:7–27
2. Yamreudeewong W, Henann NE, Fazio A, Lower DL, Cassidy TG (1995) Drug-food interactions in clinical practice. J Fam Pract 40:376–384
3. Singh BN (1999) Effects of food on clinical pharmacokinetics. Clin Pharmacokinet 37:213–255
4. Krishnaswamy K (1989) Drug metabolism and pharmacokinetics in malnourished children. Clin Pharmacokinet 17:68–88
5. Lown KS, Bailey DG, Fontana RJ, Janardan SK, Adair CH, Fortlage LA, Brown MB, Guo W, Watkins PB (1997) Grapefruit juice increases felodipine oral availability in humans by decreasing intestinal CYP3A protein expression. J Clin Investig 99:2545–2553
6. Bailey DG, Malcolm J, Arnold O, David Spence J (2002) Grapefruit juice-drug interactions. Br J Clin Pharmacol 46:101–110
7. Dresser G (2002) Fruit juices inhibit organic anion transporting polypeptide–mediated drug uptake to decrease the oral availability of fexofenadine. Clin Pharmacol Ther 71:11–20
8. Takanaga H, Ohnishi A, Matsuo H, Sawada Y (1998) Inhibition of vinblastine efflux mediated by P-glycoprotein by grapefruit juice components in caco-2 cells. Biol Pharm Bull 21:1062–1066
9. Soldner A, Christians U, Susanto M, Wacher V, Silverman J, Benet L (1999) Grapefruit juice exerts stimulatory effects on P-glycoprotein. Clin Pharmacol Ther 65:205–205
10. Eshelman FN, Spyker DA (1978) Pharmacokinetics of amoxicillin and ampicillin: crossover study of the effect of food. Antimicrob Agents Chemother 14:539–543
11. Staniforth DH, Lillystone RJ, Jackson D (1982) Effect of food on the bioavailability and tolerance of clavulanic acid/amoxycillin combination. J Antimicrob Chemother 10:131–139
12. Lode H, Stahlmann R, Koeppe P (1979) Comparative pharmacokinetics of cephalexin, cefaclor, cefadroxil, and CGP 9000. Antimicrob Agents Chemother 16:1–6
13. Barbhaiya RH, Shukla UA, Gleason CR, Shyu WC, Pittman KA (1990) Comparison of the effects of food on the pharmacokinetics of cefprozil and cefaclor. Antimicrob Agents Chemother 34:1210–1213
14. Shukla UA, Pittman KA, Barbhaiya RH (1992) Pharmacokinetic interactions of cefprozil with food, propantheline, metoclopramide, and probenecid in healthy volunteers. J Clin Pharmacol 32:725–731
15. Finn A, Straughn A, Meyer M, Chubb J (1987) Effect of dose and food on the bioavailability of cefuroxime axetil. Biopharm Drug Dispos 8:519–526
16. Sommers DK, Wyk M, Moncrieff J, Schoeman HS (1984) Influence of food and reduced gastric acidity on the bioavailability of bacampicillin and cefuroxime axetil. Br J Clin Pharmacol 18:535–539
17. Hughes GS, Heald DL, Barker KB, Patel RK, Spillers CR, Watts KC, Batts DH, Euler AR (1989) The effects of gastric pH and food on the pharmacokinetics of a new oral cephalosporin, cefpodoxime proxetil. Clin Pharmacol Ther 46:674–685
18. Nakashima M, Uematsu T, Takiguchi Y, Kanamaru M (1987) Phase I study of cefixime, a new oral cephalosporin. J Clin Pharmacol 27:425–431
19. Faulkner RD, Bohaychuk W, Haynes JD, Desjardins RE, Yacobi A, Silber BM (1988) The pharmacokinetics of cefixime in the fasted and fed state. Eur J Clin Pharmacol 34:525–528
20. Li JT, Hou F, Lu H, Li TY, Li H (1997) Phase I clinical trial of cefditoren pivoxil (ME 1207): pharmacokinetics in healthy volunteers. Drugs Exp Clin Res 23:145–150
21. Barr WH, Lin C-C, Radwanski E, Lim J, Symchowicz S, Affrime M (1991) The pharmacokinetics of ceftibuten in humans. Diagn Microbiol Infect Dis 14:93–100
22. Hovi T, Heikinheimo M (1985) Effect of concomitant food intake on absorption kinetics of erythromycin in healthy volunteers. Eur J Clin Pharmacol 28:231–233

23. DiSanto AR, Chodos DJ (1981) Influence of study design in assessing food effects on absorption of erythromycin base and erythromycin stearate. Antimicrob Agents Chemother 20:190–196

24. Clayton D, Leslie A (1981) The bioavailability of erythromycin stearate versus enteric-coated erythromycin base when taken immediately before and after food. J Int Med Res 9:470–477

25. Rutland J, Berend N, Marlin GE (1979) The influence of food on the bioavailability of new formulations of erythromycin stearate and base. Br J Clin Pharmacol 8:343–347

26. Welling PG, Huang H, Hewitt PF, Lyons LL (1978) Bioavailability of erythromycin stearate: influence of food and fluid volume. J Pharm Sci 67:764–766

27. Thompson PJ, Burgess KR, Marlin GE (1980) Influence of food on absorption of erythromycin ethyl succinate. Antimicrob Agents Chemother 18:829–831

28. Chu S-Y, Park Y, Locke C, Wilson DS, Cavanaugh JC (1992) Drug-food interaction potential of clarithromycin, a new macrolide antimicrobial. J Clin Pharmacol 32:32–36

29. Guay D. RP (2001) Pharmacokinetics and tolerability of extended-release clarithromycin. Clin Ther 23:566–577

30. Foulds G, Luke DR, Teng R, Willavize SA, Friedman H, Curatolo WJ (1996) The absence of an effect of food on the bioavailability of azithromycin administered as tablets, sachet or suspension. J Antimicrob Chemother 37:37–44

31. Bhargava V, Lenfant B, Perret C, Pascual MH, Sultan E, Montay G (2002) Lack of effect of food on the bioavailability of a new ketolide antibacterial, telithromycin. Scand J Infect Dis 34:823–826

32. Welling PG, Koch PA, Lau CC, Craig WA (1977) Bioavailability of tetracycline and doxycycline in fasted and nonfasted subjects. Antimicrob Agents Chemother 11:462–469

33. Leyden JJ (1985) Absorption of minocycline hydrochloride and tetracycline hydrochloride. J Am Acad Dermatol 12:308–312

34. Frost RW, Carlson JD, Dietz AJ Jr, Heyd A, Lettieri JT (1989) Ciprofloxacin pharmacokinetics after a standard or high-fat/high-calcium breakfast. J Clin Pharmacol 29:953–955

35. Allen A, Bygate E, Clark D, Lewis A, Pay V (2000) The effect of food on the bioavailability of oral gemifloxacin in healthy volunteers. Int J Antimicrob Agents 16:45–50

36. Lee LJ, Hafkin B, Lee ID, Hoh J, Dix R (1997) Effects of food and sucralfate on a single oral dose of 500 milligrams of levofloxacin in healthy subjects. Antimicrob Agents Chemother 41:2196–2200

37. Lettieri J, Vargas R, Agarwal V, Liu P (2001) Effect of food on the pharmacokinetics of a single oral dose of moxifloxacin 400mg in healthy male volunteers. Clin Pharmacokinet 40:19–25

38. Leroy A, Borsa F, Humbert G, Bernadet P, Fillastre JP (1987) The pharmacokinetics of ofloxacin in healthy adult male volunteers. Eur J Clin Pharmacol 31:629–630

39. Nagy J, Schipper HG, Koopmans RP, Butter JJ, Van Boxtel CJ, Kager PA (2002) Effect of grapefruit juice or cimetidine coadministration on albendazole bioavailability. Am J Trop Med Hyg 66:260–263

40. Guzzo CA, Furtek CI, Porras AG, Chen C, Tipping R, Clineschmidt CM, Sciberras DG, Hsieh JYK, Lasseter KC (2002) Safety, tolerability, and pharmacokinetics of escalating high doses of ivermectin in healthy adult subjects. J Clin Pharmacol 42:1122–1133

41. Castro N, Medina R, Sotelo J, Jung H (2000) Bioavailability of praziquantel increases with concomitant administration of food. Antimicrob Agents Chemother 44:2903–2904

42. Cuong BT, Binh VQ, Dai B, Duy DN, Lovell CM, Rieckmann KH, Edstein MD (2006) Does gender, food or grapefruit juice alter the pharmacokinetics of primaquine in healthy subjects? Br J Clin Pharmacol 61:682–689

43. Tulpule A, Krishnaswamy K (1982) Effect of food on bioavailability of chloroquine. Eur J Clin Pharmacol 23:271–273

44. Crevoisier C, Handschin J, Barré J, Roumenov D, Kleinbloesem C (1997) Food increases the bioavailability of mefloquine. Eur J Clin Pharmacol 53:135–139

45. Falloon J, Sargent S, Piscitelli SC, Bechtel C, LaFon SW, Sadler B, Walker RE, Kovacs JA, Polis MA, Davey RT, Lane HC, Masur H (1999) Atovaquone suspension in HIV-infected

volunteers: pharmacokinetics, pharmacodynamics, and TMP-SMX interaction study. Pharmacotherapy 19:1050–1056

46. McGehee RF, Smith CB, Wilcox C, Finland M (1968) Comparative studies of antibacterial activity in vitro and absorption and excretion of lincomycin and clinimycin. Am J Med Sci 256:279–292

47. Borgia M, Longo A, Lodola E (1989) Relative bioavailability of fosfomycin and of trometamol after administration of single dose by oral route of fosfomycin trometamol in fasting conditions and after a meal. Int J Clin Pharmacol Ther Toxicol 27:411–417

48. Bergogne-Berezin E, Muller-Serieys C, Joly-Guillou ML, Dronne N (1987) Trometamol-fosfomycin (Monuril) bioavailability and food-drug interaction. Eur Urol 13(Suppl 1):64–68

49. Stockis A, Allemon AM, De Bruyn S, Gengler C (2002) Nitazoxanide pharmacokinetics and tolerability in man using single ascending oral doses. Int J Clin Pharmacol Ther 40:213–220

50. Melander A, Kahlmeter G, Kamme C, Ursing B (1977) Bioavailability of metronidazole in fasting and non-fasting healthy subjects and in patients with Crohn's disease. Eur J Clin Pharmacol 12:69–72

51. Welshman IR, Sisson TA, Jungbluth GL, Stalker DJ, Hopkins NK (2001) Linezolid absolute bioavailability and the effect of food on oral bioavailability. Biopharm Drug Dispos 22:91–97

52. Flanagan SD, Bien PA, Munoz KA, Minassian SL, Prokocimer PG (2014) Pharmacokinetics of tedizolid following oral administration: single and multiple dose, effect of food, and comparison of two solid forms of the prodrug. Pharmacotherapy 34:240–250

53. Hoppu K, Tuomisto J, Koskimies O, Simell O (1987) Food and guar decrease absorption of trimethoprim. Eur J Clin Pharmacol 32:427–429

54. Peloquin CA, Namdar R, Dodge AA, Nix DE (1999) Pharmacokinetics of isoniazid under fasting conditions, with food, and with antacids. Int J Tuberc Lung Dis 3:703–710

55. Peloquin CA, Namdar R, Singleton MD, Nix DE (1999) Pharmacokinetics of rifampin under fasting conditions, with food, and with antacids. Chest 115:12–18

56. Narang PK, Lewis RC, Bianchine JR (1992) Rifabutin absorption in humans: relative bioavailability and food effect. Clin Pharmacol Ther 52:335–341

57. Peloquin CA, Bulpitt AE, Jaresko GS, Jelliffe RW, Childs JM, Nix DE (1999) Pharmacokinetics of ethambutol under fasting conditions, with food, and with antacids. Antimicrob Agents Chemother 43:568–572

58. Peloquin CA, Bulpitt AE, Jaresko GS, Jelliffe RW, James GT, Nix DE (1998) Pharmacokinetics of pyrazinamide under fasting conditions, with food, and with antacids. Pharmacotherapy 18:1205–1211

59. Peloquin CA, Zhu M, Adam RD, Singleton MD, Nix DE (2001) Pharmacokinetics of para-aminosalicylic acid granules under four dosing conditions. Ann Pharmacother 35:1332–1338

60. Zhu M, Nix DE, Adam RD, Childs JM, Peloquin CA (2001) Pharmacokinetics of cycloserine under fasting conditions and with high-fat meal, orange juice, and antacids. Pharmacotherapy 21:891–897

61. Auclair B, Nix DE, Adam RD, James GT, Peloquin CA (2001) Pharmacokinetics of ethionamide administered under fasting conditions or with orange juice, food, or antacids. Antimicrob Agents Chemother 45:810–814

62. Teo SK, Scheffler MR, Kook KA, Tracewell WG, Colburn WA, Stirling DI, Thomas SD (2000) Effect of a high-fat meal on thalidomide pharmacokinetics and the relative bioavailability of oral formulations in healthy men and women. Biopharm Drug Dispos 21:33–40

63. Nix DE, Adam RD, Auclair B, Krueger TS, Godo PG, Peloquin CA (2004) Pharmacokinetics and relative bioavailability of clofazimine in relation to food, orange juice and antacid. Tuberculosis (Edinb) 84:365–373

64. Nedelman J, Cramer JA, Robbins B, Gibiansky E, Chang CT, Gareffa S, Cohen A, Meligeni J (1997) The effect of food on the pharmacokinetics of multiple-dose terbinafine in young and elderly healthy subjects. Biopharm Drug Dispos 18:127–138

65. Daneshmend TK, Warnock DW, Ene MD, Johnson EM, Potten MR, Richardson MD, Williamson PJ (1984) Influence of food on the pharmacokinetics of ketoconazole. Antimicrob Agents Chemother 25:1–3

66. Zimmermann T, Yeates RA, Laufen H, Pfaff G, Wildfeuer A (1994) Influence of concomitant food intake on the oral absorption of two triazole antifungal agents, itraconazole and fluconazole. Eur J Clin Pharmacol 46:147–150

67. Purkins L, Wood N, Kleinermans D, Greenhalgh K, Nichols D (2003) Effect of food on the pharmacokinetics of multiple-dose oral voriconazole. Br J Clin Pharmacol 56:17–23

68. Courtney R, Wexler D, Radwanski E, Lim J, Laughlin M (2003) Effect of food on the relative bioavailability of two oral formulations of posaconazole in healthy adults. Br J Clin Pharmacol 57:218–222

69. Krishna G, Ma L, Martinho M, O'Mara E (2012) Single-dose phase I study to evaluate the pharmacokinetics of posaconazole in new tablet and capsule formulations relative to oral suspension. Antimicrob Agents Chemother 56:4196–4201

70. Krishna G, Moton A, Ma L, Medlock MM, McLeod J (2008) Pharmacokinetics and absorption of posaconazole oral suspension under various gastric conditions in healthy volunteers. Antimicrob Agents Chemother 53:958–966

71. Kersemaekers WM, Dogterom P, Xu J, Marcantonio EE, de Greef R, Waskin H, van Iersel ML (2015) Effect of a high-fat meal on the pharmacokinetics of 300-milligram posaconazole in a solid oral tablet formulation. Antimicrob Agents Chemother 59:3385–3389

72. Schmitt-Hoffmann A, Desai A, Kowalski D, Pearlman H, Yamazaki T, Townsend R (2016) Isavuconazole absorption following oral administration in healthy subjects is comparable to intravenous dosing, and is not affected by food, or drugs that alter stomach pH. Int J Clin Pharmacol Ther 54:572–580

73. Aoyagi N, Ogata H, Kaniwa N, Ejima A (1982) Effect of food on the bioavailability of griseofulvin from microsize and PEG ultramicrosize (GRIS-PEG) plain tablets. J Pharmacobiodyn 5:120–124

74. Shyu WC, Knupp CA, Pittman KA, Dunkle L, Barbhaiya RH (1991) Food-induced reduction in bioavailability of didanosine. Clin Pharmacol Ther 50:503–507

75. Damle BD, Yan J-H, Behr D, O'Mara E, Nichola P, Kaul S, Knupp C (2002) Effect of food on the oral bioavailability of didanosine from encapsulated enteric-coated beads. J Clin Pharmacol 42:419–427

76. Shelton MJ, Portmore A, Blum MR, Sadler BM, Reichman RC, Morse GD (1994) Prolonged, but not diminished, zidovudine absorption induced by a high-fat breakfast. Pharmacotherapy 14:671–677

77. Moore KHP, Shaw S, Laurent AL, Lloyd P, Duncan B, Morris DM, O'Mara MJ, Pakes GE (1999) Lamivudine/Zidovudine as a combined formulation tablet: bioequivalence compared with lamivudine and zidovudine administered concurrently and the effect of food on absorption. J Clin Pharmacol 39:593–605

78. Kaul S, Christofalo B, Raymond RH, Stewart MB, Macleod CM (1998) Effect of food on the bioavailability of stavudine in subjects with human immunodeficiency virus infection. Antimicrob Agents Chemother 42:2295–2298

79. Chittick GE, Gillotin C, McDowell JA, Lou Y, Edwards KD, Prince WT, Stein DS (1999) Abacavir: absolute bioavailability, bioequivalence of three oral formulations, and effect of food. Pharmacotherapy 19:932–942

80. Custodio JM, Yin X, Hepner M, Ling KH, Cheng A, Kearney BP, Ramanathan S (2014) Effect of food on rilpivirine/emtricitabine/tenofovir disoproxil fumarate, an antiretroviral single-tablet regimen for the treatment of HIV infection. J Clin Pharmacol 54:378–385

81. Lamorde M, Byakika-Kibwika P, Tamale WS, Kiweewa F, Ryan M, Amara A, Tjia J, Back D, Khoo S, Boffito M, Kityo C, Merry C (2012) Effect of food on the steady-state pharmacokinetics of tenofovir and emtricitabine plus efavirenz in Ugandan adults. AIDS Res Treat 2012:105980

82. Kaul S, Ji P, Lu M, Nguyen KL, Shangguan T, Grasela D (2010) Bioavailability in healthy adults of efavirenz capsule contents mixed with a small amount of food. Am J Health Syst Pharm 67:217–222

83. Schöller-Gyüre M, Boffito M, Pozniak AL, Leemans R, Kakuda TN, Woodfall B, Vyncke V, Peeters M, Vandermeulen K, Hoetelmans RMW (2008) Effects of different meal composi- tions and fasted state on the oral bioavailability of etravirine. Pharmacotherapy 28:1215–1222

84. Crauwels HM, van Heeswijk RP, Buelens A, Stevens M, Boven K, Hoetelmans RM (2013) Impact of food and different meal types on the pharmacokinetics of rilpivirine. J Clin Pharmacol 53:834–840

85. Lamorde M, Walimbwa S, Byakika-Kibwika P, Katwere M, Mukisa L, Sempa JB, Else L, Back DJ, Khoo SH, Merry C (2015) Steady-state pharmacokinetics of rilpivirine under different meal conditions in HIV-1-infected Ugandan adults. J Antimicrob Chemother 70:1482–1486

86. Yeh KC, Deutsch PJ, Haddix H, Hesney M, Hoagland V, WD J, Justice SJ, Osborne B, Sterrett AT, Stone JA, Woolf E, Waldman S (1998) Single-dose pharmacokinetics of indinavir and the effect of food. Antimicrob Agents Chemother 42:332–338

87. Kakuda TN, Falcon RW (2006) Effect of food and ranitidine on saquinavir pharmacokinetics and gastric pH in healthy volunteers. Pharmacotherapy 26:1060–1068

88. Hugen PW, Burger DM, Koopmans PP, Stuart JW, Kroon FP, van Leusen R, Hekster YA (2002) Saquinavir soft-gel capsules (Fortovase) give lower exposure than expected, even after a high-fat breakfast. Pharm World Sci 24:83–86

89. Kaeser B, Charoin JE, Gerber M, Oxley P, Birnboeck H, Saiedabadi N, Banken L (2005) Assessment of the bioequivalence of two nelfinavir tablet formulations under fed and fasted conditions in healthy subjects. Int J Clin Pharmacol Ther 43:154–162

90. Saah AJ, Winchell GA, Nessly ML, Seniuk MA, Rhodes RR, Deutsch PJ (2001) Pharmacokinetic profile and tolerability of indinavir-ritonavir combinations in healthy vol- unteers. Antimicrob Agents Chemother 45:2710–2715

91. Veldkamp AI, van Heeswijk RP, Mulder JW, Meenhorst PL, Schreij G, van der Geest S, Lange JM, Beijnen JH, Hoetelmans RM (2001) Steady-state pharmacokinetics of twice-daily dosing of saquinavir plus ritonavir in HIV-1-infected individuals. J Acquir Immune Defic Syndr 27:344–349

92. Falcoz C, Jenkins JM, Bye C, Hardman TC, Kenney KB, Studenberg S, Fuder H, Prince WT (2002) Pharmacokinetics of GW433908, a prodrug of amprenavir, in healthy male volun- teers. J Clin Pharmacol 42:887–898

93. Lamorde M, Byakika-Kibwika P, Boffito M, Nabukeera L, Mayito J, Ogwal-Okeng J, Tjia J, Back D, Khoo S, Ryan M, Merry C (2012) Steady-state pharmacokinetics of lopinavir plus ritonavir when administered under different meal conditions in HIV-infected Ugandan adults. J Acquir Immune Defic Syndr 60:295–298

94. Oki T, Usami Y, Nakai M, Sagisaka M, Ito H, Nagaoka K, Mamiya N, Yamanaka K, Utsumi M, Kaneda T (2004) Pharmacokinetics of lopinavir after administration of Kaletra in healthy Japanese volunteers. Biol Pharm Bull 27:261–265

95. Sevinsky H, Tao X, Wang R, Ravindran P, Sims K, Xu X, Jariwala N, Bertz R (2015) A randomized trial in healthy subjects to assess the bioequivalence of an atazanavir/cobici- stat fixed-dose combination tablet versus administration as separate agents. Antivir Ther 20:493–500

96. Sekar V, Kestens D, Spinosa-Guzman S, De Pauw M, De Paepe E, Vangeneugden T, Lefebvre E, Hoetelmans RMW (2007) The effect of different meal types on the pharmacokinetics of darunavir (TMC114)/ritonavir in HIV-negative healthy volunteers. J Clin Pharmacol 47:479–484

97. Brainard DM, Friedman EJ, Jin B, Breidinger SA, Tillan MD, Wenning LA, Stone JA, Chodakewitz JA, Wagner JA, Iwamoto M (2011) Effect of low-, moderate-, and high-fat meals on raltegravir pharmacokinetics. J Clin Pharmacol 51:422–427

98. Song I, Borland J, Chen S, Patel P, Wajima T, Peppercorn A, Piscitelli SC (2012) Effect of food on the pharmacokinetics of the integrase inhibitor dolutegravir. Antimicrob Agents Chemother 56:1627–1629

99. Shiomi M, Matsuki S, Ikeda A, Ishikawa T, Nishino N, Kimura M, Irie S (2014) Effects of a protein-rich drink or a standard meal on the pharmacokinetics of elvitegravir, cobici-

stat, emtricitabine and tenofovir in healthy Japanese male subjects: a randomized, three-way crossover study. J Clin Pharmacol 54:640–648

100. Sun DQ, Wang HS, Ni MY, Wang BJ, Guo RC (2007) Pharmacokinetics, safety and tolerance of single- and multiple-dose adefovir dipivoxil in healthy Chinese subjects. Br J Clin Pharmacol 63:15–23

101. Zhou XJ, Lloyd DM, Chao GC, Brown NA (2006) Absence of food effect on the pharmacokinetics of telbivudine following oral administration in healthy subjects. J Clin Pharmacol 46:275–281

102. Anonymous. Food and drug administration center for drug evaluation and research clinical pharmacology and biopharmaceutics review: Application Number 201917Orig1s000. Telaprevir. 22 November 2010. Available from http://www.accessdata.fda.gov/drugsatfda_docs/nda/2011/201917Orig1s000ClinPharmR.pdf. Accessed 19 March 2017

103. German P, Mathias A, Brainard D, Kearney BP (2016) Clinical pharmacokinetics and pharmacodynamics of ledipasvir/sofosbuvir, a fixed-dose combination tablet for the treatment of hepatitis C. Clin Pharmacokinet 55:1337–1351

104. Mogalian E, German P, Kearney BP, Yang CY, Brainard D, Link J, McNally J, Han L, Ling J, Mathias A (2017) Preclinical pharmacokinetics and first-in-human pharmacokinetics, safety, and tolerability of velpatasvir, a pangenotypic HCV NS5A inhibitor, in healthy subjects. Antimicrob Agents Chemother. https://doi.org/10.1128/AAC.02084-16

105. Wills RJ, Rodriguez LC, Choma N, Oakes M (1987) Influence of a meal on the bioavailability of rimantadine.HCl. J Clin Pharmacol 27:821–823

106. Brown F, Banken L, Saywell K, Arum I (1999) Pharmacokinetics of valganciclovir and ganciclovir following multiple oral dosages of valganciclovir in HIV- and CMV-seropositive volunteers. Clin Pharmacokinet 37:167–176

107. Gill KS, Wood MJ (1996) The clinical pharmacokinetics of famciclovir. Clin Pharmacokinet 31:1–8

108. Cronk GA, Wheatley WB, Fellers GF, Albright H (1960) The relationship of food intake to the absorption of potassium alpha-phenoxyethyl penicillin and potassium phenoxymethyl penicillin from the gastrointestinal tract. Am J Med Sci 241:125–131

109. Welling PG, Huang H, Koch PA, Craig WA, Madsen PO (1977) Bioavailability of ampicillin and amoxicillin in fasted and nonfasted subjects. J Pharm Sci 66:549–552

110. Ali HM, Farouk AM (1980) The effect of sudanese diet on the bioavailability of ampicillin. Int J Pharm 6:301–306

111. Neu HC (1974) Antimicrobial activity and human pharmacology of amoxicillin. J Infect Dis 129:S123–S131

112. Anonymous. MOXATAG (amoxicillin extended-release tablets) [package insert]. Germantown, MD; Middlebrook Pharmaceuticals, Inc. Revised December 2008. Available from: https://dailymed.nlm.nih.gov/dailymed/drugInfo.cfm?setid=a6c0e721-2021-4983-a305-49bdf8f135a9. Accessed 20 March 2017

113. Anonymous. AUGMENTIN (amoxicillin/clavulanate potassium) [package insert]. Research Triangle Park, NC; Glaxo Smith Kline. Revised May 2002. Available from: https://www.accessdata.fda.gov/drugsatfda_docs/label/2008/050575s037550597s044050725s025050726s019lbl.pdf. Accessed 20 March 2017

114. Anonymous. AUGMENTIN-ES (amoxicillin/clavulanate potassium) [package insert]. Research Triangle Park, NC; Glaxo Smith Kline. Revised November 2006. Available from: https://www.accessdata.fda.gov/drugsatfda_docs/label/2009/050755s014lbl.pdf. Accessed 20 March 2017

115. Anonymous. DYNAPEN (dicloxacillin) [package insert]. North Wales, PA; Teva Pharmaceuticals, Inc. Revised June 2016. Available from: https://www.drugs.com/pro/dicloxacillin.html. Accessed 20 March 2017

116. Gower PE, Dash CH (1969) Cephalexin: human studies of absorption and excretion of a new cephalosporin antibiotic. Br J Pharmacol 37:738–747

117. Tetzlaff TR, McCracken GH, Thomas ML (1978) Bioavailability of cephalexin in children: relationship to drug formulations and meals. J Pediatr 92:292–294

118. Sourgens H, Derendorf H, Schifferer H (1997) Pharmacokinetic profile of cefaclor. Int J Clin Pharmacol Ther 35:374–380
119. Glynne A, Goulbourn RA, Ryden R (1978) A human pharmacology study of cefaclor. J Antimicrob Chemother 4:343–348
120. Oguma T, Yamada H, Sawaki M, Narita N (1991) Pharmacokinetic analysis of the effects of different foods on absorption of cefaclor. Antimicrob Agents Chemother 35:1729–1735
121. Anonymous. CECLOR (cefaclor) [package insert]. Indianapolis, IN; Eli Lilly and Company. Revised March 2003. Available from: https://www.accessdata.fda.gov/drugsatfda_docs/label /2004/50521slr027,50522slr027_cefaclor_lbl.pdf. Accessed 20 March 2017
122. Ginsburg CM, McCracken GH, Petruska M, Olson K (1985) Pharmacokinetics and bactericidal activity of cefuroxime axetil. Antimicrob Agents Chemother 28:504–507
123. Fassbender M, Lode H, Schaberg T, Borner K, Koeppe P (1993) Pharmacokinetics of new oral cephalosporins, including a new carbacephem. Clin Infect Dis 16:646–653
124. Borin MT, Driver MR, Forbes KK (1995) Effect of timing of food on absorption of cefpodoxime proxetil. J Clin Pharmacol 35:505–509
125. Guay DRP (2000) Pharmacodynamics and pharmacokinetics of cefdinir, an oral extended spectrum cephalosporin. Pediatr Infect Dis J 19:S141–S146
126. Anonymous. OMNICEF (cefdinir) [package insert]. Chicago, IL; Abbott Laboratories. Revised June 2007. Available from: https://www.accessdata.fda.gov/drugsatfda_docs/label/ 2008/050739s015,050749s021lbl.pdf. Accessed 20 March 2017
127. Anonymous. SPECTRACEF (cefditoren pivoxil) [package insert]. Stamford, CT; Purdue Pharmaceutical Products, LP. Revised December 2005. Available from: https://www.access-data.fda.gov/drugsatfda_docs/label/2005/021222s009lbl.pdf. Accessed 20 March 2017
128. Kearns GL, Young RA (1994) Ceftibuten pharmacokinetics and pharmacodynamics. Clin Pharmacokinet 26:169–189
129. Bechtol LDBC, Perkal MB (1979) The influence of food on the absorption of erythromycin esters and enteric-coated erythromycin in single-dose studies. Curr Ther Res 25:618–625
130. Hirsch HA, Finland M (1959) Effect of food on the absorption of erythromycin propionate, erythromycin stearate and triacetyloleandomycin. Am J Med Sci 237:693–709
131. Clapper WE, Mostyn M, Meade GH (1960) An evaluation of erythromycin stearate and propionyl erythromycin in normal and hospitalized subjects. Antibiotic Med Clin Ther (New York) 7:91–96
132. Coyne TC, Shum SHU, Chun AHC, Jeansonne L, Shirkey HC (1978) Bioavailability of erythromycin ethylsuccinate in pediatric patients. J Clin Pharmacol 18:194–202
133. Kanazawa S, Ohkubo T, Sugawara K (2001) The effects of grapefruit juice on the pharmacokinetics of erythromycin. Eur J Clin Pharmacol 56:799–803
134. Cheng KL, Nafziger AN, Peloquin CA, Amsden GW (1998) Effect of grapefruit juice on clarithromycin pharmacokinetics. Antimicrob Agents Chemother 42:927–929
135. Anonymous. BIAXIN (clarithromycin) [package insert]. Chicago, IL; Abbott Laboratories. Revised May 2003. Available from: https://www.accessdata.fda.gov/drugsatfda_docs/label/2 012/050662s044s050,50698s026s030,050775s015s019lbl.pdf. Accessed 20 March 2017
136. Hopkins S (1991) Clinical toleration and safety of azithromycin. Am J Med 91:S40–S45
137. Agouridas C, Denis A, Auger JM, Benedetti Y, Bonnefoy A, Bretin F, Chantot JF, Dussarat A, Fromentin C, D'Ambrieres SG, Lachaud S, Laurin P, Le Martret O, Loyau V, Tessot N (1998) Synthesis and antibacterial activity of ketolides (6-O-methyl-3-oxoerythromycin derivatives): a new class of antibacterials highly potent against macrolide-resistant and -susceptible respiratory pathogens. J Med Chem 41:4080–4100
138. Anonymous. Ketek (telithromycin) [package insert]. Bridgewater, NJ, USA; Sanofi-Aventis. Revised December 2015. Available from http://products.sanofi.us/ketek/ketek.pdf. Accessed 27 January 2016
139. FDA (2016) FDA Drug Shortages. Available from http://www.accessdata.fda.gov/scripts/ drugshortages/dsp_ActiveIngredientDetails.cfm?AI=Telithromycin+%28Ketek%29+Tablet s&source=govdelivery&st=d&tab=tabs-4&utm_medium=email&utm_source=govdelivery. Accessed 11 March 2016
140. Jonas M, Cunha BA (1982) Minocycline. Ther Drug Monit 4:115–146
141. Neuvonen PJ (1976) Interactions with the absorption of tetracyclines. Drugs 11:45–54

142. Von Wittenau S (1968) Some pharmacokinetic aspects of doxycycline metabolism in man. Chemotherapy 13:41–50
143. Meyer FP, Specht H, Quednow B, Walther H (1989) Influence of milk on the bioavailability of doxycycline — new aspects. Infection 17:245–246
144. Neuvonen PJ, Penttilä O. (1974) Effect of oral ferrous sulphate on the half-life of doxycycline in man. Eur J Clin Pharmacol 7:361–363
145. Allen JC (1976) Minocycline. Ann Intern Med 85:482–487
146. Cunha BA, Sibley CM, Ristuccia AM (1982) Doxycycline. Ther Drug Monit 4:115
147. Anonymous. DECLOMYCIN (demeclocycline) [package insert]. Manati, Puerto Rico; Mova Pharmaceuticals, Corp. Available from: https://www.accessdata.fda.gov/drugsatfda_docs/label/2008/050261s100lbl.pdf. Accessed 20 March 2017
148. Shah A (1999) Oral bioequivalence of three ciprofloxacin formulations following single-dose administration: 500 mg tablet compared with 500 mg/10 mL or 500 mg/5 mL suspension and the effect of food on the absorption of ciprofloxacin oral suspension. J Antimicrob Chemother 43:49–54
149. Anonymous. CIPRO (ciprofloxacin) [package insert]. Whippany, NJ; Bayer Healthcare Pharmaceuticals, Inc. Revised July 2016. Available from: https://www.accessdata.fda.gov/drugsatfda_docs/label/2016/019537s086lbl.pdf. Accessed 20 March 2017
150. Stass H, Kubitza D (2001) Effects of iron supplements on the oral bioavailability of moxifloxacin, a novel 8-methoxyfluoroquinolone, in humans. Clin Pharmacokinet 40:57–62
151. Anonymous. AVELOX (moxifloxacin) [package insert]. Whippany, NJ; Bayer Healthcare Pharmaceuticals, Inc. Revised July 2016. Available from: https://www.merck.com/product/usa/pi_circulars/a/avelox/avelox_pi.pdf. Accessed 20 March 2017
152. Anonymous. LEVAQUIN (levofloxacin) [package insert]. Titusville, NJ; Janssen Pharmaceuticals, Inc. Revised September 2008. Available from: https://www.accessdata.fda.gov/drugsatfda_docs/label/2013/020634s065,020635s071,021721s032lbl.pdf. Accessed 20 March 2017
153. Kawakami J, Matsuse T, Kotaki H, Seino T, Fukuchi Y, Orimo H, Sawada Y, Iga T (1994) The effect of food on the interaction of ofloxacin with sucralfate in healthy volunteers. Eur J Clin Pharmacol 47:67–69
154. Verho M, Malerczyk V, Dagrosa E, Korn A (1986) The effect of food on the pharmacokinetics of ofloxacin. Curr Med Res Opin 10:166–171
155. Kalager T, Digranes A, Bergan T, Rolstad T (1986) Ofloxacin: serum and skin blister fluid pharmacokinetics in the fasting and non-fasting state. J Antimicrob Chemother 17:795–800
156. Dudley MN, Marchbanks CR, Flor SC, Beals B (1991) The effect of food or milk on the absorption kinetics of ofloxacin. Eur J Clin Pharmacol 41:569–571
157. Anonymous. FLOXIN (ofloxacin) [package insert]. Titusville, NJ; Ortho-McNeil-Janssen Therapeutics, Inc. Revised September 208. Available from: https://www.accessdata.fda.gov/drugsatfda_docs/label/2008/019735s059lbl.pdf. Accessed 20 March 2017
158. Ledergerber B, Bettex JD, Joos B, Flepp M, Luthy R (1985) Effect of standard breakfast on drug absorption and multiple-dose pharmacokinetics of ciprofloxacin. Antimicrob Agents Chemother 27:350–352
159. Shah A, Liu MC, Vaughan D, Heller AH (1999) Oral bioequivalence of three ciprofloxacin formulations following single-dose administration: 500 mg tablet compared with 500 mg/10 mL or 500 mg/5 mL suspension and the effect of food on the absorption of ciprofloxacin oral suspension. J Antimicrob Chemother 43(Suppl A):49–54
160. Martin SJ, Meyer JM, Chuck SK, Jung R, Messick CR, Pendland SL (1998) Levofloxacin and sparfloxacin: new quinolone antibiotics. Ann Pharmacother 32:320–336
161. Ballow C, Lettieri J, Agarwal V, Liu P, Stass H, Sullivan JT (1999) Absolute bioavailability of moxifloxacin. Clin Ther 21:513–522
162. Anonymous. FACTIVE (gemifloxacin mesylate) [package insert]. Waltham, MA; Oscient Pharmaceuticals Corporation. Revised May 2007. Available from: https://www.accessdata.fda.gov/drugsatfda_docs/label/2007/021158s007lbl.pdf. Accessed 20 March 2017
163. Hoogkamer JFW, Kleinbloesem CH (1995) The effect of milk consumption on the pharmacokinetics of fleroxacin and ciprofloxacin in healthy volunteers. Drugs 49:346–348

164. Neuvonen PJ, Kivistö KT, Lehto P (1991) Interference of dairy products with the absorption of ciprofloxacin. Clinical Pharmacology & Therapeutics 50:498–502

165. Minami R, Inotsume N, Nakano M, Sudo Y, Higashi A, Matsuda I (1993) Effect of milk on absorption of norfloxacin in healthy volunteers. J Clin Pharmacol 33:1238–1240

166. Kivisto KT, Ojala-Karlsson P, Neuvonen PJ (1992) Inhibition of norfloxacin absorption by dairy products. Antimicrob Agents Chemother 36:489–491

167. Neuvonen PJ, Kivisto KT (1992) Milk and yoghurt do not impair the absorption of ofloxacin. Br J Clin Pharmacol 33:346–348

168. Stass H, Kubitza D (2001) Effects of dairy products on the oral bioavailability of moxifloxacin, a novel 8-methoxyfluoroquinolone, in healthy volunteers. Clin Pharmacokinet 40:33–38

169. Neuhofel AL, Wilton JH, Victory JM, Hejmanowski LG, Amsden GW (2002) Lack of bioequivalence of ciprofloxacin when administered with calcium-fortified orange juice: a new twist on an old interaction. J Clin Pharmacol 42:461–466

170. Aminimanizani A, Beringer P, Jelliffe R (2001) Comparative pharmacokinetics and pharmacodynamics of the newer fluoroquinolone antibacterials. Clin Pharmacokinet 40:169–187

171. Amsden GW, Whitaker A-M, Johnson PW (2003) Lack of bioequivalence of levofloxacin when coadministered with a mineral-fortified breakfast of juice and cereal. J Clin Pharmacol 43:990–995

172. Yamaguchi H (2002) Pharmacokinetic role of P-glycoprotein in oral bioavailability and intestinal secretion of grepafloxacin in vivo. J Pharmacol Exp Ther 300:1063–1069

173. Takanaga H, Ohnishi A, Yamada S, Matsuo H, Morimoto S, Shoyama Y, Ohtani H, Sawada Y (2000) Polymethoxylated flavones in orange juice are inhibitors of P-glycoprotein but not cytochrome P450 3A4. J Pharmacol Exp Ther 293:230–236

174. Fuhr U, Anders EM, Mahr G, Sorgel F, Staib AH (1992) Inhibitory potency of quinolone antibacterial agents against cytochrome P450IA2 activity in vivo and in vitro. Antimicrob Agents Chemother 36:942–948

175. Carbó M, Segura J, De la Torre R, Badenas JM, Camí J (1989) Effect of quinolones on caffeine disposition. Clin Pharmacol Ther 45:234–240

176. Healy DP, Polk RE, Kanawati L, Rock DT, Mooney ML (1989) Interaction between oral ciprofloxacin and caffeine in normal volunteers. Antimicrob Agents Chemother 33:474–478

177. Staib AH, Harder S, Mieke S, Beer C, Stille W, Shah P (1987) Gyrase-inhibitors impair caffeine elimination in man. Methods Find Exp Clin Pharmacol 9:193–198

178. Wright DH, Pietz SL, Konstantinides FN, Rotschafer JC (2000) Decreased in vitro fluoroquinolone concentrations after admixture with an enteral feeding formulation. J Parenter Enter Nutr 24:42–48

179. Mueller BA, Brierton DG, Abel SR, Bowman L (1994) Effect of enteral feeding with ensure on oral bioavailabilities of ofloxacin and ciprofloxacin. Antimicrob Agents Chemother 38:2101–2105

180. Yuk JH, Nightingale CH, Sweeney KR, Quintiliani R, Lettieri JT, Frost RW (1989) Relative bioavailability in healthy volunteers of ciprofloxacin administered through a nasogastric tube with and without enteral feeding. Antimicrob Agents Chemother 33:1118–1120

181. Burkhardt O, Stass H, Thuss U, Borner K, Welte T (2005) Effects of enteral feeding on the oral bioavailability of moxifloxacin in healthy volunteers. Clin Pharmacokinet 44:969–976

182. Healy DP, Brodbeck MC, Clendening CE (1996) Ciprofloxacin absorption is impaired in patients given enteral feedings orally and via gastrostomy and jejunostomy tubes. Antimicrob Agents Chemother 40:6–10

183. Deppermann KM, Lode H (1993) Fluoroquinolones: interaction profile during enteral absorption. Drugs 45(Suppl 3):65–72

184. Lange H, Eggers R, Bircher J (1988) Increased systemic availability of albendazole when taken with a fatty meal. Eur J Clin Pharmacol 34:315–317

185. Dayan AD (2003) Albendazole, mebendazole and praziquantel. Review of non-clinical toxicity and pharmacokinetics. Acta Trop 86:141–159

186. Winstanley PA, Orme ML (1989) The effects of food on drug bioavailability. Br J Clin Pharmacol 28:621–628

187. Anonymous. MINTEZOL (thiabendazole) [package insert]. West Point, PA; Merck & Co., Inc. Revised June 2003. Available from: https://www.accessdata.fda.gov/drugsatfda_docs/label/2003/16096slr033,16097slr026_mintezol_lbl.pdf. Accessed 20 March 2017

188. Anonymous. STROMECTOL (ivermectin) [package insert]. Whitehouse Station, NJ; Merck & Co., Inc. Revised October 2003. Available from: https://www.accessdata.fda.gov/drugsatfda_docs/label/2009/050742s026lbl.pdf. Accessed 20 March 2017

189. Anonymous. BILTRICIDE (praziquantel) [package insert]. Wayne, NJ; Bayer Healthcare Pharmaceuticals, Inc. Revised August 2010. Available from: https://www.accessdata.fda.gov/drugsatfda_docs/label/2010/018714s012lbl.pdf. Accessed 20 March 2017

190. Castro N, Jung H, Medina R, Gonzalez-Esquivel D, Lopez M, Sotelo J (2002) Interaction between grapefruit juice and praziquantel in humans. Antimicrob Agents Chemother 46:1614–1616

191. Anonymous. QUALAQUIN (quinine sulfate) [package insert]. Philadelphia, PA; AR Scientific, Inc. Revised June 2008. Available from: https://www.accessdata.fda.gov/drugsatfda_docs/label/2008/021799s008lbl.pdf. Accessed 20 March 2017

192. Anonymous. PLAQUENIL (hydroxychloroquine) [package insert]. Bridgewater, NJ; Sanofi-Aventis, LLC. Revised October 2006. Available from: https://www.accessdata.fda.gov/drugsatfda_docs/label/2007/009768s041lbl.pdf. Accessed 20 March 2017

193. Anonymous. DARAPRIM (pyrimethamine) [package insert]. Greenville, NC; DSM Pharmaceuticals, Inc. Revised March 2003. Available from: https://www.accessdata.fda.gov/drugsatfda_docs/label/2003/08578slr016_daraprim_lbl.pdf. Accessed 20 March 2017

194. Anonymous. FANSIDAR (sulfadoxine and pyrimethamine) [package insert]. Nutley, NJ; Roche Pharmaceuticals, Inc. Revised February 2004. Available from: https://www.accessdata.fda.gov/drugsatfda_docs/label/2004/18557slr015_fansidar_lbl.pdf. Accessed 20 March 2017

195. Anonymous. COARTEM (artemether/lumefantrine) [package insert]. East Hanover, NJ; Novartis Pharmaceuticals Corporation. Revised March 2015. Available from: https://www.pharma.us.novartis.com/sites/www.pharma.us.novartis.com/files/coartem.pdf. Accessed 20 March 2017

196. Anonymous. LARIAM (mefloquine hydrochloride) [package insert]. Nutley, NJ; Roche Pharmaceuticals, Inc. Revised August 2008. Available from: https://www.accessdata.fda.gov/drugsatfda_docs/label/2009/019591s026s028lbl.pdf. Accessed 20 March 2017

197. Freeman CD, Klutman NE, Lamp KC, Dall LH, Strayer AH (1998) Relative bioavailability of atovaquone suspension when administered with an enteral nutrition supplement. Ann Pharmacother 32:1004–1007

198. Anonymous. MEPRON (atovaquone) [package insert]. Research Triangle Park, NC; GalxoSmithKline . Revised May 2008. Available from: https://www.accessdata.fda.gov/drugsatfda_docs/label/2008/020500s010lbl.pdf. Accessed 20 March 2017

199. Anonymous. MALARONE (atovaquone and proguanil hydrochloride) [package insert]. Research Triangle Park, NC; GalxoSmithKline . Revised July 2016. Available from: https://www.gsksource.com/pharma/content/dam/GlaxoSmithKline/US/en/Prescribing_Information/Malarone/pdf/MALARONE.PDF Accessed 20 March 2017

200. Anonymous. CLEOCIN HCL (clindamycin hydrochloride) [package insert]. Kalamazoo, MI; Pharmacia and Upjohn Company. Revised 2009. Available from: https://www.accessdata.fda.gov/drugsatfda_docs/label/2009/050162s085lbl.pdf. Accessed 20 March 2017

201. Anonymous. DIFICID (fidaxomicin) [package insert]. Whitehouse Station, NJ; Merck & Co., Inc. Revised December 2015. Available from: https://www.merck.com/product/usa/pi_circulars/d/dificid/dificid_pi.pdf. Accessed 20 March 2017

202. Anonymous. MONUROL (fosfomycin tromethamine) [package insert]. St. Louis, MO; Forest Pharmaceuticals, Inc. Revised July 2007. Available from http://www.accessdata.fda.gov/drugsatfda_docs/label/2008/050717s005lbl.pdf. Accessed 11 May 2015

203. Anonymous. ALINIA (nitazoxanide) [package insert]. Tampa, FL; Romark Pharmaceuticals, LC. Revised June 2005. Available from: https://www.accessdata.fda.gov/drugsatfda_docs/label/2005/021818lbl.pdf. Accessed 20 March 2017

204. D'Arcy PF (1985) Nitrofurantoin. Ann Pharmacother 19:540–547
205. Bates TR, Sequeira JA, Tembo AV (1974) Effect of food on nitrofurantoin absorption. Clin Pharmacol Ther 16:63–68
206. Rosenberg HA, Bates TR (1976) The influence of food on nitrofurantoin bioavailability. Clini PharmacolTher 20:227–232
207. Anonymous. TINDAMAX (tinidazole) [package insert]. Mumbai, India; Unique Pharmaceuticals Laboratories. Revised May 2007. Available from: https://www.rising-pharma.com/Files/Prescribing-Info/Package%20Insert-Tinidazole%20Tablets-250mg-500mg.pdf. Accessed 20 March 2017
208. Antal EJ, Hendershot PE, Batts DH, Sheu W-P, Hopkins NK, Donaldson KM (2001) Linezolid, a novel oxazolidinone antibiotic: assessment of monoamine oxidase inhibition using pressor response to oral tyramine. J Clin Pharmacol 41:552–562
209. Flanagan S, Bartizal K, Minassian SL, Fang E, Prokocimer P (2013) In vitro, in vivo, and clinical studies of tedizolid to assess the potential for peripheral or central monoamine oxidase interactions. Antimicrob Agents Chemother 57:3060–3066
210. Anonymous. XIFAXAN (rifaximin) [package insert]. Morrisville, NJ; Salix Pharmaceuticals, Inc. Revised 2010. Available from: https://www.accessdata.fda.gov/drugsatfda_docs/label/2010/022554lbl.pdf. Accessed 20 March 2017
211. Anonymous. Sulfadiazine [package insert]. New Hyde Park, NY; Eon Labs, Inc. Revised March 2016. Available from: https://dailymed.nlm.nih.gov/dailymed/drugInfo.cfm?setid=10549cba-9c15-4d2e-a68c-5afbc178591d. Accessed 20 March 2017
212. Anonymous. BACTRIM (sulfamethoxazole and trimethoprim) [package insert]. Philadelphia, PA; Mutual Pharmaceutical Company, Inc. Revised June 2013. Available from: https://www.accessdata.fda.gov/drugsatfda_docs/label/2013/017377s068s073lbl.pdf. Accessed 20 March 2017
213. Anonymous. VANCOCIN (vancomycin hydrochloride) [package insert]. Exton, PA; ViroPharma, Inc. Revised December 2011. Available from: https://www.accessdata.fda.gov/drugsatfda_docs/label/2011/050606s028lbl.pdf. Accessed 20 March 2017
214. Melander A, Danielson K, Hanson A, Jansson L, Rerup C, Scherstén B, Thulin T, Wåhlin E (2009) Reduction of isoniazid bioavailability in normal men by concomitant intake of food. Acta Med Scand 200:93–97
215. Self TH, Chrisman CR, Baciewicz AM, Bronze MS (1999) Isoniazid drug and food interactions. Am J Med Sci 317:304–311
216. Smith CK (1978) Isoniazid and reaction to cheese. Ann Intern Med 88:520
217. Baciewicz AM, Self TH (1985) Isoniazid interactions. South Med J 78:714–718
218. Hauser MJ, Baier H (1982) Interactions of isoniazid with foods. Drug Intell Clin Pharm 16:617–618
219. Siegler D (1974) Effect of meals on rifampicin absorption. Lancet 304:197–198
220. Ameer B, Polk RE, Kline BJ, Grisafe JP (1982) Effect of food on ethambutol absorption. Clin Pharm 1:156–158
221. van Heeswijk RP, Dannemann B, Hoetelmans RM (2014) Bedaquiline: a review of human pharmacokinetics and drug-drug interactions. J Antimicrob Chemother 69:2310–2318
222. Anonymous. SIRTURO (bedaquiline) [package insert]. Titusville, NJ; Janssen Therapeutics, LP. Revised December 2012. Available from: https://www.accessdata.fda.gov/drugsatfda_docs/label/2012/204384s000lbl.pdf. Accessed 20 March 2017
223. Holdiness MR (1984) Clinical pharmacokinetics of the antituberculosis drugs. Clin Pharmacokinet 9:511–544
224. Anonymous. PASER (aminosalicyclic acid delayed-release granules) [package insert]. Princeton, NJ; Jacobus Pharmaceutical Company, Inc. Revised July 1996. Available from: https://dailymed.nlm.nih.gov/dailymed/drugInfo.cfm?setid=6f2753dc-5f0c-4f49-9335-e519afb69ba6. Accessed 20 March 2017
225. Anonymous. CYCLOSERINE (cycloserine) [package insert]. West Lafayette, IN; The Chao Center for Industrial Pharmacy and Contract Manufacturing. Revised October 2011.

Available from: https://dailymed.nlm.nih.gov/dailymed/drugInfo.cfm?setid=10727aa8-a03e-4b61-9a80-a024cb4a2d28. Accessed 20 March 2017

226. Venkatesan K (1989) Clinical pharmacokinetic considerations in the treatment of patients with leprosy. Clin Pharmacokinet 16:365–386

227. Anonymous. TRECATOR (ethionamide) [package insert]. Philadelphia, PA; Wyeth Pharmaceuticals, Inc. Revised December 2005. Available from: https://www.accessdata.fda.gov/drugsatfda_docs/label/2006/013026s024lbl.pdf. Accessed 20 March 2017

228. Zvada SP, Van Der Walt JS, Smith PJ, Fourie PB, Roscigno G, Mitchison D, Simonsson US, McIlleron HM (2010) Effects of four different meal types on the population pharmacokinetics of single-dose rifapentine in healthy male volunteers. Antimicrob Agents Chemother 54:3390–3394

229. Chan SL, Yew WW, Porter JH, McAdam KP, Allen BW, Dickinson JM, Ellard GA, Mitchison DA (1994) Comparison of Chinese and Western rifapentines and improvement of bioavailability by prior taking of various meals. Int J Antimicrob Agents 3:267–274

230. Anonymous. PRIFTIN (rifapentine) [package insert]. Bridgewater, NJ; Sanofi-Aventis, LLC. Revised December 2014. Available from: http://products.sanofi.us/priftin/priftin.pdf. Accessed 20 March 2017

231. Anonymous. DAPSONE (dapsone) [package insert]. Princeton, NJ; Jacobus Pharmaceutical Company, Inc. Revised July 2011. Available from: https://dailymed.nlm.nih.gov/dailymed/drugInfo.cfm?setid=0792169d-c6f9-4af0-93ae-b75d710c47a9. Accessed 20 March 2017

232. Teo SK, Colburn WA, Tracewell WG, Kook KA, Stirling DI, Jaworsky MS, Scheffler MA, Thomas SD, Laskin OL (2004) Clinical pharmacokinetics of thalidomide. Clin Pharmacokinet 43:311–327

233. Anonymous. THALOMID (thalidomide) [package insert]. Summit, NJ; Celgene Corporation. Revised January 2017. Available from: https://www.celgene.com/content/uploads/thalomid-pi.pdf. Accessed 20 March 2017

234. Holdiness MR (1989) Clinical pharmacokinetics of clofazimine. A review. Clin Pharmacokinet 16:74–85

235. Schaad-Lanyi Z, Dieterle W, Dubois JP, Theobald W, Vischer W (1987) Pharmacokinetics of clofazimine in healthy volunteers. Int J Lepr Other Mycobact Dis 55:9–15

236. Anonymous. LAMPRENE (clofazimine) [package insert]. East Hanover, NJ; Novartis Pharmaceuticals Corporation. Revised May 2006. Available from: https://www.drugs.com/pro/lamprene.html. Accessed 20 March 2017

237. Anonymous. ANCOBON (flucytosine) [package insert]. Bridgewater, NJ; Valeant Pharmaceuticals, LLC. Revised November 2013. Available from: https://dailymed.nlm.nih.gov/dailymed/drugInfo.cfm?setid=aea0df00-a88c-4a16-abcf-750f3ff2004e. Accessed 20 March 2017

238. Anonymous. LAMISIL (terbinafine) [package insert]. East Hanover, NJ; Novartis Pharmaceuticals Corportation. Revised January 2017. Available from: https://www.pharma.us.novartis.com/sites/www.pharma.us.novartis.com/files/Lamisil_tablets.pdf. Accessed 20 March 2017

239. Lelawongs P, Barone JA, Colaizzi JL, Hsuan AT, Mechlinski W, Legendre R, Guarnieri J (1988) Effect of food and gastric acidity on absorption of orally administered ketoconazole. Clin Pharm 7:228–235

240. Mannisto PT, Mantyla R, Nykanen S, Lamminsivu U, Ottoila P (1982) Impairing effect of food on ketoconazole absorption. Antimicrob Agents Chemother 21:730–733

241. Wishart JM (1987) The influence of food on the pharmacokinetics of itraconazole in patients with superficial fungal infection. J Am Acad Dermatol 17:220–223

242. Barone JA, Koh JG, Bierman RH, Colaizzi JL, Swanson KA, Gaffar MC, Moskovitz BL, Mechlinski W, Van de Velde V (1993) Food interaction and steady-state pharmacokinetics of itraconazole capsules in healthy male volunteers. Antimicrob Agents Chemother 37:778–784

243. Lange D, Pavao JH, Wu J, Klausner M (1997) Effect of a cola beverage on the bioavailability of itraconazole in the presence of H2Blockers. J Clin Pharmacol 37:535–540

244. Jaruratanasirikul S, Kleepkaew A (1997) Influence of an acidic beverage (Coca-Cola) on the absorption of itraconazole. Eur J Clin Pharmacol 52:235–237

245. Kawakami M, Suzuki K, Ishizuka T, Hidaka T, Matsuki Y, Nakamura H (1998) Effect of grapefruit juice on pharmacokinetics of itraconazole in healthy subjects. Int J Clin Pharmacol Ther 36:306–308

246. Penzak SR, Gubbins PO, Gurley BJ, Wang PL, Saccente M (1999) Grapefruit juice decreases the systemic availability of itraconazole capsules in healthy volunteers. Ther Drug Monit 21:304–309

247. Gubbins PO, McConnell SA, Gurley BJ, Fincher TK, Franks AM, Williams DK, Penzak SR, Saccente M (2004) Influence of grapefruit juice on the systemic availability of itraconazole oral solution in healthy adult volunteers. Pharmacotherapy 24:460–467

248. Sansone-Parsons A, Krishna G, Calzetta A, Wexler D, Kantesaria B, Rosenberg MA, Saltzman MA (2006) Effect of a nutritional supplement on posaconazole pharmacokinetics following oral administration to healthy volunteers. Antimicrob Agents Chemother 50:1881–1883

249. Krishna G, Ma L, Vickery D, Yu X, Wu I, Power E, Beresford E, Komjathy S (2009) Effect of varying amounts of a liquid nutritional supplement on the pharmacokinetics of posaconazole in healthy volunteers. Antimicrob Agents Chemother 53:4749–4752

250. Wiederhold NP (2016) Pharmacokinetics and safety of posaconazole delayed-release tablets for invasive fungal infections. Clin Pharmacol 8:1–8

251. Anonymous. NOXAFIL (posaconazole) [package insert]. Whitehouse Station, NJ; Merck & Co., Inc. Revised March 2017. Available from: https://www.merck.com/product/usa/pi_circulars/n/noxafil/noxafil_pi.pdf. Accessed 20 March 2017

252. Kraft WK, Chang PS, van Iersel ML, Waskin H, Krishna G, Kersemaekers WM (2014) Posaconazole tablet pharmacokinetics: lack of effect of concomitant medications altering gastric pH and gastric motility in healthy subjects. Antimicrob Agents Chemother 58:4020–4025

253. Yamazaki T, Desai A, Goldwater R, Han D, Howieson C, Akhtar S, Kowalski D, Lademacher C, Pearlman H, Rammelsberg D, Townsend R (2017) Pharmacokinetic effects of isavuconazole coadministration with the cytochrome P450 enzyme substrates bupropion, repaglinide, caffeine, dextromethorphan, and methadone in healthy subjects. Clin Pharmacol Drug Dev 6:54–65

254. Bijanzadeh M, Mahmoudian M, Salehian P, Khazainia T, Eshghi L, Khosravy A (1990) The bioavailability of griseofulvin from microsized and ultramicrosized tablets in nonfasting volunteers. Indian J Physiol Pharmacol 34:157–161

255. Crounse RG (1961) Human pharmacology of griseofulvin: the effect of fat intake on gastrointestinal absorption11from the Department of Dermatology, University of Miami Medical School, Miami, Florida. J Investig Dermatol 37:529–533

256. Hartman NR, Yarchoan R, Pluda JM, Thomas RV, Marczyk KS, Broder S, Johns DG (1990) Pharmacokinetics of 2′, 3′-dideoxyadenosine and 2′, 3′-dideoxyinosine in patients with severe human immunodeficiency virus infection. Clin Pharmacol Ther 47:647–654

257. Knupp CA, Shyu WC, Dolin R, Valentine FT, McLaren C, Martin RR, Pittman KA, Barbhaiya RH (1991) Pharmacokinetics of didanosine in patients with acquired immunodeficiency syndrome or acquired immunodeficiency syndrome—related complex. Clin Pharmacol Ther 49:523–535

258. McGowan JJ, Tomaszewski JE, Cradock J, Hoth D, Grieshaber CK, Broder S, Mitsuya H (1990) Overview of the preclinical development of an antiretroviral drug, 2′,3′-dideoxyinosine. Clin Infect Dis 12:S513–S521

259. Damle BD, Kaul S, Behr D, Knupp C (2002) Bioequivalence of two formulations of didanosine, encapsulated enteric-coated beads and buffered tablet, in healthy volunteers and HIV-infected subjects. J Clin Pharmacol 42:791–797

260. Anonymous. VIDEX EC (didanosine) [package insert]. Princeton, NJ; Bristol-Myers Squibb. Revised August 2015. Available from: https://packageinserts.bms.com/pi/pi_videx_ec.pdf. Accessed 20 March 2017

261. Klecker RW, Collins JM, Yarchoan R, Thomas R, Jenkins JF, Broder S, Myers CE (1987) Plasma and cerebrospinal fluid pharmacokinetics of 3′-azido-3′-deoxythymidine: a novel pyrimidine analog with potential application for the treatment of patients with AIDS and related diseases. Clin Pharmacol Ther 41:407–412

262. Sahai J, Gallicano K, Garber G, McGilveray I, Hawley-Foss N, Turgeon N, Cameron DW (1992) The effect of a protein meal on zidovudine pharmacokinetics in HIV- infected patients. Br J Clin Pharmacol 33:657–660

263. Unadkat JD, Collier AC, Crosby SS, Cummingst D, Opheimt KE, Coreyt L (1990) Pharmacokinetics of oral zidovudine (azidothymidine) in patients with AIDS when administered with and without a high-fat meal. AIDS 4:229–232

264. Lotterer E, Ruhnke M, Trautmann M, Beyer R, Bauer FE (1991) Decreased and variable systemic availability of zidovudine in patients with AIDS if administered with a meal. Eur J Clin Pharmacol 40:305–308

265. Anonymous. ZERIT (stavudine) [package insert]. Princeton, NJ; Bristol-Myers Squibb. Revised December 2012. Available from: https://www.accessdata.fda.gov/drugsatfda_docs/label/2008/020412s029,020413s020lbl.pdf. Accessed 20 March 2017

266. Foster RH, Faulds D (1998) Abacavir. Drugs 55:729–736

267. Yuen GJ, Lou Y, Thompson NF, Otto VR, Allsup TL, Mahony WB, Hutman HW (2001) Abacavir/Lamivudine/Zidovudme as a combined formulation tablet: bioequivalence compared with each component administered concurrently and the effect of food on absorption. J Clin Pharmacol 41:277–288

268. Weller S, Chen S, Borland J, Savina P, Wynne B, Piscitelli SC (2014) Bioequivalence of a dolutegravir, abacavir, and lamivudine fixed-dose combination tablet and the effect of food. J Acquir Immune Defic Syndr 66:393–398

269. Anonymous. EMTRIVA (emtricitabine) [package insert]. Foster City, CA; Gilead Sciences, Inc. Revised November 2012. Available from: http://www.gilead.com/~/media/files/pdfs/medicines/hiv/emtriva/emtriva_pi.pdf. Accessed 20 March 2017

270. Anonymous. TRUVADA (emtricitabine/tenofovir disoproxil fumarate) [package insert]. Foster City, CA; Gilead Sciences, Inc. Revised April 2016. Available from: http://www.gilead.com/~/media/files/pdfs/medicines/hiv/truvada/truvada_pi.pdf. Accessed 20 March 2017

271. Anonymous. VIREAD (tenofovir disoproxil fumarate) [package insert]. Foster City, CA; Gilead Sciences, Inc. Revised February 2016. Available from: http://gilead.com/~/media/files/pdfs/medicines/liver-disease/viread/viread_pi.pdf. Accessed 20 March 2017

272. Barditch-Crovo P, Deeks SG, Collier A, Safrin S, Coakley DF, Miller M, Kearney BP, Coleman RL, Lamy PD, Kahn JO, McGowan I, Lietman PS (2001) Phase I/II trial of the pharmacokinetics, safety, and antiretroviral activity of tenofovir disoproxil fumarate in human immunodeficiency virus-infected adults. Antimicrob Agents Chemother 45:2733–2739

273. Anonymous. VEMLIDY (tenofovir alafenamide) [package insert]. Foster City, CA; Gilead Sciences, Inc. Revised November 2016. Available from: https://www.gilead.com/~/media/files/pdfs/medicines/liver-disease/vemlidy/vemlidy_pi.pdf?la=en. Accessed 20 March 2017

274. Anonymous. SUSTIVA (efavirenz) [package insert]. Princeton, NJ; Bristol-Myers Squibb. Revised January 2017. Available from: https://packageinserts.bms.com/pi/pi_sustiva.pdf. Accessed 20 March 2017

275. Anonymous. ATRIPLA (efavirenz/emtricitabine/tenofovir disoproxil fumarate) [package insert]. Foster City, CA; Gilead Sciences, Inc. Revised February 2016. Available from: http://packageinserts.bms.com/pi/pi_atripla.pdf. Accessed 20 March 2017

276. Beach JW (1998) Chemotherapeutic agents for human immunodeficiency virus infection: mechanism of action, pharmacokinetics, metabolism, and adverse reactions. Clin Ther 20:2–25; discussion 1

277. Anonymous. INTELENCE (etravirine) [package insert]. Titusville, NJ; Janssen Pharmaceuticals, Inc. Revised August 2014. Available from: http://www.intelence.com/shared/product/intelence/prescribing-information.pdf. Accessed 20 March 2017

278. Anonymous, Hoetelmans R, Van Heeswijk R, Kestens D, et al (2005) Effect of food on and multiple-dose pharmacokinetics of TMC278 as an oral tablet formulation. Third International AIDS Conference, Rio de Janeiro, Brazil, July 2005. Abstract TuPe3.1 B10

279. Williams GC, Sinko PJ (1999) Oral absorption of the HIV protease inhibitors: a current update. Adv Drug Deliv Rev 39:211–238

280. Anonymous. CRIXIVAN (indinavir) [package insert]. Whitehouse Station, NJ; Merck & Co., Inc. Revised September 2016. Available from: https://www.merck.com/product/usa/pi_circulars/c/crixivan/crixivan_pi.pdf. Accessed 20 March 2017

281. Penzak SR, Acosta EP, Turner M, Edwards DJ, Hon YY, Desai HD, Jann MW (2002) Effect of seville orange juice and grapefruit juice on indinavir pharmacokinetics. J Clin Pharmacol 42:1165–1170

282. Shelton MJ, Wynn HE, Hewitt RG, DiFrancesco R (2001) Effects of grapefruit juice on pharmacokinetic exposure to indinavir in HIV-positive subjects. J Clin Pharmacol 41:435–442

283. Noble S, Faulds D (1996) Saquinavir. Drugs 52:93–112

284. Kupferschmidt HHT, Fattinger KE, Ha HR, Follath F, Krähenbühl S (2003) Grapefruit juice enhances the bioavailability of the HIV protease inhibitor saquinavir in man. Br J Clin Pharmacol 45:355–359

285. Anonymous. INVIRASE (saquinavir mesylate) [package insert]. Nutley, NJ; Roche Pharmaceuticals, Inc. Revised April 2010. Available from: https://www.accessdata.fda.gov/drugsatfda_docs/label/2010/020628s032,021785s009lbl.pdf. Accessed 20 March 2017

286. Kenyon CJ, Brown F, McClelland GR, Wilding IR (1998) The use of pharmacoscintigraphy to elucidate food effects observed with a novel protease inhibitor (saquinavir). Pharm Res 15:417–422

287. Perry CM, Benfield P (1997) Nelfinavir. Drugs 54:81–87

288. Anonymous. LEXIVA (fosamprenavir calcium) [package insert]. Research Triangle Park, NC; GlaxoSmithKline. Revised September 2009. Available from: https://www.accessdata.fda.gov/drugsatfda_docs/label/2009/021548s021,022116s005lbl.pdf. Accessed 20 March 2017

289. Anonymous. KALETRA (lopinavir and ritonavir) [package insert]. Chicago, IL; AbbVie, Inc. Revised November 2016. Available from: http://www.rxabbvie.com/pdf/kaletratabpi.pdf. Accessed 20 March 2017

290. Klein CE, Chiu YL, Awni W, Zhu T, Heuser RS, Doan T, Breitenbach J, Morris JB, Brun SC, Hanna GJ (2007) The tablet formulation of lopinavir/ritonavir provides similar bioavailability to the soft-gelatin capsule formulation with less pharmacokinetic variability and diminished food effect. J Acquir Immune Defic Syndr 44:401–410

291. Anonymous. REYATAZ (atazanavir) [package insert]. Princeton, NJ; Bristol-Myers Squibb. Revised September 2016. Available from: https://packageinserts.bms.com/pi/pi_reyataz.pdf. Accessed 20 March 2017

292. Anonymous. APTIVUS (tipranavir) [package insert]. Ridgefield, CT; Boehringer Ingelheim Pharmaceuticals, Inc. Revised September 2016. Available from: http://docs.boehringer-ingelheim.com/Prescribing%20Information/PIs/Aptivus/10003515%20US%2001.pdf?DMW_FORMAT=pdf. Accessed 20 March 2017

293. Kakuda TN, Van De Casteele T, Petrovic R, Neujens M, Salih H, Opsomer M, Hoetelmans RM (2014) Bioequivalence of a darunavir/cobicistat fixed-dose combination tablet versus single agents and food effect in healthy volunteers. Antivir Ther 19:597–606

294. Kiser JJ, Bumpass JB, Meditz AL, Anderson PL, Bushman L, Ray M, Predhomme JA, Rower J, Mawhinney S, Brundage R (2010) Effect of antacids on the pharmacokinetics of raltegravir in human immunodeficiency virus-seronegative volunteers. Antimicrob Agents Chemother 54:4999–5003

295. Cottrell ML, Hadzic T, Kashuba AD (2013) Clinical pharmacokinetic, pharmacodynamic and drug-interaction profile of the integrase inhibitor dolutegravir. Clin Pharmacokinet 52:981–994

296. Song I, Borland J, Arya N, Wynne B, Piscitelli S (2015) Pharmacokinetics of dolutegravir when administered with mineral supplements in healthy adult subjects. J Clin Pharmacol 55:490–496

297. Patel P, Song I, Borland J, Patel A, Lou Y, Chen S, Wajima T, Peppercorn A, Min SS, Piscitelli SC (2011) Pharmacokinetics of the HIV integrase inhibitor S/GSK1349572 co-administered with acid-reducing agents and multivitamins in healthy volunteers. J Antimicrob Chemother 66:1567–1572

298. Ramanathan S, Mathias AA, German P, Kearney BP (2011) Clinical pharmacokinetic and pharmacodynamic profile of the HIV integrase inhibitor elvitegravir. Clin Pharmacokinet 50:229–244

299. Anonymous. STRIBILD (elvitegravir, cobicistat, emtricitabine, tenofovir disoproxil fumurate) [package insert]. Foster City, CA; Gilead Sciences, Inc. Revised September 2016. Available from: http://www.gilead.com/~/media/files/pdfs/medicines/hiv/stribild/stribild_pi_old.pdf. Accessed 20 March 2017

300. Anonymous. GENVOYA (elvitegravir, cobicistat, emtricitabine, tenofovir alefenamide) [package insert]. Foster City, CA; Gilead Sciences, Inc. Revised February 2017. Available from: https://www.gilead.com/~/media/files/pdfs/medicines/hiv/genvoya/genvoya_pi.pdf?la=en. Accessed 20 March 2017

301. Cattaneo D, Baldelli S, Minisci D, Meraviglia P, Clementi E, Galli M, Gervasoni C (2016) When food can make the difference: the case of elvitegravir-based co-formulation. Int J Pharm 512:301–304

302. Anonymous. Food and drug administration center for drug evaluation and research clinical pharmacology and biopharmaceutics review: Application Number 22–128. Maraviroc. 19 December 2006. Available from https://www.accessdata.fda.gov/drugsatfda_docs/nda/2007/022128s000_ClinPharmR.pdf. Accessed 19 March 2017

303. Anonymous. HEPSERA (adefovir) [package insert]. Foster City, CA; Gilead Sciences, Inc. Revised November 2012. Available from: https://www.accessdata.fda.gov/drugsatfda_docs/label/2012/021449s020lbl.pdf. Accessed 20 March 2017

304. Anonymous. BARACLUDE (entecavir) [package insert]. New York, NY; Bristol-Myers Squibb. Revised August 2015. Available from: http://packageinserts.bms.com/pi/pi_baraclude.pdf. Accessed 20 March 2017

305. Anonymous. TYZEKA (telbivudine) [package insert]. East Hanover, NJ; Novartis Pharmaceuticals, Corp. Revised January 2013. Available from: https://www.accessdata.fda.gov/drugsatfda_docs/label/2013/022011s013lbl.pdf. Accessed 20 March 2017

306. Li L, Koo SH, Limenta LM, Han L, Hashim KB, Quek HH, Lee EJ (2009) Effect of dietary purines on the pharmacokinetics of orally administered ribavirin. J Clin Pharmacol 49:661–667

307. Chang MH, Gordon LA, Fung HB (2012) Boceprevir: a protease inhibitor for the treatment of hepatitis C. Clin Ther 34:2021–2038

308. Anonymous. VICTRELIS (boceprevir) [package insert]. Whitehouse Station, NJ; Merck & Co, Inc. Revised January 2017. Available from: https://www.merck.com/product/usa/pi_circulars/v/victrelis/victrelis_pi.pdf. Accessed 20 March 2017

309. Kiang TK, Wilby KJ, Ensom MH (2013) Telaprevir: clinical pharmacokinetics, pharmacodynamics, and drug-drug interactions. Clin Pharmacokinet 52:487–510

310. Anonymous. INCIVEK (telapravir) [package insert]. Cambridge, MA; Vertex Pharmaceuticals, Inc. Revised December 2012. Available from: http://www.accessdata.fda.gov/drugsatfda_docs/label/2012/201917s007lbl.pdf. Accessed 20 March 2017

311. Anonymous. OLYSIO (simeprevir) [package insert]. Titusville, NJ; Janssen Therapeutics. Revised February 2017. Available from: https://www.olysio.com/shared/product/olysio/prescribing-information.pdf. Accessed 20 March 2017

312. Anonymous. VIEKIRA PAK (ombitasvir, paritaprevir, ritonavir, and dasabuvir) [package insert]. Chicago, IL; AbbVie, Inc. Revised February 2017. Available from: http://www.rxabbvie.com/pdf/viekirapak_pi.pdf. Accessed 20 March 2017

313. King JR, Zha J, Khatri A, Dutta S, Menon RM (2017) Clinical pharmacokinetics of dasabuvir. Clin Pharmacokinet. https://doi.org/10.1007/s40262-017-0519-3

314. Anonymous. ZEPATIER (elbasvir and grazoprevir) [package insert]. Whitehouse Station, NJ; Merck & Co, Inc. Revised February 2017. Available from: https://www.merck.com/product/usa/pi_circulars/z/zepatier/zepatier_pi.pdf. Accessed 20 March 2017

315. Aoki FY, Sitar DS (1988) Clinical pharmacokinetics of amantadine hydrochloride. Clin Pharmacokinet 14:35–51
316. Anonymous. FLUMADINE (rimantadine) [package insert]. Saint Louis, MO; Forest Pharmaceuticals, Inc. Revised April 2010. Available from: https://www.accessdata.fda.gov/drugsatfda_docs/label/2010/019649s015lbl.pdf. Accessed 20 March 2017
317. Acosta EP, Fletcher CV (1997) Valacyclovir. Ann Pharmacother 31:185–191
318. Anonymous. ZOVIRAX (acyclovir) [package insert]. Research Triangle Park, NC; GlaxoSmithKline. Revised June 2005. Available from: https://www.accessdata.fda.gov/drugsatfda_docs/label/2005/018828s030,020089s019,019909s020lbl.pdf. Accessed 20 March 2017
319. Jung D, Dorr A (1999) Single-dose pharmacokinetics of valganciclovir in HIV- and CMV-seropositive subjects. J Clin Pharmacol 39:800–804
320. Pescovitz MD, Rabkin J, Merion RM, Paya CV, Pirsch J, Freeman RB, O'Grady J, Robinson C, To Z, Wren K, Banken L, Buhles W, Brown F (2000) Valganciclovir results in improved oral absorption of ganciclovir in liver transplant recipients. Antimicrob Agents Chemother 44:2811–2815
321. Fuller RK (1979) Disulfiram for the treatment of alcoholism. Ann Intern Med 90:901
322. Adams WL (1995) Interactions between alcohol and other drugs. Int J Addict 30:1903–1923
323. Kannangara DW, Gallagher K, Lefrock JL (1984) Disulfiram-like reactions with newer cephalosporins: cefmenoxime. Am J Med Sci 287:45–47
324. Campbell B, Taylor JT, Haslett WL (1967) Anti-alcohol properties of metronidazole in rats. Exp Biol Med 124:191–195
325. Edwards JA, Price J (1967) Metronidazole and human alcohol dehydrogenase. Nature 214:190–191
326. Kline SS, Mauro VF, Forney RB, Freimer EH, Somani P (1987) Cefotetan-induced disulfiram-type reactions and hypoprothrombinemia. Antimicrob Agents Chemother 31:1328–1331
327. Lassman HB, Hubbard JW, Chen BL, Puri SK (1992) Lack of interaction between cefpirome and alcohol. J Antimicrob Chemother 29:47–50
328. McMahon FG, Noveck RJ (1982) Lack of disulfiram-like reactions with ceftizoxime. J Antimicrob Chemother 10:129–133
329. Kitson TM (1987) The effect of cephalosporin antibiotics on alcohol metabolism: a review. Alcohol 4:143–148
330. Uri JV, Parks DB (1983) Disulfiram-like reaction to certain cephalosporins. Ther Drug Monit 5:219–224
331. Azarnoff DL (1974) Drug interactions: clinical significance. Clin Pharmacol Ther 16:986–988
332. Heelon MW, White M (1998) Disulfiram-cotrimoxazole reaction. Pharmacotherapy 18:869–870
333. Anonymous. NORVIR (ritonavir) [package insert]. Chicago, IL; AbbVie, Inc. Revised December 2016. Available from: http://www.rxabbvie.com/pdf/norvirtab_pi.pdf. Accessed 20 March 2017

Chapter 5
Drug-Cytokine Interactions

Kerry B. Goralski, Matthew A. Ladda, and Jenna O. McNeil

5.1 Introduction

Drug disposition is the general term describing what the body does to a drug and is governed by the processes of drug absorption, distribution, metabolism, and elimination. Evidence of altered drug disposition during infection dates back some 50 years from observations of impaired quinine metabolism in humans with malaria or enhanced cerebrospinal fluid accumulation of rifampin and ethambutol in cases of meningitis [1–3]. While the mechanisms were not known at the time, pioneering preclinical work carried out in the mid-1970s solidified the idea of drug-cytokine interactions [4–6]. The traditionally described drug-cytokine interactions referred to reduced hepatic cytochrome P450 (CYP) metabolism that occurred following exposure to mediators of the innate immune response. It is now established that drug transporters and possibly drug receptors are regulated by cytokines [7–10]. Further, the effects of cytokines on drug disposition are not liver specific but involve the brain, intestine, kidney, placenta, and immune and cancer cells [7, 11–18]. Herein the term "drug-cytokine interaction" will refer to any interaction between a cytokine

K.B. Goralski (✉)
Department of Pharmacology, Dalhousie University, Halifax, NS, Canada

College of Pharmacy, Faculty of Health Professions, Dalhousie University, Halifax, NS, Canada
e-mail: kerry.goralski@dal.ca

M.A. Ladda
College of Pharmacy, Faculty of Health Professions, Dalhousie University, Halifax, NS, Canada
e-mail: ladda@Dal.Ca

J.O. McNeil
Department of Family Medicine, Dalhousie University, Halifax, NS, Canada
e-mail: jennamcneil@Dal.Ca

© Springer International Publishing AG 2018
M.P. Pai et al. (eds.), *Drug Interactions in Infectious Diseases: Mechanisms and Models of Drug Interactions*, Infectious Disease,
https://doi.org/10.1007/978-3-319-72422-5_5

Table 5.1 List of abbreviations

ABC: ATP-binding cassette transporter	*IFN*: interferon
ABCB1: p-glycoprotein	*IL*: interleukin
ABCC1-4: multidrug resistance proteins 1-4	*LPS*: lipopolysaccharide
ABCG2: breast cancer resistance protein	*M3G*; morphine-3-glucuronide
AhR: aryl hydrocarbon receptor	*M6G*; morphine-6-glucuronide
BBB: blood-brain barrier	*NF-κB*: nuclear factor kappa B
CD: Crohn's disease	*NO*: nitric oxide
CL: clearance	*PolyIC*: polyinosinic-polycytidylic acid
CNS: central nervous system	*PPI*: proton pump inhibitor
CSF: cerebrospinal fluid	*PXR*: pregnane-x-receptor
CYP: cytochrome P450 enzyme	*SLC*: solute carrier transporter
ET-1: endothelin-1	*TNF*: tumor necrosis factor
HIV: human immunodeficiency virus	*UC*: ulcerative colitis
IBD: inflammatory bowel disease	

and drug-metabolizing enzyme, drug transporter, or receptor that leads to altered drug disposition and/or drug response. To assist the reader, a complete list of abbreviations used in this chapter is provided in Table 5.1.

Over the past 35 years, the understanding of drug-cytokine interactions has greatly expanded. It is currently appreciated that inflammatory conditions including bacterial and viral infections, surgical procedures, inflammatory diseases of the central nervous system (CNS), cancer and autoimmune diseases, and cytokine therapies alter drug disposition processes (Fig. 5.1 and Table 5.2) [13, 19–22]. These positive primary stimuli trigger the signaling of inflammatory cytokines, interleukins 1 and 6 (IL-1 and IL-6), tumor necrosis factor (TNF), and interferons (IFNs). The inflammatory cytokines (primary mediators) bind to cell surface receptors in target organs and activate intracellular signaling cascades that increase or decrease transcription factors to regulate CYP and drug transporter gene transcription, protein levels, and corresponding metabolic and transport activity [12, 13, 20–22]. A second mechanism involves production of nitric oxide (NO) by nitric oxide synthase, which affects drug metabolism and transport through transcriptional or posttranslational mechanisms [12, 23–26]. The end result is typically a loss in drug metabolism and transport, but there are instances where enhanced metabolic or transport activity occurs. This ultimately depends on the target organ, the nature and duration of the primary inflammatory stimuli, and the CYP or transporter involved. In recent years, it has also become apparent that some anti-cytokine therapies have the ability to induce or restore CYP function to normal levels through the blockade of inflammatory cytokine signaling in chronic inflammatory conditions. This chapter provides an overview of organ-specific drug-cytokine interactions and the specific infectious and inflammatory conditions that may lead to drug-cytokine interactions in humans. For the purpose of clarifying nomenclature, italicized upper case (e.g., *CYP3A* and *ABCB1*) and lower case (*cyp3a* and *abcb1*) abbreviations specifically refer to CYP or drug transporter gene or mRNA in humans and rodents,

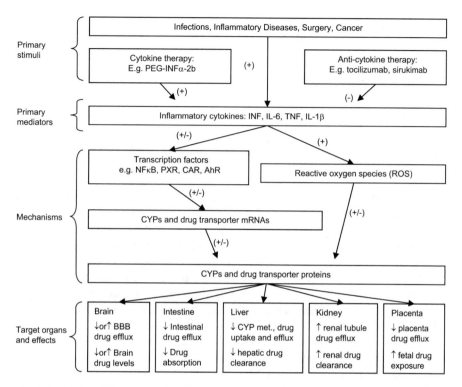

Fig. 5.1 Outline of the proposed pathways and target organs of drug-cytokine interactions in infections and inflammatory diseases. (+) or ↑ symbols and (−) or ↓ symbols denote activation or inhibition, respectively. A complete list of abbreviations is provided in Table 5.1

Table 5.2 Organisms and inflammatory stimuli known to alter CYP metabolism, drug transporter function, drug disposition, or drug effectiveness in humans

Organisms and inflammatory stimuli	Examples
Viruses	Hepatitis [66, 205, 241], influenza [67, 68], adenovirus [67], herpes simplex [242], meningitis [2, 3], HIV [73, 137], hepatitis C [85, 215]
Bacteria	*Helicobacter pylori* [10, 117, 119, 121]
Parasites	*Plasmodium falciparum* [1, 243], *Schistosoma mansoni* [244], *Opisthorchiasis viverrini* [245]
Inflammatory stimuli	Vaccines [75], IFNα, β and γ [23, 57], PEG-IFNα2b [33, 208, 209], IL-1α, -1β,-2 and -6 [23, 57], TNF [23, 57], LPS [246, 247]
Inflammatory conditions	Tissue injury/trauma [139], surgical stress [248], cancer [77, 201], IBD [106, 110], CNS diseases [136–138, 140], heart failure [74], sepsis [249, 250]
Anti-cytokine therapies	Adalimumab [230], sarilumab [233], tocilizumab [69], basiliximab [224], sirukumab [234]

A complete list of abbreviations is provided in Table 5.1

respectively. In all other instances, non-italicized upper case (e.g., CYP3A or ABCB1 for humans) or lower case (e.g., cyp3a or abcb1 for rodents) abbreviations are used.

5.2 Drug Metabolism and Drug Transport

Drug metabolism and transport are integrated processes that dictate drug disposition in the body and provide protection against drugs and chemicals (Fig. 5.2) [27, 28]. Both CYP enzymes and transport proteins are sites of drug-cytokine interactions and should be considered with respect to altered drug disposition in conditions where inflammation is present.

The CYP, a gene superfamily of heme-containing enzymes, has a major role in phase I drug detoxification [22, 29]. The highest concentrations of drug-metabolizing CYPs including CYP1A1/CYP1A2, CYP2B6, CYP2C9, CYP2C19, CYP2D6, CYP2E1, and CYP3A4 (Table 5.3) are found in the liver and intestine with lower amounts in other major organs. The CYP3A enzymes are particularly important with regard to drug interactions as they comprise 30–50% of CYP content in the liver and metabolize 50–60% of clinically used drugs [29]. The effects of inflammatory stimuli on the phase II conjugation enzymes including N-acetyltransferases, UDP-glucuronosyltransferases, and sulfotransferases have been described but remain poorly defined relative to CYPs [30–33].

Drug transporters are a collection of membrane proteins that exist in all major organs where they contribute to organ physiology and drug disposition [34–38]. Intestinal transporters mediate dietary nutrient and drug absorption into the mesenteric circulation (Fig. 5.2a). Hepatic transporters are critical for cholesterol transport, bile secretion, and biliary drug elimination (Fig. 5.2b). Renal tubule transporters mediate solute reabsorption and urinary drug elimination (Fig. 5.2c). Brain capillary endothelial transporters control the uptake of nutrients into the brain while simultaneously preventing harmful compounds from accumulating in the CNS (Fig. 5.2d). Transporters are grouped into the solute carrier (SLC) and the ATP-binding cassette (ABC) superfamilies [35–38]. Drug uptake into cells is primarily mediated by the SLC22 family (organic cation and organic anion transporters) and the SLC01 family (organic anion transporting polypeptides) [39–41]. Additional

---

Fig. 5.2 (continued) and efflux via ABCB1, ABCC2, ABCG2, and ABCC11 across the canalicular membrane contributes to the biliary elimination of drugs and their metabolites (**b**). The ABC transporters ABCC3 and ABCC4 can transport metabolites drugs and metabolites from the hepatocyte back into the sinusoidal blood for distribution to the circulation and other organs. In the kidney proximal tubules, various drug uptake (SLC22s and SLCOs) and efflux (ABCB1, ABCC2, ABCC4, and ALC22A4, 5) transporters contribute renal tubule secretion of drugs and metabolites and their subsequent excretion in the urine (**c**). In the brain capillary endothelium SLCs (e.g., SLCO1A2 and SLCO2B1) can help deliver certain drugs from the blood to the brain, whereas ABC transporters (ABCB and ABCG2) protect the brain by pumping drugs and metabolites from the capillary endothelial cells into the blood (**d**)

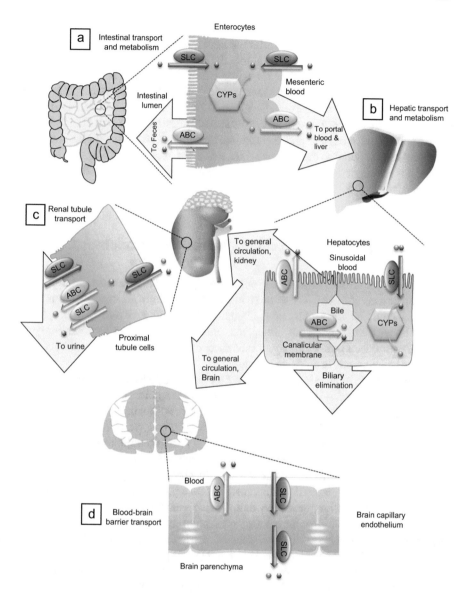

Fig. 5.2 *The role of CYPs and drug transport proteins in absorption, metabolism, and excretion of drugs.* The absorption, tissue distribution, and excretion of drugs (black hexagons) and metabolites (orange circles) are determined by the combined action of solute carrier (SLC) transporters (red ovals), cytochrome P450 enzymes (blue hexagons), and ATP-binding cassette (ABC) transporters (green ovals). SLC transporters including SLCOs, SLC15A1, and SLC16A mediate drug uptake into the intestinal enterocytes following oral administration. This can be followed intracellular metabolism by CYPs and ABC transporter-mediated efflux of the metabolite or parent drug into the intestinal lumen for elimination in the feces or into the mesenteric blood for delivery to the liver via the portal circulation (**a**). In the liver, drug uptake by SLCOs, SLC22s, and SLC10A1 across the sinusoidal membrane, intracellular CYP metabolism or phase II conjugation not shown),

nutrient transporters including SLC1 (amino acid transporters), SLC15 (oligopeptide transporters), and SLC16 (monocarboxylic acid transporters) mediate the cellular uptake of drugs that structurally resemble the natural transported ligands [42, 43]. The ABC transporters (Table 5.4) including ABCB1, multidrug resistance proteins (ABCC1-4), and breast cancer resistant protein (ABCG2) are the primary mediators of drug transport out of cells (efflux) [34]. ABCB1 is the best-understood ABC transporter. It exists in the apical membrane of intestinal enterocytes, the biliary membrane of hepatocytes, and the luminal membrane of renal tubules, where it mediates drug efflux into the intestine, bile, and urine, respectively [44, 45]. Further, ABCB1 is an important blood-brain barrier efflux transporter that limits drug accumulation in CNS [46].

5.3 Cytokines and the Acute Inflammatory Response

Cytokines are a diverse superfamily of secreted proteins that function in immunity and metabolism [47, 48]. These molecules are secreted from monocytes, macrophages, T-cells, and mast cells and nonhematopoietic cells such as adipocytes, fibroblasts, hepatocytes, epithelial cells, and chondrocytes [49]. Cytokines are not normally produced constitutively, rather their expression and secretion occurs in response to infectious or injurious stimuli.

Drug-cytokine interactions have been commonly associated with acute inflammation, a generalized immune response that provides a potent early defense against primary infection or tissue injury in order to counteract the source of the disturbance and restore homeostasis [50]. Stressors including infections, trauma, and surgery activate the innate immune response leading to local inflammation and systemic responses, which can alter drug disposition in humans (Table 5.2). The primary immune sensors, host tissue macrophages and blood monocytes, contain pattern recognition receptors, including the transmembrane toll-like receptors and cytosolic nucleotide-binding oligomerization domain-like receptors, which bind conserved pathogen-associated molecular patterns present on infecting microorganisms (e.g., lipopolysaccharide, LPS), virulence factors, particulate irritants, and endogenous molecular indicators of cell stress or cell death [51, 52]. This sets in motion a signaling cascade leading to enhanced expression and release of pro-inflammatory cytokines (e.g., TNF, IL-1, and IL-6), chemoattractant molecules, prostaglandins, histamine, bradykinin, complement proteins, NO, and proteolytic factors [51, 53]. Locally, these mediators enhance vascular permeability and recruit immune cells into the infected or injured tissue for removal of invading pathogens and/or damaged tissue and contribute to wound healing [50, 51]. With increased severity of tissue insult, greater quantities of inflammatory mediators are secreted into the circulation [50]. This allows for activation of their respective receptors in target organs, which produces physiologic changes that define the systemic inflammatory response: fever, appetite suppression, activation of the hypothalamic-pituitary adrenal axis, muscle protein catabolism, production of hepatic acute phase proteins, and

Table 5.3 Drug-metabolizing CYP enzymes and representative substrates

Enzyme	Drug class
CYP1A1/ CYP1A2	*Analgesics*: acetaminophen *Anticancer*: erlotinib, tamoxifen *Methylxanthines*: theophylline, caffeine
CYP2B6	*Antidepressants*: bupropion *Anticancer*: cyclophosphamide, ifosfamide, tamoxifen
CYP2C9	*Antiviral*: ritonavir *Anti-inflammatories*: celecoxib, ibuprofen, indomethacin, naproxen *Anticancer*: idarubicin, cyclophosphamide *Anticoagulants*: warfarin *Antidiabetics*: glipizide, glibenclamide
CYP2C19	*Anticancer*: cyclophosphamide *Proton pump inhibitors*: omeprazole, pantoprazole, rabeprazole *Antiplatelet*: clopidogrel
CYP2D6	*Analgesics*: codeine, fentanyl, hydrocodone, oxycodone, *Anticancer*: cyclophosphamide, idarubicin, tamoxifen
CYP2E1	*Antibiotics*: dapsone *Analgesics*: acetaminophen *Other*: chlorzoxazone, ethanol
CYP3A4	*Antibiotics*: clarithromycin, erythromycin, metronidazole *Antifungals*: fluconazole, itraconazole, ketoconazole, miconazole *Antivirals*: delavirdine, efavirenz, indinavir, nelfinavir, ritonavir, saquinavir *Anti-inflammatories*: cortisol, hydrocortisone, methylprednisolone, prednisolone, prednisone *Analgesics/sedatives*: fentanyl, midazolam, triazolam *Anticancer*: doxorubicin, etoposide, vinblastine, vincristine *Direct-acting HCV antivirals:* daclatasvir, dasabuvir, elbasvir, grazoprevir, paritaprevir, simeprevir

altered drug disposition [50, 54–56]. The pro-inflammatory cytokines IL-1β, IL-6, IFNα/β/γ, and TNF appear to be particularly important mediators, which link the immune/inflammatory response with altered drug disposition in mammals. When applied individually to cells, cytokines regulate CYPs and transporters with a certain amount of redundancy [7, 20, 23, 57–59]. Thus, in humans, the overall effects on drug disposition are likely due to the collective and redundant actions of the multiple cytokines that are released upon immune stimulation.

5.4 Drug-Cytokine Interactions and the Liver

The first step in hepatic drug elimination is SLC-mediated passage of the drug from the sinusoids into the hepatocyte where the drug may undergo metabolism by CYPs and conjugating enzymes (Fig. 5.2a). Canalicular ABC transporters then mediate drug or metabolite secretion into bile, whereas sinusoidal ABC transporters deliver drugs or metabolites back into circulation (Fig. 5.2b). Cytokine-mediated losses in

Table 5.4 Common ABC drug efflux transporters and representative substrates

Transporter	Tissues	Substrates
ABCB1	Intestine, kidney, liver, brain, placenta, cancer cells	*Antibiotics*: erythromycin, levofloxacin, rifampin, sparfloxacin *Antifungals*: ketoconazole *Antivirals*: amprenavir, indinavir, nelfinavir, ritonavir, saquinavir *Analgesics*: morphine *Anticancer*: anthracyclines, anthracenes, epipodophyllotoxins, taxanes, vinca alkaloids *Anti-inflammatories*: dexamethasone, prednisolone, cortisol *Direct-acting HCV antivirals*: daclatasvir, dasabuvir, ledipasvir, ombitasvir, paritaprevir, sofosbuvir
ABCC1	Ubiquitous, low in the liver	*Antivirals*: indinavir, ritonavir, saquinavir *Anticancer*: anthracenes, anthracyclines, cisplatin, epipdodphyllines, flutamide, methotrexate, vinca alkaloids
ABCC2	Liver, gut, kidney, brain, placenta, gall bladder	*Antibiotics*: ceftriaxone, rifampin *Antivirals*: indinavir, ritonavir, saquinavir *Analgesics*: acetaminophen, diclofenac *Anticancer*: cisplatin, doxorubicin, etoposide, methotrexate, vinblastine, vincristine
ABCC3	Liver, gut, brain, kidney, lung, prostate, gall bladder, prostate, placenta,	*Analgesics*: acetaminophen *Anticancer*: etoposide, leucovorin, methotrexate, teniposide
ABCG2	Placenta, liver, kidney, intestine, brain	*Antiviral*: azidothymidine, lamivudine *Anticancer*: adriamycin, daunorubicin, doxorubicin, etoposide, flavopiridol, irinotecan, methotrexate, mitoxantrone, topotecan *Direct-acting HCV antivirals*: ombitasvir, sofosbuvir

A complete list of abbreviations is provided in Table 5.1

hepatic CYP metabolism, drug uptake, or efflux transporter function are established and may reduce drug clearance, increase plasma drug levels, and enhance drug efficacy and/or toxicity.

The pioneering work related to hepatic drug-cytokine interactions was carried out in the 1970s by several research groups. A seminal observation published in 1972 identified that polyinosinic-polycytidylic acid (PolyIC), a molecule that mimics double-stranded viral RNA, induces an interferon (IFN) response and suppresses hepatic CYP metabolism in vitro and in vivo [60]. Morahan and coworkers correctly speculated that the reduction in hepatic metabolism involved the inhibition of RNA and protein synthesis, but the connection between the immune response and reduced hepatic drug metabolism would have to wait for an unexpected observation by Drs. Renton and Mannering in 1975. Based on previously reported cell culture studies, these investigators had reason to believe that an IFN-inducing agent, tilorone, would potentiate the induction of hepatic cyp metabolism by barbiturates and polycyclic aromatic hydrocarbons [5, 61]. Instead, the opposite was observed; tilorone administered to rats transiently reduced their total hepatic cyp protein content and microsomal cyp enzyme activity. The loss in hepatic metabolic activity was pharmacologically relevant as it elevated plasma levels of hexobarbital and lengthened barbiturate-induced sleeping time. This pivotal finding led to the hypothesis that IFN or some aspect of the IFN induction mechanism mediates the loss in cyp-mediated drug metabolism [5]. The hypothesis was quickly reinforced by a follow-up study in which a diverse array of IFN-inducing agents including an RNA virus (Mengo), fungal mycophage (statolon), PolyIC, E. coli LPS, and an attenuated bacteria (B. pertussis) vaccine similarly inhibited hepatic cyp metabolism [4]. Around the same time, investigations of Corynebacterium parvum (C. parvum) and Bacillus Calmette-Guerin as immunotherapeutic agents for cancer therapy demonstrated similar immune-mediated reductions in hepatic drug metabolism [6, 62]. These two studies extended the concept that immune stimulation alters drug pharmacokinetics and pharmacodynamics in animals and provided a potential explanation for enhanced toxicity of short-acting barbiturates and hematopoietic toxicity of chemotherapeutic agents in rodents treated with C. parvum [63, 64]. Subsequently, it was shown that irradiation or splenectomy could block the C. parvum-mediated reduction of hepatic drug-metabolizing activity identifying for the first time that monocytes and macrophages, which release cytokines during the inflammatory response, are cellular mediators of the interaction [65].

In the 35 years, hence, the effects of cytokines on hepatic cyp metabolism have been firmly established in animals [20–22, 66]. The human situation is more complicated, due to inherent variability in drug disposition, polypharmacy, and underlying diseases. Nonetheless, the clinical reality of drug-cytokine interactions was recognized early on through observations that asthmatic children previously controlled on theophylline experienced reduced clearance, higher steady-state peak concentrations, and toxicity of theophylline during febrile viral illness [67, 68]. In addition, many human studies and/or case reports support the notion that specific hepatic drug-cytokine interactions may arise in clinically applicable situations. These include impaired theophylline elimination and attainment of toxic theophylline levels in recipients of influenza vaccine; increased half-life and decreased clearance of midazolam in critical illness; decreased clearance of cyclosporine, carbamazepine, and omeprazole in patients following allogeneic bone marrow transplants, temporal lobectomy, and spinal cord injuries, respectively; decreased metabolism of omeprazole and erythromycin in advanced cancer; decreased dextromethorphan metabolism in patients with active HIV infection; and decreased caffeine and mephenytoin metabolism in patients with heart failure and altered simvastatin metabolism in patients with rheumatoid arthritis following treatment with the IL-6 receptor antibody tocilizumab [69–77].

The question of whether alterations in drug disposition in humans are explained by hepatic drug-cytokine interactions has been addressed using the HepaRG hepatoma cell line and primary hepatocytes or liver microsomes prepared from human donors (Table 5.5) [23, 25, 57, 78–83]. The general effect of IL-6, TNF, IFNγ, IL-1β, and IL-2 is to differentially reduce the basal levels of mRNA, protein, and/or

Table 5.5 Summary of the documented or proposed effects of cytokines or inflammatory diseases on drug metabolism and transport in humans

Inflammatory mediator or disease/model	General effect on CYPs or transporters	Documented or proposed effect on drug disposition in vivo	References
Liver			
LPS, IL-6, TNF, IFNα,γ IL-1β, IL-2 and IL-4 in primary hepatocytes, liver microsomes	↓ CYP1A1/CYP1A2, CYP2B6, CYP2C9/19, CYP2E1, and CYP3A4	↓ hepatic drug CL (documented for LPS and IFNα, proposed for other cytokines)	[23, 25, 57, 79–83]
	↑ CYP2E1 by IL-4	↑ hepatic CL of CYP2E1 substrates (proposed)	[80]
3D hepatocyte Kupffer cell culture	↓ CYP3A4 by IL-6 and reversed by tocilizumab	↓ hepatic drug CL and recovery with IL-6 antibody treatment (proposed)	[19]
LPS, IL-6, TNF, IL-1β In primary hepatocytes or liver slices	↓ sinusoidal uptake transporters SLC10A, SLC22A, SLCO1B1, 1B3 and 2B1, ↓ canalicular efflux transporters ABCB11, ABCC2 and ABCG2 ↓ sinusoidal efflux transporters ABCC4	↓ hepatic drug CL (proposed)	[58, 59, 97]
Gastrointestinal tract			
CD and UC/intestinal biopsies	↓ ABCB1 and ABCG2 in inflamed regions of intestine vs. noninflamed regions or vs. healthy controls. ABCB1 and ABCG2 return to control levels in UC remission	↑ oral drug bioavailability (proposed)	[106–110]
H. pylori + IL1βT/T and T/C genotypes	↑ inflammatory response Unknown effect on CYPs or transporters	↑ efficacy of triple therapy in CYP2C19 rapid metabolizers (documented)	[10, 117, 121, 125]
Or H. pylori cagA+/vacA s1	↑ inflammatory response Unknown effect on CYPs or transporters	↑ efficacy of triple therapy (documented)	
Kidney			
Human data not available	Proposed ↓ in proximal tubule uptake and ↑ in proximal tubule ABCB1 efflux transporters based on animal data	↑ renal elimination of ABCB1 substrates and ↓ proximal tubule drug accumulation (proposed based on animal data)	[17, 86, 125–127]

(continued)

Table 5.5 (continued)

Inflammatory mediator or disease/model	General effect on CYPs or transporters	Documented or proposed effect on drug disposition in vivo	References
Brain			
IL-1β, IL-6 and TNF Human brain capillary endothelial cells	↓ ABCG2 with all cytokines, ↓ ABCB1 with IL-6, ↑ABCB1 with TNF	↑ brain penetration of ABCG2 substrates and ↑↓ ABCB1 substrates (proposed)	[161]
Meningitis	Proposed ↓ in BBB ABCB1	↑ CSF levels of rifampin and ethambutol (documented)	[2, 3]
Postmortem brain from HIV⁻, HIVE⁻HIVE⁺ subjects	ABCB1 ↓ in brain capillaries of HIVE⁻ and HIVE⁺ ABCB1 ↑ in astrocytes and microglia of HIVE⁺ comparedHIVE⁻andHIV⁻	↑ brain penetration of antiretrovirals but ↓ into infected glia (proposed)	[137]
Acute head injury	Proposed ↓ in BBB ABC transporters	↑ CSF levels of M3G and M6G (documented)	[139]
Parkinson's disease	Proposed ↓ in BBB ABCB1	↑ midbrain penetration of ¹¹C-verapamil (documented)	[136]
Placenta			
TNF, IL-1β and Il-6 in term placental trophoblasts	↓ apical ABCB1, ABCG2 ↑ or ↔ basolateral ABCC1,4	↑ or ↓fetal drug exposure depending on stage of pregnancy and complications (proposed)	[174]
LPS and Poly-IC in 1st or 3rd trimester placental trophoblasts	↓ apical ABCB1, ABCG2 by LPS in 1st but not 3rd trimester ↓ apical ABCB1 by Poly-IC in 3rd but not 1st trimester		[181]
Placenta from preterm labor with inflammation	↑ ABCB1, ABCG2		[188]

Definition of table symbols: HIV⁻, not infected with HIV; HIVE⁻, HIV infected without encephalitis; HIVE⁺, HIV infected with encephalitis; ↑, increased compared to controls; ↓, decreased compared to controls; and ↔, unchanged compared to controls. A complete list of abbreviations is provided in Table 5.1

activities of hepatic CYP1A, CYP2B, CYP2C, CYP2E, and CYP3A [23, 25, 57, 80–83]. One exception is a substantial IL-4-mediated induction of CYP2E1 mRNA and protein in primary human hepatocytes or human hepatoma cells [57, 80]. The cytokine-mediated reductions in CYP mRNA, protein, and/or activity typically have ranged between 40 and 90% and occur 1–4 days after cytokine treatment. The degree of reduction depends on the cytokine and CYP in question, and the effect may increase or decrease in magnitude upon simultaneous exposure to multiple cytokines, as would occur during a systemic inflammatory response [23, 79, 84]. Reduced CYP mRNA levels and subsequent reductions in protein or enzyme activity are usually observed indicating regulation at the level of gene transcription. Donato and colleagues identified that NO contributes to a portion (50%) of the total loss of hepatic CYP1A1/CYP1A2 activity after INFγ stimulation [25]. The effect of NO is independent of transcriptional regulation and only affects specific CYP isoforms [79]. Binding of pathogen-associated molecular patterns (e.g., LPS) and cytokines directly to their cognate receptors on hepatocytes and activation of intracellular signaling mechanisms regulate CYPs. However, these effects are also augmented through interactions with Kupffer cells and peripheral blood mononuclear cells. For instance, the IL-2, but not the IL-1- or IL-6-mediated reduction in hepatocyte CYP3A4 activity, was of greater magnitude and sustained for a longer period of time when hepatocytes were co-cultured with Kupffer cells [82]. This indirect effect could occur following IL-2-mediated release of IL-1 and IL-6 from Kupffer cells, which in turn activate their hepatocyte receptors. In a second more recent study, ANC28.1, a monoclonal antibody against human T-cell CD28, did not directly affect CYP1A2, CYP2B6, and CYP3A4 mRNA or activity when applied directly to human hepatocyte/Kupffer cell co-cultures [84]. However, when plasma from ANC28.1-treated blood was applied to the same co-cultures, suppression of CYP1A2, CYP2B6, and CYP3A4 mRNA and activity was observed and attributed to cytokine release from leukocytes in human whole blood [84]. However, there are limitations in predicting clinically significant drug interactions from primary hepatocyte cultures given that their metabolic function changes over a short period of time. The exciting recent development of a three-dimensional perfusable human hepatocyte/Kupffer cell co-culture model offers a more physiologically relevant model to further evaluate complex drug-cytokine interactions over longer periods [19]. As proof of principle, Long et al. have used this model to demonstrate IL-6-mediated downregulation of CYP3A4 expression and function [19]. The observation of reduced metabolism of the CYP3A4 probe midazolam in humans with chronic hepatitis C infection compared to healthy controls supports the potential clinical relevance of cytokine-mediated drug interactions in humans [85].

Equally important is the consideration that inflammatory cytokines impair hepatic drug elimination through suppression of drug transporter function. Activation of the innate immune response in rats by the administration of LPS, inflammatory cytokines, or the IL-6 inducer turpentine reduced the hepatic mRNA, protein expression, and function of the canalicular abcb1 drug efflux transporter [86–88]. In rodents, the loss of hepatic abcb1 manifests as reduced biliary clearance and increased hepatic accumulation and/or plasma levels of its substrates, digoxin, doxorubicin, and [99m]-Tc-sestamibi [14, 86, 89, 90]. Further, inflammation and competitive inhibitors of abcb1 reduce biliary drug elimination in an additive fashion in rats indicating the combination of polypharmacy and inflammation is a situation of potential clinical concern [14]. Cytokine effects on hepatic drug transport are potentially broad as reductions in sinusoidal bile salt (slc10a1), organic anion (slco1a1 and slco1a5) and organic cation (slc22a1) uptake transporters, other bile canalicular efflux transporters including bile salt export protein (abcb11) and multidrug-resistant protein 2 (abcc2), and sinusoidal efflux transporters (abcc3) occur following treatment of rodents with LPS, turpentine, IL-6, and IL-1 [86, 91–96].

Initial evidence for human cytokine-hepatic drug transporter interactions stems from the finding that LPS decreased the expression of the bile salt uptake transporter (*SLC10A*) and the canalicular efflux transporter *ABCC2* in liver slices [97]. The change in *SLC10A* was inversely correlated with IL-1β and TNF production by the liver slices indicating the effect was likely cytokine mediated [97]. The regulatory link has been further evaluated in primary human hepatocytes isolated from hepatic tissue from individuals with primary and secondary tumors [58, 59]. IL-1β, IL-6, and TNF globally reduced the expression of sinusoidal organic cation (*SLC22A*), organic anion (*SLCO1B1, 1B3,* and *2B1*), and bile acid uptake transporters (*SLC10A*) and differentially reduced drug (*ABCC2, ABCC4,* and *ABCG2*) and bile salt (*ABCB11*) efflux transporters (Table 5.5). The loss of drug transporter mRNA expression occurred 8–48 h after cytokine treatments. For select transporters, corresponding reductions in protein and transporter activity were shown. Studies with human hepatocyte/parenchymal cell co-cultures also suggest that the anti-HCV therapy IFN-α2b could contribute to drug interactions through its combined regulatory effects on hepatic uptake (SLC22A6) and efflux transporters (ABCC2) as well as CYPs and the phase II metabolic enzymes [33]. Limited data supports that cytokine effects on hepatic drug and bile acid transporters are probable in humans with a potential impact on impaired drug and bile acid elimination caused by inflammation or treatment with cytokine therapies [98].

5.5 Drug-Cytokine Interactions in the Gastrointestinal System

The intestine is the primary site of absorption for orally administered drugs. Intestinal SLC transporters facilitate drug absorption, whereas enterocyte CYP3A metabolism and ABCB1 and ABCC2 efflux transporters provide barriers against drug absorption (Fig. 5.2a). Reductions in intestinal abcb1 and abcc2 mRNA, protein expression and/or function, and cyp3a expression and metabolism occur in rodents with bacterial infection, colitis, and chronic kidney disease [16, 99–102]. Chronic treatment of mice with IL-2 lowered intestinal abcb1 protein and increased oral bioavailability of digoxin providing evidence of cytokine involvement [103]. Theoretically, such changes to intestinal metabolism or efflux transport could induce variability in oral drug absorption. In humans, many inflammatory conditions affect the gastrointestinal system and therefore have the potential to increase local cytokine concentrations and modify drug absorption. Two relatively common gastrointestinal conditions in which the evidence supports the possibility of drug-cytokine interactions are inflammatory bowel disease (IBD) and *Helicobacter pylori* (*H. pylori*) infection.

IBD is a term used to encompass a group of autoimmune disorders affecting the GI tract, of which Crohn's disease (CD) and ulcerative colitis (UC) are the most prevalent. In these diseases, the expression of IL-1β, IL-4, IL-5, IL-8, IL-10, IL-12, TNF, and IFNγ can be elevated [104, 105]. Several studies have documented differential dysregulation of genes involved in intestinal drug detoxification and drug efflux in humans with IBD. Sizable reductions in ABCB1 and ABCG2 mRNA and protein expression have been demonstrated in biopsies from inflamed intestinal regions of subjects newly diagnosed with UC compared to noninflamed sections, treatment refractory patients, or healthy mucosa of control patients [106]. A second study demonstrated an induction of IL-6 and IL-1β combined with a 70–80% reduction in ABCB1 and ABCG2 mRNAs and proteins in inflamed colons and rectums of subjects with active UC compared to those in remission or healthy controls [107]. Similarly, ABCB1 mRNA was reduced in sigmoidal tissue in humans with active UC compared to healthy controls. In this study, the strong inverse association between ABCB1 expression and disease activity was postulated to involve time-dependent modulatory effects of IL-8, which was also elevated [108]. In a fourth study, *ABCB1* mRNA levels were reduced in inflamed colons of subjects with active UC and CD compared to controls [109]. The depression of *ABCB1* mRNA was recapitulated by treatment of intestinal biopsies with a cytokine cocktail containing TNF, IL-1β, and IFNγ [109]. In comparison, Langmann reported that *ABCB1* mRNA was reduced in the colon but not ileum of those with UC but not CD suggesting there could be disease and tissue-specific regulation of *ABCB1* [110]. These studies generally support that a reduction of intestinal ABCB1 and ABCG2 drug efflux transporters occurs during active IBD and that this effect is related to the inflammatory process (Table 5.5). A second observation was that ABCB1 and ABCG2 mRNA and protein levels in the colonic mucosa of UC subjects in remission were similar or higher than in healthy controls indicating that the intestinal barrier function afforded by these transporters returns to normal with resolution of the inflammatory process [106, 107]. While more pharmacokinetic studies are needed, some have hypothesized that reduced intestinal drug efflux may have implications for oral drug absorption, aggravating intestinal inflammation and contributing to increased rates of colorectal cancer in UC patients due to an accumulation of carcinogens [106, 107, 109]. In addition to directly affecting intestinal drug transport and metabolism, preclinical studies provide some evidence that IBD may impact systemic drug metabolism and disposition [111–116]. In mice, the dextran sulfate sodium and *citrobacter rodentium* infection models of UC affected oral drug absorption through cytokine-mediated downregulation of various drug-metabolizing hepatic cyps as well as abcb1 during peak disease [111–113, 115, 116]. Differing results have been observed in 2,4,6-trinitorbenzene sulfonic acid model of UC in mice and rat [114]. In these experiments, an upregulation of abcb1 in polymorphic blood mononuclear cells was observed in the subacute phase of the disease. The change in abcb1 was associated with elevated plasma lipopolysaccharide levels and inflammatory cytokines including TNF, IL-6, IL-17, and INFγ and a reduction in PBMC intracellular accumulation of cyclosporine and plasma cyclosporine area under the curve (AUC) [114]. It was postulated that the enhanced function of ABCB1 could be a mechanism of acquired resistance in patients with IBD; however, this idea remains to be confirmed in humans [114].

H. pylori infection is a second relevant condition in which drug-cytokine interactions occur in the gastrointestinal tract. *H. pylori* colonizes the gastric mucosa of humans with relatively high prevalence: 25% in developed countries and up to 80–95% in the developing world [10]. Infection with *H. pylori* causes chronic gastritis, which leads to gastric atrophy and metaplasia, a known risk factor for gastric cancer [117]. The current gold standard for *H. pylori* eradication is triple therapy with two antibiotics (generally amoxicillin and clarithromycin) and a proton pump inhibitor (PPI). This combination therapy reaches eradication rates of 80–90%, but individual success depends on many factors including local resistance, host genetics, bacteria virulence factors, and level of gastric acid inhibition [10, 118, 119]. Gastric acid suppression is crucial to therapeutic success as it decreases degradation of the acid sensitive antibiotics and increases antibiotic susceptibility of the bacteria [10, 117]. In this regard, a positive correlation between the level of gastric inflammation caused by *H. pylori* and the success of bacterial eradication using triple therapy has been shown. A possible explanation for this is that inflammatory mediators IL-1β and TNF that are produced in the gastric mucosa are also potent inhibitors of gastric acid secretion [10, 117]. Further insight into this relationship stems from studies of *H. pylori* virulence factors and naturally occurring genetic polymorphisms in human IL-1β, IL-1 receptor antagonist, and TNF. Of these, an IL-1β polymorphism (IL-1β -511) is associated with differences in acid inhibition in response to *H. pylori* infection [117]. *H pylori* infected individuals with the IL1β -511 T/T and T/C genotypes have significantly higher IL-1β production and elevated median intragastric pH levels compared to those with the IL-1β -511C/C allele [120]. Correspondingly, the IL-1β-511C/C allele is associated with reduced clinical effectiveness of PPI/amoxicillin/clarithromycin triple therapy in CYP2C19 extensive metabolizers [117, 121]. A second group of polymorphisms concerns the *cagA* and *vacA H. pylori* virulence factors. The *cagA*-positive *H. pylori* strains are associated with severe gastric inflammation and produce significantly higher levels of IL-1β [10]. Although there have been conflicting results, the most recent publication by Sugimoto et al. concluded that the cure rates of patients with the *cagA*-positive/*vacA* s1 *H. pylori* strains were significantly higher than those with *cagA*-negative/*vacA* s2 strains [10, 119]. This elevated cure rate is considered to be the result of higher cytokine levels in the gastric mucosa. These interesting relationships suggest a previously unrecognized and beneficial drug-cytokine interaction in which the degree of inflammation produced by infection enhances antibiotic effectiveness. This could occur by reducing antibiotic degradation and increasing bacterial susceptibility to antibiotic action as compared to what would occur in a more acidic environment. The notion that the pharmacodynamic response involves inflammation-mediated reductions in CYP metabolism and/or drug efflux transport of drugs (PPIs, macrolides, and amoxicillin) used in the triple therapy regimen is unknown but is an intriguing possibility.

5.6   Drug-Cytokine Interactions and the Kidney

The kidney proximal tubules are home to a variety of drug transporters that facilitate the secretion of potentially harmful drugs, endogenous compounds, and metabolic wastes into the urinary ultrafiltrate [34]. The first step in drug secretion is SLC22A- and SLCO-mediated drug transfer from the peritubular capillaries into the proximal tubule cells (Fig. 5.2c). The second step is ABC- and SLC-mediated drug transfer from the proximal tubule cells into the nephron lumen. As commonly seen in acute kidney injury, inflammatory cytokines alter the renal tubule expression of glucose, sodium, and urea transporters [122–124] and decrease glucose reabsorption, urine osmolarity, and the urine-to-plasma urea quotient [122–124]. In a similar fashion, alteration of renal tubule drug transporters by inflammatory cytokines would be expected to impact urinary drug elimination.

In rats, *E. coli* LPS treatment reduced the mRNA, protein, and function of the proximal tubule basolateral membrane organic cation uptake transporters slc22a1 and slc22a2 [125]. Contrasting that result and the response to inflammation seen in the liver and intestine, the expression and function of the proximal tubule apical efflux abcb1 transporter is enhanced in rodent models of *E. coli* endotoxemia and ischemia reperfusion injury [86, 126, 127]. In one study, enhanced abcb1 levels were associated with increased renal clearance of doxorubicin [86]. In another, *Klebsiella pneumoniae* endotoxin transiently reduced the renal tubule secretion of rhodamine 123 [89]. This discrepancy could be attributed to the fact that the rate-limiting step in rhodamine 123 renal elimination is tubule uptake by the slc22a1 and slc22a2 transporters. Therefore, the suppression of the basolateral membrane organic cation transporters and not a loss in abcb1 is the most likely explanation for the reduction in rhodamine 123 renal clearance [125, 128]. The reduction of proximal tubule drug uptake combined with enhanced efflux are particularly interesting observations, which could indicate that changes in kidney drug transport occur to diminish the renal proximal tubule accumulation of harmful metabolites or cytokines thereby mitigating the extent of proximal tubule damage created by endotoxemia or ischemic injury [125–127].

The effect of endotoxemia or ischemia/reperfusion on renal tubule abcb1 expression and function could be recapitulated by treatment of spontaneously immortalized rat kidney proximal tubule cells with TNF indicating direct cytokine involvement [17]. The study by Heemskerk et al. also evaluated the process by which this upregulation occurs, suspecting NO produced by renal inducible NO-synthase plays a central role [17, 126]. Interestingly, the induction of renal abcb1 by LPS occurred via a NO-dependent mechanism whereas TNF increased abcb1 de novo synthesis via TLR4 activation and nuclear factor kappa B (NF-κB) signaling without NO involvement.

Overall, the preliminary results in animal models are intriguing, some showing opposing effects on inflammatory stimuli on abcb1 compared to other tissues. Given the importance of renal drug elimination the area of drug-cytokine interactions in the kidney is an area that requires further investigation including clinical studies in 'human participants (Table 5.5).

5.7 Drug-Cytokine Interactions at the Blood-Brain Barrier

The blood-brain barrier formed by brain capillary endothelial cells limits paracellular and transcellular diffusion of macromolecules and hydrophilic drugs into the brain [129]. Uptake transporters that facilitate passage of nutrients, hormones, and some drugs into the brain are integral components of the blood-brain barrier (Fig. 5.2d). ABC efflux transporter systems simultaneously limit the passage of potentially harmful and therapeutic substances into the CNS [129, 130]. Neuroinflammatory conditions, including meningitis, acute traumatic brain injury, multiple sclerosis, Parkinson's and Alzheimer's diseases, and HIV-related encephalopathy, are associated with altered blood-brain permeability [131–134]. Investigation into whether these inflammatory conditions affect drug transporter function at the blood-brain barrier interface and impact brain accumulation and pharmacological effects of CNS drugs has garnered particular interest [12, 13, 135–140].

The primary focus has been on ABCB1, a CNS-protecting blood-brain barrier efflux transporter [141, 142]. The CNS-protective function of ABCB1 is exemplified by animal studies in which its absence, mutation, or blockade is associated with increased CNS accumulation and/or toxicity of anticonvulsants, antidepressants, antineoplastics, antiretrovirals, antipsychotics, calcineurin inhibitors, calcium channel blockers, glucocorticoids, and opioids, among other medications [45, 46, 143–148]. Similarly, humans with polymorphisms in *ABCB1*, or those receiving ABCB1 inhibitors, may exhibit significantly altered CNS pharmacological responses [148–150].

Expression and/or function of blood-brain barrier abcb1 are decreased in animals with CNS bacterial and/or fungal infections and following stroke or neuronal injury [14, 21, 151–153]. Decreased activity of this transporter promotes enhanced accumulation of abcb1 substrates in the CNS, which may alter pharmacological or toxicological responses [14, 151–153]. Similarly, some have reported that blood-brain barrier abcb1 function is reduced by inflammatory and infectious stimuli that originate in compartments peripheral to the CNS such as in the circulation or the highly vascular peritoneal cavity [90, 154, 155]. In contrast, others found increased abcb1 expression and function in cerebral capillaries isolated from rats exposed to a painful peripheral inflammatory stimulus or to particulate irritants [156, 157]. A recent study has suggested that this increase in abcb1 activity is the result of alterations of abcb1 trafficking at the blood-brain barrier [158]. These contrasting results most likely indicate that the direction and degree of change in ABCB1 activity depend upon the particular inflammatory stimulus studied, the anatomical site in which the inflammatory response was generated, and the particular drug substrate [13]. Differential effects could also relate to short- versus long-term exposure to inflammatory stimuli such as would be observed during an acute versus chronic inflammatory response. In support of this idea, abcb1 activity in isolated rat brain capillaries was decreased after short-term exposure to LPS, TNF, and endothelin-1 (ET-1), whereas prolonged exposure to TNF and ET-1 produced a biphasic response with an initial decrease in abcb1 function and then an increase in both abcb1 expression and activity [26, 159, 160]. The effects of inflammatory stimuli on blood-brain barrier transport are not restricted to ABCB1. There are reports of differential regulation of ABCB1, ABCC2, ABCC4, and ABCG2 by multiple inflammatory mediators including TNF, IL-6, and IL-β in human or rodent capillary endothelial and glial cells [24, 26, 156, 161–163]. These collective findings indicate modified CNS drug accumulation through an alteration in blood-brain barrier efflux transport processes during episodes of infection and inflammation is probable.

Altered ABCB1 expression and/or activity or enhanced CNS drug levels during meningitis, HIV infection, Parkinson's disease, epilepsy, and after acute traumatic brain injury suggest that a regulatory link exists between inflammation and blood-brain barrier transport in humans (Table 5.5) [2, 3, 136, 137, 139, 164]. Further investigations are needed, however, to delineate if drug-cytokine interactions involving blood-brain barrier transporters impact CNS drug efficacy and toxicity in humans. In this regard, it should be recognized that alterations in blood-brain barrier ABCB1 activity during inflammatory or infectious conditions might have either positive or negative consequences depending on the drug in question and the therapeutic goals for the patient. For example, a transient reduction in blood-brain barrier ABCB1 activity could improve the CNS delivery of neuroprotectant agents in diseases such as ischemic or hemorrhagic stroke or antibiotics in meningitis with the potential for enhanced therapeutic efficacy. This is a plausible explanation for the historical observations that patients with meningitis had higher cerebrospinal fluid concentrations of the antituberculosis agents, ethambutol, and rifampin [2, 3]. In contrast, chronic diseases such as epilepsy appear to increase levels of blood-brain barrier ABCB1 potentially reducing CNS drug levels resulting in drug resistance and even pharmacotherapeutic failure.

A second finding of note is competitive drug interactions at the blood-brain barrier may amplify the pharmacological effects of ABCB1 substrate drugs during episodes of CNS inflammation [152, 153, 165]. In animals with CNS infection or ischemia, the administration of competitive inhibitors of abcb1 enhances the CNS concentration and efficacy of itraconazole, rifampin, and tacrolimus above that of the disease process alone, suggesting a beneficial drug-immune system interaction [152, 165]. Conversely, near-maximal inhibition of abcb1 activity during CNS inflammation could contribute to neurotoxicity. A novel clinical example of such a situation is implied from data derived from critically ill patients with acute inflammatory brain injury. These patients can receive upward of 30 drugs concomitantly, including ABCB1 substrates (e.g., dexamethasone, morphine, and ranitidine) and inhibitors (e.g., amiodarone and diltiazem) as part of their routine care. Such polypharmacy during an acute neuroinflammatory reaction could place these patients at risk for drug interactions involving blood-brain barrier ABCB1 [139].

Inflammatory conditions can also alter blood-brain permeability via abcb1-independent mechanisms. An experiment in which mice were concomitantly treated with the antibiotic colistin and abcb1 inhibitors found no statistically significant increase in brain-to-plasma concentration of colistin, suggesting that colistin is not a substrate of abcb1 [166]. However, when mice were concomitantly treated with

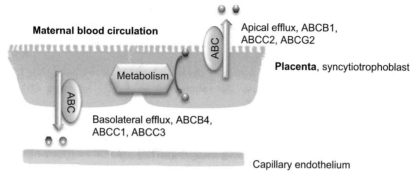

Fig. 5.3 Select drug efflux transporters in the human placenta at term. ATP-binding cassette transporters (ABC) exist in the apical and basolateral membranes of the syncytiotrophoblasts where they mediate drug (black hexagons) or metabolite (orange circles) efflux into the maternal blood circulation or in the direction of the fetal circulation. The green ovals represent drug efflux transporters. For simplicity only those transporters that are discussed in this chapter are shown on the diagram

colistin and LPS, an increase in brain concentration of colistin was noted. This was likely due to increased drug penetration through paracellular transport. Cytokines such as TNF have been shown to increase blood-barrier permeability by dysregulating the tight junction complexes that normally restrict paracellular transport [167–169]. Decreasing barrier function can lead to indiscriminate increases in CNS concentrations of therapeutic and toxic agents alike. As such, patients experiencing inflammatory conditions may be more sensitive to the neurological and psychoactive effects of a wide variety of medications.

5.8   Drug-Cytokine Interactions During Pregnancy

In pregnancy, the placenta is an additional location of many drug transporters [170]. Placental transporters are expressed on the apical (facing maternal blood) and basolateral (facing fetal capillaries) surfaces of syncytiotrophoblasts and facilitate the exchange of drugs and endogenous compounds between the maternal and fetal circulations (Fig. 5.3) [171–173]. On the apical surface of the syncytiotrophoblasts, ABCB1 and ABCG2 are the most abundant of the ABC transporters and are prominent contributors to the efflux of drugs and metabolites from the fetoplacental space [174]. This is exemplified by several fold increases in the fetal exposure to digoxin, saquinavir, and Taxol in mice with complete deficiency in *abcb1* and nitrofurantoin and glyburide in *abcg2*-deficient mice [175, 176]. There are similar relevant examples of human placental drug transport of clinically used medications. In term human placentas ex vivo, inhibition of ABCB1 led to enhanced maternal-placental transfer of indinavir, vinblastine, and saquinavir, whereas inhibition of ABCG2 enhanced the maternal-placental transfer of glyburide [177–180]. Opposing the actions of the apical transporters are the transporters located on the basolateral side of the syncytiotrophoblasts including ABCB4, ABCC1, and ABCC3 [174]. While the human data remains limited, the above studies identify that placental drug transporters are likely to be important determinants of fetal drug exposure and fetal development and safety (Table 5.5).

The role of cytokines in the regulation of placental drug transport has gained considerable attention as the placenta is a source of TNF, IL-1β, and Il-6 production, and there are reports that the levels of these cytokines are aberrantly increased in pregnancy complications with associated inflammation such as placental insufficiency/fetal growth restriction, preeclampsia, chorioamnionitis, and gestational diabetes as well as a number of unrelated comorbid conditions [11, 181–184]. Evseenko et al. evaluated the effect of TNF, IL-1β, and IL-6 on the prominent ABC transporters in primary trophoblasts from term human placentas [174]. TNF and IL-1β but not IL-6 significantly decreased *ABCB1* and *ABCG2* mRNA by >40% after 12 h and corresponding ABCB1 and ABCG2 protein levels by 50% after 48 h. With respect to the basolateral transporters, ABCB4 mRNA and protein were specifically and significantly increased following IL-6 but not TNF and IL-1β. Comparatively, TNF, IL-6, and IL-1β significantly increased the expression of *ABCC1* mRNA but not ABCC1 protein. The combined depression of the maternal facing ABCB1 and ABCG2 apical transporters and increased or static expression of the fetal circulation facing ABCC1 and ABCB4 basolateral transporters indicate that placental exposure to inflammatory cytokines may decrease fetal protection from drugs and enhance active transport of drugs to the fetus in conditions in which those cytokines are aberrantly elevated. Animal studies do provide some evidence that this occurs [8, 90, 185]. For instance, treatment of near-term rats with LPS dose-dependently increased plasma TNF and IL-6 levels while reducing the placental mRNA and protein expression of abcb1 and abcg2 among other uptake and efflux transporters [186]. The functional outcome was an increase in the fetal-to-maternal concentration of glyburide, an agent that is used for treating gestational diabetes in humans and is primarily restricted from accessing the fetal circulation by ABCG2 [186, 187]. Supporting the likelihood that such an alteration could occur in humans, treatment with LPS has been shown to decrease *ABCB1* and *ABCG2* mRNA and protein levels in first-trimester human placental explants. Of note, no change in *ABCB1* or *ABCG2* mRNA levels was noted when the experiment was conducted in third-trimester human placental explants, suggesting that the influence of inflammatory processes on the human placenta in regards to drug disposition may gestational age dependent [181]. In a second study, placental samples from preterm pregnancies diagnosed with chorioamnionitis, a bacterial infection of the fetal chorion and amnion membranes, displayed increased *TNF*, *IL-1β*, and *IL-6* expression by 2.5- to 3.0-fold and reduced ABCG2 mRNA and protein expression by 50%, compared to control placental samples [11]. However, not all studies are in agreement. Mason et al. found increased ABCB1 and ABCG2 protein expression in placentas of women with preterm labor with inflammation compared to those without inflammation [188]. These diverging findings suggest the effect of inflammation on placental drug transporters may depend on the stage of pregnancy and the type of complication.

5.9   Drug-Cytokine Interactions and Cancer

The role of inflammation in the pathophysiology of cancer is becoming increasingly accepted. It is now thought that inflammatory components are present in the microenvironment of most, if not all, tumors, and the level of inflammation appears to correlate with the severity of the cancer [189]. An interesting connection is that IL-6, a potent regulator of drug disposition, is produced by tumor cells and its concentration in circulation increases with many cancers [18, 21, 190–192]. Increasing evidence indicates that the elevated IL-6 levels that occur in malignancy affect pharmacological responses to chemotherapy drugs by activating multidrug resistance mechanisms in tumor cells and/or through alteration of host drug disposition.

Multidrug resistance occurs when tumors develop cross-resistance to a number of structurally and mechanistically unrelated drugs. This has become an increasing problem in the field of oncology [193]. It can arise through tumor cell modifications including the inhibition of apoptosis, activation of DNA repair mechanisms, and decreased intracellular chemotherapeutic drug accumulation due to suppression of SLC drug uptake transporters or increased levels of the ABC drug efflux transporters [194–196]. Previous research has shown that autocrine production of IL-6 by breast, osteosarcoma, and ovarian cancer cells caused the cells to develop resistance to the cytotoxic effects of doxorubicin, paclitaxel, or cisplatin [190, 191, 197]. Depending on the cell, different mechanisms were implicated. These included IL-6 induction of ABCB1 efflux (breast and ovarian cancer), inhibition of apoptosis (osteosarcoma and ovarian cancer), and increased glutathione-s-transferase (ovarian cancer) [190, 191, 197]. In mice inoculated with Engelbreth-Holm-Swarm (EHS) sarcoma, *abcb1a*, but not *abcg2*, *abcc2*, *abcc3*, and *abcb4* mRNA, increased significantly in the developing tumor in association with increasing levels of plasma and intratumoral IL-6 [18]. Contrary to the effects observed in tumor cells, mice bearing extrahepatic tumors displayed widespread repressed hepatic uptake (*slc10a1*, *slco1b1*), sinusoidal efflux (*abcc3*), and biliary efflux (*abcb4*, *abcc2*, *abcg2*, and *abcb11*) transporters [18]. This indicates the possibility of drug-cytokine interactions involving IL-6-mediated induction of multidrug-resistant efflux transport in tumors in vivo and/or reduction in hepatic drug transport capacity. However, the effects of these drug transporter gene expression changes on chemotherapy resistance or biliary drug elimination were not investigated. Although IL-6 is a predominantly elevated inflammatory cytokine in cancer, this does not imply other cytokines are not involved in regulation of specific transporters. For example, Mosaffa et al. demonstrated that IL-1β and TNF increased ABCG2 efflux transport of mitoxantrone to a greater degree in mitoxantrone-sensitive versus mitoxantrone-resistant MCF7 breast cancer cells [198]. Consistent with the study by Sharma et al., IL-6 did not regulate ABCG2 mRNA or function in human breast cancer cells [18, 198]. More recently, IL-1β and TNF have been shown to increase mitoxantrone accumulation in cervical cancer HeLa cells, and IL-1β to increase mitoxantrone accumulation in the gastric cancer cell line EPG85-257, suggesting a tissue-dependent effect [199].

In humans, the main enzyme responsible for inter-patient variability in anticancer drug metabolism is CYP3A4 as it metabolizes many important chemotherapeutic drugs including the taxanes, vinca alkaloids, camptothecins, cyclophosphamide, etoposide, tamoxifen, imatinib, and gefitinib [21, 200]. The reported inverse associations between CYP3A4 metabolic activities and the inflammatory mediators IL-6 and C-reactive protein in patients with advanced cancer suggest that cytokine-CYP interactions may contribute to the clinically observed variations [18, 76, 200]. Supporting this, when breast, melanoma, and EHS sarcoma tumors were introduced into mice, circulating levels of IL-6 increased, but IL-1β and TNF levels were unchanged. With the three tumors, the increased plasma IL-6 concentration corresponded with activation of the hepatic acute phase response, a precipitous drop in hepatic cyp3a11 (mouse equivalent of human CYP3A4) mRNA and protein levels and extended sedation by midazolam, a cyp3a11-specific substrate [18, 200]. Direct evidence that cytokines mechanistically link malignancy and human CYP3A4 metabolism stems from the demonstration of reduced expression of a human CYP3A4 transgene in the livers of mice with extrahepatic tumors [200]. Further, the similar reduction in *cyp3a11* mRNA using several cancer models would argue that the tumor-derived inflammation and suppressed hepatic drug metabolism is a common occurrence among malignancies [18, 192]. Less is known regarding the effect of malignancy on other hepatic CYPs. Two studies have reported that a proportion of patients with advanced cancer displayed a CYP2C19 poor metabolizer phenotype despite having a rapid metabolizer genotype [77, 201]. However, an association between the reduced CYP2C19 metabolism and circulating levels of IL-1α/β, IL-6, TNF, or C-reactive protein was not demonstrated [201]. While some individuals with advanced cancer may have decreased ability to metabolize CYP2C19 substrates like cyclophosphamide, the involvement of cytokines in this interaction remains inconclusive.

It is worth reflecting upon the apparent differential activation of multidrug resistance in cancer cells versus a loss of drug metabolism and transport capacity in the liver caused by inflammatory cytokines. This presents a potential situation of double jeopardy, whereby reduced hepatic elimination of chemotherapeutic agents could pose problems for host toxicity at the same time that the drugs are becoming less effective against the tumors because of activation of cellular multidrug resistance. As chemotherapeutic drugs have a very narrow therapeutic window and individual variability in response is so vast, a comprehensive understanding of how inflammatory cytokines alter the disposition of these agents at the level of the cancer cell and the host is an essential area for continued study.

5.10   Drug-Cytokine Interactions and IFN Therapy

Proof of principle that IFN could be a source of cytokine-drug interactions stems from clinical pharmacokinetic studies of conventional forms of recombinant INFα in humans with chronic hepatitis B or C or metastatic disease. Two studies of human subjects with chronic active hepatitis B indicated variable (5–63%) reductions in hepatic CYP-mediated drug metabolism after a single high dose (4.5–18 × 10^6 units) of IFNα and after chronic IFNα (6 × 10^6 units for 4 weeks) treatment [202, 203]. The effects of IFNα appeared dose-dependent as treatment of subjects with chronic active hepatitis C for 1 month with a lower IFNα dose (3 × 10^6 units/3 times per week) did not reduce metabolism of CYP1A2 or CYP3A substrates [204]. A fourth study showed a trend toward higher CYP2D6 and CYP3A4 enzyme activities in individuals with chronic active hepatitis C who responded to 1 month of IFNα (3 × 10^6 units/3 times per week)/ribavirin (600 mg/twice daily) combination therapy compared to nonresponders [205]. A particularly informative study of individuals with high-risk melanoma examined the effect of high-dose IFNα-2b (a glycosylated form of IFNα) on the pharmacokinetics of a drug cocktail containing substrates for CYP1A2 (caffeine), CYP2C19 (mephenytoin), CYP2D6 (debrisoquine), CYP2E1 (chlorzoxazone), and CYP2C8/CYP2C9, CYP3A4/CYP3A5, CYP2E1, and N-acetyltransferase (dapsone) [206]. One day after a single IFNα-2b dose, the metabolism of the CYP1A2 and CYP2D6 probe substrates was lowered by 20% and 10%, respectively. After 4 weeks of chronic IFNα-2b treatment, the magnitudes of the reductions in CYP1A2 (53%) and CYP2D6 (25%) metabolism increased. Further, metabolism of the CYP2C19 substrate mephenytoin was now reduced by 25%. The metabolism of the CYP2E1 substrate, chlorzoxazone, and the CYP2C8/9, CYP3A4/5, CYP2E1, and N-acetyltransferase substrate dapsone was not altered indicating that IFNα-2b differently effects CYP metabolism in humans.

IFNα has largely been replaced by pegylated IFNα (PEG-IFNα-2a or 2b) in clinical practice due to the latter's improved pharmacokinetic characteristics [207]. As such, it is important to ascertain whether the differences in pharmacokinetics, dosing, and selected pegylated moiety of IFN impact its effect on CYP activity. PEG-IFNα-2b treatment significantly downregulated CYP1A2, UGT2B7, and SLC22A7 protein and mRNA in freshly prepared co-cultures of human primary hepatocytes and non-parenchymal cells after 2–3 days [33]. Consistent with the reduction in CYP1A2 in cell cultures, a study in healthy human subjects found a 20% decrease in the metabolism of the CYP1A2 probe theophylline after four doses of PEG-IFNα-2a (180 mcg/weekly), and no difference in CYP2C9, CYP2C19, CYP2D6, or CYP3A4 probe metabolism [208]. In regard to PEG-IFNα-2b, Gupta et al. found metabolism of the CYP2D6 substrate dextromethorphan was increased by 67% in human subjects with chronic hepatitis C after 1 month of therapy, though no difference was noted in the metabolism of the CYP3A4 substrate midazolam. Supporting an increase in CYP2D6 metabolic activity, a 2-month treatment regimen of PEG-IFNα-2b decreased the terminal elimination of the CYP2D6 substrate fluoxetine from 47 to 33 h in patients with chronic hepatitis C. A minor statistically significant increase in CYP2C8/9 activity and decrease in CYP1A2 activity was found, though the magnitude of change for these enzymes was small and unlikely to be clinically significant [209]. These studies suggest that differences in CYP induction and inhibition may exist for PEG-IFNα-2a and PEG-IFNα-2b. Furthermore, the observation that INFα increased the activity of certain CYPs in individuals with chronic hepatitis C suggests that a normalization of CYP activity may occur over time with successful antiviral therapy.

Over the past 3 years, the clinical management of chronic hepatitis C has entered a new era with the introduction of the second-generation direct-acting antiviral agents including sofosbuvir, simeprevir, daclatasvir, ledipasvir, elbasvir/grazoprevir, and paritaprevir/ritonavir/ombitasvir fixed-dose combinations [210]. Combinations of direct-acting antivirals have emerged as recommended therapies for hepatitis C given their improved efficacy and safety profile in comparison to traditional PEG-INFα/ribavirin-containing regimens [211, 212]. As they become more universally available, the first-line use of IFN-free therapies will undoubtedly lessen the concern for IFNα-mediated drug interactions. However, as the direct-acting antiviral agents are so new, their safety profiles and interactions including those involving cytokines are not completely established [210, 213]. The direct-acting antivirals are metabolized primarily by CYP3A4 and/or transported by ABCB1 [213, 214], which are sensitive to regulation by various pro-inflammatory cytokines. Thus, altered disposition of direct-acting antivirals due to cytokine-mediated changes in hepatic drug metabolism and transport during active and resolution phases of hepatitis C infection is possible and should be investigated. Supporting this, hepatitis C patients receiving single or multiple oral simeprevir doses (200 mg/day) displayed an approximate two fold higher simeprevir C_{max}, AUC, and $t_{1/2}$ versus healthy patients [215].

5.11   Drug-Cytokine Interactions and Immunosuppression Therapy

An untoward effect of immunosuppression therapy is increased susceptibility to opportunistic viral and bacterial infections [216–218]. A number of human studies and case reports support a potential association between opportunistic infections and the disposition of the low therapeutic index calcineurin inhibitors, cyclosporine, and tacrolimus. For instance, higher blood cyclosporine levels have been reported in lung transplant recipients with cytomegalovirus infection compared to uninfected patients [219]. Tacrolimus blood levels over time as measured by AUC were higher in adult kidney transplant patients presenting with infections compared to those without [220]. In a third example, adult kidney transplant patients with hepatitis C infection required 25% lower daily doses of cyclosporine or tacrolimus to maintain target blood levels of these drugs compared to uninfected patients [221].

However, this observation could be confounded by hepatitis C-mediated reductions in hepatic CYP3A4 function. In addition to a potential pharmacokinetic interaction, clinical findings of renal impairment after low-dose cyclosporine therapy in patients with human immunodeficiency virus (HIV) and autoimmune diseases [222, 223] suggest a potential interaction between immune responses and calcineurin inhibitors that may augment nephrotoxicity produced by lower doses of those drugs. Presumably, these pharmacokinetic changes could be due to altered intestinal, hepatic, and/or renal CYP metabolism and transport; however, this remains to be determined. The use of anti-cytokine immunosuppressive therapies following organ transplantation represents an additional situation where drug-cytokine interactions could arise. Supporting this, a retrospective analysis identified pediatric kidney transplant patients receiving the IL-2 receptor antibody basiliximab required lower initial cyclosporine levels to achieve therapeutic target concentrations and prevent toxicity compared to controls [224]. However, after 4–7 weeks of basiliximab treatment, higher cyclosporine doses were required to maintain desired concentrations. The mechanisms responsible for these pharmacokinetic changes were not determined but were postulated to involve IL-2-mediated changes in hepatic CYP metabolism of cyclosporine [225].

5.12   Drug-Cytokine Interactions Involving Anti-cytokine Therapy of Immune-Mediated Inflammatory Diseases

Immune-mediated inflammatory diseases are a broad array of conditions with diverse clinical presentations that share common inflammatory pathways and therapeutic goals: gain control of the inflammation, prevent tissue damage, improve quality of life, and, if possible, achieve long-term remission [226]. Common examples of these diverse diseases are rheumatoid arthritis, IBD, and psoriasis. The etiology of these diseases remains unknown but substantial advances have been made in identification of many cytokines involved in the underlying pathophysiology [226]. This has led to the rapid development and approval of many antibody therapies for immune-mediated inflammatory diseases, cancer, and immunosuppression following organ transplantation [227, 228].

Approved biologic agents for inflammatory diseases include inhibitors of TNF (e.g., infliximab, etanercept, adalimumab), IL-6 (e.g., tocilizumab), and IL-1β (e.g., canakinumab) signaling. With the emergence of these new anti-cytokine drugs comes the potential of new drug interactions. As pro-inflammatory cytokines reduce the extent of drug metabolism and elimination through suppression of hepatic CYPs and drug transporters, it is reasonable to hypothesize that when anti-cytokine agents are administered to patients experiencing chronic inflammatory conditions, a relative induction in these same enzymes should initially occur [229]. In theory, such interactions would increase the clearance of CYP-metabolized medications requiring increased dosing of the CYP-metabolized drug over time in order to maintain its therapeutic efficacy. A retrospective analysis of the effects of the basiliximab on the dosing requirements of the CYP3A4-metabolized drug cyclosporine in pediatric kidney transplant recipients and case report of a possible drug-drug interaction between the adalimumab and CYP2D6 and 1A2-metabolized duloxetine support this idea [224, 230]. Arguing against this hypothesis, two clinical studies found no effect of etanercept on the pharmacokinetics or pharmacodynamics of digoxin, an ABCB1 substrate nor warfarin, a CYP2C9-metabolized drug [231, 232]. However, these studies were completed in healthy volunteers whom would not have had pre-existing inflammation. The strongest support for the idea that anti-cytokine therapy alters metabolism drug in a clinical setting comes from recent prospective pharma-cokinetic studies of the CYP3A4 substrate simvastatin in humans with active rheu-matoid arthritis. Following treatment with IL-6 receptor antibodies sarilumab or tocilizumab, plasma simvastatin exposure was significantly reduced compared to pretreatment values [69, 233]. A similar reduction in the C_{max} and AUC of the CYP3A4 substrate midazolam was observed following treatment with the IL-6 receptor antibody sirukumab [234]. These results were consistent with the allevia-tion of IL-6-mediated suppression of CYP3A4 metabolism with time after initiating the monoclonal antibody therapies, a mechanism that is supported by studies con-ducted in three-dimensional human hepatocyte cell cultures [19]. In patients with active rheumatoid arthritis, sirukumab treatment also significantly decreased plasma exposure to the CYP2C19 substrate omeprazole and to a lesser extent S-warfarin, a CYP2C9 substrate, while increasing the exposure to the CYP1A2 substrate caffeine [234]. This implies that the pharmacokinetics of other CYP-metabolized drugs may be differentially affected by monoclonal antibody therapies. As a result, monitoring and dosing adjustments of certain CYP3A4-metabolized drugs following the initia-tion of IL-6 receptor monoclonal antibody therapies in rheumatoid arthritis have been suggested [69]. In comparison, the treatment of patients with multiple sclero-sis with the IL-2 receptor antibody daclizumab had no significant effect on the phar-macokinetics of an orally administered drug cocktail containing CYP1A2 (caffeine), CYP2C9 (warfarin), CYP2C19 (omeprazole), CYP2D6 (dextromethorphan), and CYP3A4 (midazolam) [235] indicating that such therapeutic protein-drug interac-tions depend on the cytokine pathway targeted and the inflammatory disease.

5.13   Drug-Cytokine Interactions and Vaccines

There is some evidence to support that the known interactions between drugs and cytokines may be utilized to enhance the effectiveness of vaccine adjuvants. The particular situation applies to the CYP-mediated production of calcitriol, a known immune adjuvant. Calcitriol produced and secreted by myeloid dendritic cells causes those cells to migrate from cutaneous vaccination sites into multiple second-ary lymphoid organs where they stimulate B and T lymphocyte cell responses [236]. Enioutina et al. recently showed that monophosphoryl lipid A, an LPS derivative, elicited a similar immune response. Interestingly, it was found that the mucosal adjuvant properties of monophosphoryl A directly correlated with its capacity to induce in dendritic cells, the expression of CYP27B1 the enzyme that converts vitamin D to its active form (calcitriol) [236]. Further, monophosphoryl A was unable to upregulate CYP27B1 in IFN receptor-deficient (IFNR$^{-/-}$) dendritic cells nor stimulate the migration of IFNR$^{-/-}$ dendritic cells to secondary lymphoid organs confirming that it was an IFN-CYP interaction. As recently reviewed by Pellegrino et al., a collection of older case reports have raised concern that vaccinations could trigger interactions with some drugs due to cytokine-mediated reductions in hepatic CYP metabolism [237]. However, this concern has not been borne out by clinical studies [238–240], and others have pointed out that case reports of drug interactions attributed to vaccines could simply be caused by simultaneous infections that would produce a larger cytokine response [239].

5.14   Concluding Remarks

The ability of infectious and inflammatory stimuli to alter the disposition of commonly used drugs through cytokine-mediated reductions in hepatic CYP metabolism has been recognized for some time. Based on recent discoveries, it is likely that some of the historical reports of altered drug disposition during infectious diseases also involved cytokine-mediated reductions in drug transporters.

Despite the well-recognized changes in hepatic metabolism, there is a continued need for human studies to determine the broader clinical importance of drug-cytokine interactions involving extrahepatic tissues and drug transporters. For example, do reduced intestinal ABCB1 and ABCG2 in IBD increase oral drug absorption or serve as risk factors for colorectal cancer? Is human kidney ABCB1 upregulated in inflammatory diseases and does this compensate for an inflammation-mediated loss in intestinal and hepatic drug elimination? When infectious or inflammatory complications arise in pregnancy, does altered placental drug transport increase fetal exposure to medications? Do diseases of the central nervous system or acute brain injuries impart cytokine-mediated changes to blood-brain barrier ABCB1 such that drug efficacy or disease progression is affected?

An increased awareness of how evolving therapeutic approaches including the use of anti-cytokine monoclonal antibody therapies, hepatitis C direct-acting antivirals, and pharmacogenetic approaches to personalized medicine may influence or be influenced by drug-cytokine interactions is required. For instance, does acute or chronic inflammation lead to a mismatch between genotype and phenotype, such that an individual with a rapid metabolism genotype converts to a slow metabolism phenotype? Are cytokine-mediated reductions in CYP metabolism additive with genetic poor metabolizer CYP polymorphisms? Do successful anti-hepatitis C or anti-cytokine therapies lead a normalization or induction of drug metabolism and transport over time necessitating dosage adjustments of concomitantly administered medications?

Answering these questions will elucidate the situations in which drug-cytokine interactions are most likely to occur, the pharmacological outcomes of the interactions, and who is most at risk. With this information, it will then be possible to appropriately inform and caution physicians and pharmacists about the potential positive and negative impact of infectious and inflammatory diseases on the safe and effective use of medications. Until this information becomes available, it would be correct to assume that hepatic drug elimination will be impaired in any disease state that has an inflammatory component or one that activates host defense. This especially applies to the elderly and critically ill, who may be more susceptible because they have a reduced capacity to eliminate drugs and tend to receive multiple medications concurrently. Consideration of drug-cytokine interactions in addition to other patient-specific factors such as kidney function and drug-specific factors such as therapeutic index can assist the clinician in making rationale drug therapy decisions. Knowledge of the known and suspected changes inflammation can have on drug disposition can lead to more appropriate drug and dose selection for a given patient, as well as aid the clinician in modifying therapy in the event of drug toxicity. Until further information on drug-cytokine interactions in humans becomes available, empirical dose alterations and/or rigorous patient monitoring is warranted until the infectious or inflammatory condition is resolved.

References

1. Brooks MH, Malloy JP, Bartelloni PJ, Sheehy TW, Barry KG (1969) Quinine, pyrimethamine, and sulphorthodimethoxine: clinical response, plasma levels, and urinary excretion during the initial attack of naturally acquired falciparum malaria. Clin Pharmacol Ther 10(1):85–91
2. Place VA, Pyle MM, De la Huerga J (1969) Ethambutol in tuberculous meningitis. Am Rev Respir Dis 99(5):783–785
3. Sippel JE, Mikhail IA, Girgis NI, Youssef HH (1974) Rifampin concentrations in cerebrospinal fluid of patients with tuberculous meningitis. Am Rev Respir Dis 109(5):579–580
4. Renton KW, Mannering GJ (1976) Depression of hepatic cytochrome P-450-dependent monooxygenase systems with administered interferon inducing agents. Biochem Biophys Res Commun 73(2):343–348
5. Renton KW, Mannering GJ (1976) Depression of the hepatic cytochrome P-450 monooxygenase system by administered tilorone (2,7-bis(2-(diethylamino)ethoxy)fluoren-9-one dihydrochloride). Drug Metab Dispos 4(3):223–231
6. Soyka LF, Hunt WG, Knight SE, Foster RS Jr (1976) Decreased liver and lung drug-metabolizing activity in mice treated with Corynebacterium parvum. Cancer Res 36(12):4425–4428
7. Cressman AM, Petrovic V, Piquette-Miller M (2012) Inflammation-mediated changes in drug transporter expression/activity: implications for therapeutic drug response. Expert Rev Clin Pharmacol 5(1):69–89. https://doi.org/10.1586/ecp.11.66
8. Petrovic V, Teng S, Piquette-Miller M (2007) Regulation of drug transporters during infection and inflammation. Mol Interv 7(2):99–111
9. Gonzalez-Gay MA, Gonzalez-Juanatey C, Vazquez-Rodriguez TR, Miranda-Filloy JA, Llorca J (2010) Insulin resistance in rheumatoid arthritis: the impact of the anti-TNF-alpha therapy. Ann N Y Acad Sci 1193(1):153–159. https://doi.org/10.1111/j.1749-6632.2009.05287.x. NYAS5287 [pii]

10. Sugimoto M, Furuta T, Yamaoka Y (2009) Influence of inflammatory cytokine polymorphisms on eradication rates of Helicobacter pylori. J Gastroenterol Hepatol 24(11):1725–1732. https://doi.org/10.1111/j.1440-1746.2009.06047.x. JGH6047 [pii]
11. Petrovic V, Kojovic D, Cressman A, Piquette-Miller M (2015) Maternal bacterial infections impact expression of drug transporters in human placenta. Int Immunopharmacol 26(2):349–356. https://doi.org/10.1016/j.intimp.2015.04.020
12. Miller DS (2015) Regulation of ABC transporters blood-brain barrier: the good, the bad, and the ugly. Adv Cancer Res 125:43–70. https://doi.org/10.1016/bs.acr.2014.10.002
13. Roberts DJ, Goralski KB (2008) A critical overview of the influence of inflammation and infection on P-glycoprotein expression and activity in the brain. Expert Opin Drug Metab Toxicol 4(10):1245–1264
14. Goralski KB, Hartmann G, Piquette-Miller M, Renton KW (2003) Downregulation of mdr1a expression in the brain and liver during CNS inflammation alters the in vivo disposition of digoxin. Br J Pharmacol 139(1):35–48
15. DX X, Wang JP, Sun MF, Chen YH, Wei W (2006) Lipopolysaccharide downregulates the expressions of intestinal pregnane X receptor and cytochrome P450 3a11. Eur J Pharmacol 536(1–2):162–170. https://doi.org/10.1016/j.ejphar.2006.02.029. S0014-2999(06)00197-X [pii]
16. Kalitsky-Szirtes J, Shayeganpour A, Brocks DR, Piquette-Miller M (2004) Suppression of drug-metabolizing enzymes and efflux transporters in the intestine of endotoxin-treated rats. Drug Metab Dispos 32(1):20–27. https://doi.org/10.1124/dmd.32.1.20. 32/1/20 [pii]
17. Heemskerk S, Peters JG, Louisse J, Sagar S, Russel FG, Masereeuw R (2010) Regulation of P-glycoprotein in renal proximal tubule epithelial cells by LPS and TNF-alpha. J Biomed Biotechnol 2010:525180. https://doi.org/10.1155/2010/525180
18. Sharma R, Kacevska M, London R, Clarke SJ, Liddle C, Robertson G (2008) Downregulation of drug transport and metabolism in mice bearing extra-hepatic malignancies. Br J Cancer 98(1):91–97. https://doi.org/10.1038/Sj.Bjc.6604101
19. Long TJ, Cosgrove PA, Dunn RT 2nd, Stolz DB, Hamadeh H, Afshari C, McBride H, Griffith LG (2016) Modeling therapeutic antibody-small molecule drug-drug interactions using a three-dimensional perfusable human liver coculture platform. Drug Metab Dispos 44(12):1940–1948. https://doi.org/10.1124/dmd.116.071456
20. Harvey RD, Morgan ET (2014) Cancer, inflammation, and therapy: effects on cytochrome p450-mediated drug metabolism and implications for novel immunotherapeutic agents. Clin Pharmacol Ther 96(4):449–457. https://doi.org/10.1038/clpt.2014.143
21. Morgan ET, Goralski KB, Piquette-Miller M, Renton KW, Robertson GR, Chaluvadi MR, Charles KA, Clarke SJ, Kacevska M, Liddle C, Richardson TA, Sharma R, Sinal CJ (2008) Regulation of drug-metabolizing enzymes and transporters in infection, inflammation, and cancer. Drug Metab Dispos 36(2):205–216
22. Renton KW (2005) Regulation of drug metabolism and disposition during inflammation and infection. Expert Opin Drug Metab Toxicol 1(4):629–640
23. Aitken AE, Morgan ET (2007) Gene-specific effects of inflammatory cytokines on cytochrome P450 2C, 2B6 and 3A4 mRNA levels in human hepatocytes. Drug Metab Dispos 35(9):1687–1693. https://doi.org/10.1124/dmd.107.015511. dmd.107.015511 [pii]
24. Bauer B, Hartz AM, Miller DS (2007) Tumor necrosis factor alpha and endothelin-1 increase P-glycoprotein expression and transport activity at the blood-brain barrier. Mol Pharmacol 71(3):667–675
25. Donato MT, Guillen MI, Jover R, Castell JV, Gomez-Lechon MJ (1997) Nitric oxide-mediated inhibition of cytochrome P450 by interferon-gamma in human hepatocytes. J Pharmacol Exp Ther 281(1):484–490
26. Hartz AM, Bauer B, Fricker G, Miller DS (2006) Rapid modulation of P-glycoprotein-mediated transport at the blood-brain barrier by tumor necrosis factor-alpha and lipopolysaccharide. Mol Pharmacol 69(2):462–470
27. Kim RB (2006) Transporters and drug discovery: why, when, and how. Mol Pharm 3(1):26–32

28. Petzinger E, Geyer J (2006) Drug transporters in pharmacokinetics. Naunyn Schmiedeberg's Arch Pharmacol 372(6):465–475
29. Anzenbacher P, Anzenbacherova E (2001) Cytochromes P450 and metabolism of xenobiotics. Cell Mol Life Sci 58(5–6):737–747
30. Ramirez-Alcantara V, Montrose MH (2014) Acute murine colitis reduces colonic 5-aminosalicylic acid metabolism by regulation of N-acetyltransferase-2. Am J Physiol Gastrointest Liver Physiol 306(11):G1002–G1010. https://doi.org/10.1152/ajpgi.00389.2013
31. Richardson TA, Sherman M, Kalman D, Morgan ET (2006) Expression of UDP-glucuronosyltransferase isoform mRNAs during inflammation and infection in mouse liver and kidney. Drug Metab Dispos 34(3):351–353. https://doi.org/10.1124/dmd.105.007435
32. Mimche SM, Nyagode BA, Merrell MD, Lee CM, Prasanphanich NS, Cummings RD, Morgan ET (2014) Hepatic cytochrome P450s, phase II enzymes and nuclear receptors are downregulated in a Th2 environment during Schistosoma mansoni infection. Drug Metab Dispos 42(1):134–140. https://doi.org/10.1124/dmd.113.054957
33. Chen C, Han YH, Yang Z, Rodrigues AD (2011) Effect of interferon-alpha2b on the expression of various drug-metabolizing enzymes and transporters in co-cultures of freshly prepared human primary hepatocytes. Xenobiotica 41(6):476–485. https://doi.org/10.3109/0049 8254.2011.560971
34. International Transporter C, Giacomini KM, Huang SM, Tweedie DJ, Benet LZ, Brouwer KL, Chu X, Dahlin A, Evers R, Fischer V, Hillgren KM, Hoffmaster KA, Ishikawa T, Keppler D, Kim RB, Lee CA, Niemi M, Polli JW, Sugiyama Y, Swaan PW, Ware JA, Wright SH, Yee SW, Zamek-Gliszczynski MJ, Zhang L (2010) Membrane transporters in drug development. Nat Rev 9(3):215–236. https://doi.org/10.1038/nrd3028
35. He L, Vasiliou K, Nebert DW (2009) Analysis and update of the human solute carrier (SLC) gene superfamily. Hum Genomics 3(2):195–206
36. Vasiliou V, Vasiliou K, Nebert DW (2009) Human ATP-binding cassette (ABC) transporter family. Hum Genomics 3(3):281–290
37. Dean M, Rzhetsky A, Allikmets R (2001) The human ATP-binding cassette (ABC) transporter superfamily. Genome Res 11(7):1156–1166
38. Hediger MA, Romero MF, Peng JB, Rolfs A, Takanaga H, Bruford EA (2004) The ABCs of solute carriers: physiological, pathological and therapeutic implications of human membrane transport proteinsIntroduction. Pflugers Arch 447(5):465–468
39. Stieger B, Hagenbuch B (2014) Organic anion-transporting polypeptides. Curr Top Membr 73:205–232. https://doi.org/10.1016/B978-0-12-800223-0.00005-0
40. Koepsell H, Schmitt BM, Gorboulev V (2003) Organic cation transporters. Rev Physiol Biochem Pharmacol 150:36–90
41. Rizwan AN, Burckhardt G (2007) Organic anion transporters of the SLC22 family: biopharmaceutical, physiological, and pathological roles. Pharm Res 24(3):450–470
42. Tsuji A, Tamai I (1996) Carrier-mediated intestinal transport of drugs. Pharm Res 13(7):963–977
43. Yang CY, Dantzig AH, Pidgeon C (1999) Intestinal peptide transport systems and oral drug availability. Pharm Res 16(9):1331–1343
44. Tsuruoka S, Sugimoto KI, Fujimura A, Imai M, Asano Y, Muto S (2001) P-glycoprotein-mediated drug secretion in mouse proximal tubule perfused in vitro. J Am Soc Nephrol 12(1):177–181
45. van Asperen J, van Tellingen O, Beijnen JH (2000) The role of mdr1a P-glycoprotein in the biliary and intestinal secretion of doxorubicin and vinblastine in mice. Drug Metab Dispos 28(3):264–267
46. Schinkel AH, Smit JJ, van Tellingen O, Beijnen JH, Wagenaar E, van Deemter L, Mol CA, van der Valk MA, Robanus-Maandag EC, te Riele HP et al (1994) Disruption of the mouse mdr1a P-glycoprotein gene leads to a deficiency in the blood-brain barrier and to increased sensitivity to drugs. Cell 77(4):491–502

47. Steinke JW, Borish L (2006) 3. Cytokines and chemokines. J Allergy Clin Immunol 117(2 Suppl Mini-Primer):S441–S445
48. Commins SP, Borish L, Steinke JW (2010) Immunologic messenger molecules: cytokines, interferons, and chemokines. J Allergy Clin Immunol 125(2 Suppl 2):S53–S72. https://doi. org/10.1016/j.jaci.2009.07.008
49. Oppenheim JJ, Feldmann M (2000) Introduction to the role of cytokines in innate host defense and adaptive immunity. In: Oppenheim JJ, Feldmann M, Durum SK, Hirano T, Vilcek J, Nicola NA Cytokine reference: a compendium of cytokines and other mediators of host defense. Ligands., vol 1. 1st edn. Academic Press, San Diego
50. Medzhitov R (2010) Inflammation 2010: new adventures of an old flame. Cell 140(6):771–776. https://doi.org/10.1016/j.cell.2010.03.006
51. Medzhitov R (2008) Origin and physiological roles of inflammation. Nature 454(7203): 428–435. https://doi.org/10.1038/nature07201. nature07201 [pii]
52. Barton GM (2008) A calculated response: control of inflammation by the innate immune system. J Clin Invest 118(2):413–420. https://doi.org/10.1172/JCI34431
53. Medzhitov R Inflammation (2010) New adventures of an old flame. Cell 140(6):771–776
54. Baumann H, Gauldie J (1994) The acute phase response. Immunol Today 15(2):74–80
55. Gruys E, Toussaint MJ, Niewold TA, Koopmans SJ (2005) Acute phase reaction and acute phase proteins. J Zhejiang Univ Sci B 6(11):1045–1056. https://doi.org/10.1631/jzus.2005.B1045
56. Ramadori G, Christ B (1999) Cytokines and the hepatic acute-phase response. Semin Liver Dis 19(2):141–155. https://doi.org/10.1055/s-2007-1007106
57. Abdel-Razzak Z, Loyer P, Fautrel A, Gautier JC, Corcos L, Turlin B, Beaune P, Guillouzo A (1993) Cytokines down-regulate expression of major cytochrome P-450 enzymes in adult human hepatocytes in primary culture. Mol Pharmacol 44(4):707–715
58. Le Vee M, Gripon P, Stieger B, Fardel O (2008) Down-regulation of organic anion trans- porter expression in human hepatocytes exposed to the proinflammatory cytokine interleu- kin 1beta. Drug Metab Dispos 36(2):217–222. https://doi.org/10.1124/dmd.107.016907. dmd.107.016907 [pii]
59. Vee ML, Lecureur V, Stieger B, Fardel O (2009) Regulation of drug transporter expression in human hepatocytes exposed to the proinflammatory cytokines tumor necrosis factor-alpha or interleukin-6. Drug Metab Dispos 37(3):685–693. https://doi.org/10.1124/dmd.108.023630. dmd.108.023630 [pii]
60. Morahan PS, Munson AE, Regelson W, Commerford SL, Hamilton LD (1972) Antiviral activity and side effects of polyriboinosinic-cytidylic acid complexes as affected by molecu- lar size. Proc Natl Acad Sci U S A 69(4):842–846
61. Nebert DW, Friedman RM (1973) Stimulation of aryl hydrocarbon hydroxylase induction in cell cultures by interferon. J Virol 11(2):193–197
62. Farquhar D, Loo TL, Gutterman JU, Hersh EM, Luna MA (1976) Inhibition of drug- metabolizing enzymes in the rat after Bacillus Calmette-Guerin treatment. Biochem Pharmacol 25(13):1529–1535
63. Castro JE (1974) The effect of Corynebacterium parvum on the structure and function of the lymphoid system in mice. Eur J Cancer 10(2):115–120
64. Foster RS Jr (1976) The immunostimulant Corynibacterium parvum and hematopoietic tox- icity of chemotherapy. Surg Forum 27(62):140–142
65. Soyka LF, Stephens CC, MacPherson BR, Foster RS Jr (1979) Role of mononuclear phago- cytes in decreased hepatic drug metabolism following administration of Corynebacterium parvum. Int J Immunopharmacol 1(2):101–112
66. Aitken AE, Richardson TA, Morgan ET (2006) Regulation of drug-metabolizing enzymes and transporters in inflammation. Annu Rev Pharmacol Toxicol 46:123–149
67. Chang KC, Bell TD, Lauer BA, Chai H (1978) Altered theophylline pharmacokinetics during acute respiratory viral illness. Lancet 1(8074):1132–1133
68. Kraemer MJ, Furukawa CT, Koup JR, Shapiro GG, Pierson WE, Bierman CW (1982) Altered theophylline clearance during an influenza B outbreak. Pediatrics 69(4):476–480

69. Schmitt C, Kuhn B, Zhang X, Kivitz AJ, Grange S (2011) Disease-drug-drug interaction involving tocilizumab and simvastatin in patients with rheumatoid arthritis. Clin Pharmacol Ther 89(5):735–740. https://doi.org/10.1038/clpt.2011.35

70. Chen YL, Le Vraux V, Leneveu A, Dreyfus F, Stheneur A, Florentin I, De Sousa M, Giroud JP, Flouvat B, Chauvelot-Moachon L (1994) Acute-phase response, interleukin-6, and alteration of cyclosporine pharmacokinetics. Clin Pharmacol Ther 55(6):649–660

71. Gidal BE, Reiss WG, Liao JS, Pitterle ME (1996) Changes in interleukin-6 concentrations following epilepsy surgery: potential influence on carbamazepine pharmacokinetics. Ann Pharmacother 30(5):545–546

72. Shelly MP, Mendel L, Park GR (1987) Failure of critically ill patients to metabolise midazolam. Anaesthesia 42(6):619–626

73. O'Neil WM, Gilfix BM, Markoglou N, Di Girolamo A, Tsoukas CM, Wainer IW (2000) Genotype and phenotype of cytochrome P450 2D6 in human immunodeficiency virus-positive patients and patients with acquired immunodeficiency syndrome. Eur J Clin Pharmacol 56(3):231–240

74. Frye RF, Schneider VM, Frye CS, Feldman AM (2002) Plasma levels of TNF-alpha and IL-6 are inversely related to cytochrome P450-dependent drug metabolism in patients with congestive heart failure. J Card Fail 8(5):315–319. doi:S1071916402004232 [pii]

75. Renton KW, Gray JD, Hall RI (1980) Decreased elimination of theophylline after influenza vaccination. Can Med Assoc J 123(4):288–290

76. Rivory LP, Slaviero KA, Clarke SJ (2002) Hepatic cytochrome P450 3A drug metabolism is reduced in cancer patients who have an acute-phase response. Br J Cancer 87(3):277–280. https://doi.org/10.1038/sj.bjc.6600448

77. Williams ML, Bhargava P, Cherrouk I, Marshall JL, Flockhart DA, Wainer IW (2000) A discordance of the cytochrome P450 2C19 genotype and phenotype in patients with advanced cancer. Br J Clin Pharmacol 49(5):485–488

78. Rubin K, Janefeldt A, Andersson L, Berke Z, Grime K, Andersson TB (2015) HepaRG cells as human-relevant in vitro model to study the effects of inflammatory stimuli on cytochrome P450 isoenzymes. Drug Metab Dispos 43(1):119–125. https://doi.org/10.1124/dmd.114.059246

79. Aitken AE, Lee CM, Morgan ET (2008) Roles of nitric oxide in inflammatory downregulation of human cytochromes P450. Free Radic Biol Med 44(6):1161–1168. https://doi.org/10.1016/j.freeradbiomed.2007.12.010. S0891-5849(07)00812-X [pii]

80. Lagadic-Gossmann D, Lerche C, Rissel M, Joannard F, Galisteo M, Guillouzo A, Corcos L (2000) The induction of the human hepatic CYP2E1 gene by interleukin 4 is transcriptional and regulated by protein kinase C. Cell Biol Toxicol 16(4):221–233

81. Muntane-Relat J, Ourlin JC, Domergue J, Maurel P (1995) Differential effects of cytokines on the inducible expression of CYP1A1, CYP1A2, and CYP3A4 in human hepatocytes in primary culture. Hepatology 22(4 Pt 1):1143–1153. doi:S0270913995003569 [pii]

82. Sunman JA, Hawke RL, LeCluyse EL, Kashuba AD (2004) Kupffer cell-mediated IL-2 suppression of CYP3A activity in human hepatocytes. Drug Metab Dispos 32(3):359–363. https://doi.org/10.1124/dmd.32.3.359. 32/3/359 [pii]

83. Yang J, Hao C, Yang D, Shi D, Song X, Luan X, Hu G, Yan B (2010) Pregnane X receptor is required for interleukin-6 mediated down-regulation of cytochrome P450 3A4 in human hepatocytes. Toxicol Lett. https://doi.org/10.1016/j.toxlet.2010.06.003. S0378-4274(10)01542-0 [pii]

84. Czerwinski M, Kazmi F, Parkinson A, Buckley DB (2015) Anti-CD28 monoclonal antibody-stimulated cytokines released from blood suppress CYP1A2, CYP2B6, and CYP3A4 in human hepatocytes in vitro. Drug Metab Dispos 43(1):42–52. https://doi.org/10.1124/dmd.114.060186

85. Morcos PN, Moreira SA, Brennan BJ, Blotner S, Shulman NS, Smith PF (2013) Influence of chronic hepatitis C infection on cytochrome P450 3A4 activity using midazolam as an in vivo probe substrate. Eur J Clin Pharmacol 69(10):1777–1784. https://doi.org/10.1007/s00228-013-1525-5

86. Hartmann G, Vassileva V, Piquette-Miller M (2005) Impact of endotoxin-induced changes in P-glycoprotein expression on disposition of doxorubicin in mice. Drug Metab Dispos 33(6):820–828. https://doi.org/10.1124/dmd.104.002568. dmd.104.002568 [pii]

87. Piquette-Miller M, Pak A, Kim H, Anari R, Shahzamani A (1998) Decreased expression and activity of P-glycoprotein in rat liver during acute inflammation. Pharm Res 15(5):706–711

88. Sukhai M, Yong A, Kalitsky J, Piquette-Miller M (2000) Inflammation and interleukin-6 mediate reductions in the hepatic expression and transcription of the mdr1a and mdr1b Genes. Mol Cell Biol Res Commun 4(4):248–256. https://doi.org/10.1006/mcbr.2001.0288. S1522472401902880 [pii]

89. Ando H, Nishio Y, Ito K, Nakao A, Wang L, Zhao YL, Kitaichi K, Takagi K, Hasegawa T (2001) Effect of endotoxin on P-glycoprotein-mediated biliary and renal excretion of rhodamine-123 in rats. Antimicrob Agents Chemother 45(12):3462–3467. https://doi.org/10.1128/AAC.45.12.3462-3467.2001

90. Wang JH, Scollard DA, Teng S, Reilly RM, Piquette-Miller M (2005) Detection of P-glycoprotein activity in endotoxemic rats by 99mTc-sestamibi imaging. J Nucl Med 46(9):1537–1545. doi:46/9/1537 [pii]

91. Cherrington NJ, Slitt AL, Li N, Klaassen CD (2004) Lipopolysaccharide-mediated regulation of hepatic transporter mRNA levels in rats. Drug Metab Dispos: Biol Fate Chem 32(i):734–741. doi:32/7/734 [pii]

92. Geier A, Dietrich CG, Voigt S, Kim SK, Gerloff T, Kullak-Ublick GA, Lorenzen J, Matern S, Gartung C (2003) Effects of proinflammatory cytokines on rat organic anion transporters during toxic liver injury and cholestasis. Hepatology 38(2):345–354. https://doi.org/10.1053/jhep.2003.50317. S0270913903005330 [pii]

93. Hartmann G, Cheung AK, Piquette-Miller M (2002) Inflammatory cytokines, but not bile acids, regulate expression of murine hepatic anion transporters in endotoxemia. J Pharmacol Exp Ther 303(1):273–281. https://doi.org/10.1124/jpet.102.039404

94. Sukhai M, Yong A, Pak A, Piquette-Miller M (2001) Decreased expression of P-glycoprotein in interleukin-1beta and interleukin-6 treated rat hepatocytes. Inflamm Res 50(7):362–370

95. Teng S, Piquette-Miller M (2005) The involvement of the pregnane X receptor in hepatic gene regulation during inflammation in mice. J Pharmacol Exp Ther 312(2):841–848. https://doi.org/10.1124/jpet.104.076141. jpet.104.076141 [pii]

96. Siewert E, Dietrich CG, Lammert F, Heinrich PC, Matern S, Gartung C, Geier A (2004) Interleukin-6 regulates hepatic transporters during acute-phase response. Biochem Biophys Res Commun 322(1):232–238. https://doi.org/10.1016/j.bbrc.2004.07.102. S0006-291X(04)01616-X [pii]

97. Elferink MG, Olinga P, Draaisma AL, Merema MT, Faber KN, Slooff MJ, Meijer DK, Groothuis GM (2004) LPS-induced downregulation of MRP2 and BSEP in human liver is due to a posttranscriptional process. Am J Physiol Gastrointest Liver Physiol 287(5):G1008–G1016. https://doi.org/10.1152/ajpgi.00071.2004. 00071.2004 [pii]

98. Fardel O, Le Vee M (2009) Regulation of human hepatic drug transporter expression by pro-inflammatory cytokines. Expert Opin Drug Metab Toxicol 5(12):1469–1481. https://doi.org/10.1517/17425250903304056

99. Buyse M, Radeva G, Bado A, Farinotti R (2005) Intestinal inflammation induces adaptation of P-glycoprotein expression and activity. Biochem Pharmacol 69(12):1745–1754. https://doi.org/10.1016/j.bcp.2005.03.025. S0006-2952(05)00199-1 [pii]

100. Iizasa H, Genda N, Kitano T, Tomita M, Nishihara K, Hayashi M, Nakamura K, Kobayashi S, Nakashima E (2003) Altered expression and function of P-glycoprotein in dextran sodium sulfate-induced colitis in mice. J Pharm Sci 92(3):569–576. https://doi.org/10.1002/jps.10326

101. Masubuchi Y, Enoki K, Horie T (2008) Down-regulation of hepatic cytochrome P450 enzymes in rats with trinitrobenzene sulfonic acid-induced colitis. Drug Metab Dispos 36(3):597–603. https://doi.org/10.1124/dmd.107.018754. dmd.107.018754 [pii]

102. Naud J, Michaud J, Boisvert C, Desbiens K, Leblond FA, Mitchell A, Jones C, Bonnardeaux A, Pichette V (2007) Down-regulation of intestinal drug transporters in chronic renal failure in rats. J Pharmacol Exp Ther 320(3):978–985. https://doi.org/10.1124/jpet.106.112631. jpet.106.112631 [pii]

103. Veau C, Faivre L, Tardivel S, Soursac M, Banide H, Lacour B, Farinotti R (2002) Effect of interleukin-2 on intestinal P-glycoprotein expression and functionality in mice. J Pharmacol Exp Ther 302(2):742–750

104. Niessner M, Volk BA (1995) Altered Th1/Th2 cytokine profiles in the intestinal mucosa of patients with inflammatory bowel disease as assessed by quantitative reversed transcribed polymerase chain reaction (RT-PCR). Clin Exp Immunol 101(3):428–435

105. Sawa Y, Oshitani N, Adachi K, Higuchi K, Matsumoto T, Arakawa T (2003) Comprehensive analysis of intestinal cytokine messenger RNA profile by real-time quantitative polymerase chain reaction in patients with inflammatory bowel disease. Int J Mol Med 11(2): 175–179

106. Gutmann H, Hruz P, Zimmermann C, Straumann A, Terracciano L, Hammann F, Lehmann F, Beglinger C, Drewe J (2008) Breast cancer resistance protein and P-glycoprotein expression in patients with newly diagnosed and therapy-refractory ulcerative colitis compared with healthy controls. Digestion 78(2–3):154–162. https://doi.org/10.1159/000179361. 000179361 [pii]

107. Englund G, Jacobson A, Rorsman F, Artursson P, Kindmark A, Ronnblom A (2007) Efflux transporters in ulcerative colitis: decreased expression of BCRP (ABCG2) and Pgp (ABCB1). Inflamm Bowel Dis 13(3):291–297. https://doi.org/10.1002/ibd.20030

108. Ufer M, Hasler R, Jacobs G, Haenisch S, Lachelt S, Faltraco F, Sina C, Rosenstiel P, Nikolaus S, Schreiber S, Cascorbi I (2009) Decreased sigmoidal ABCB1 (P-glycoprotein) expression in ulcerative colitis is associated with disease activity. Pharmacogenomics 10(12):1941–1953. https://doi.org/10.2217/pgs.09.128

109. Blokzijl H, Vander Borght S, Bok LI, Libbrecht L, Geuken M, van den Heuvel FA, Dijkstra G, Roskams TA, Moshage H, Jansen PL, Faber KN (2007) Decreased P-glycoprotein (P-gp/MDR1) expression in inflamed human intestinal epithelium is independent of PXR protein levels. Inflamm Bowel Dis 13(6):710–720. https://doi.org/10.1002/ibd.20088

110. Langmann T, Moehle C, Mauerer R, Scharl M, Liebisch G, Zahn A, Stremmel W, Schmitz G (2004) Loss of detoxification in inflammatory bowel disease: dysregulation of pregnane X receptor target genes. Gastroenterology 127(1):26–40. doi:S0016508504007140 [pii]

111. Kawauchi S, Nakamura T, Miki I, Inoue J, Hamaguchi T, Tanahashi T, Mizuno S (2014) Downregulation of CYP3A and P-glycoprotein in the secondary inflammatory response of mice with dextran sulfate sodium-induced colitis and its contribution to cyclosporine A blood concentrations. J Pharmacol Sci 124(2):180–191

112. Kusunoki Y, Ikarashi N, Hayakawa Y, Ishii M, Kon R, Ochiai W, Machida Y, Sugiyama K (2014) Hepatic early inflammation induces downregulation of hepatic cytochrome P450 expression and metabolic activity in the dextran sulfate sodium-induced murine colitis. Eur J Pharm Sci 54:17–27. https://doi.org/10.1016/j.ejps.2013.12.019

113. Kusunoki Y, Ikarashi N, Matsuda S, Matsukawa Y, Kitaoka S, Kon R, Tajima M, Wakui N, Ochiai W, Machida Y, Sugiyama K (2015) Expression of hepatic cytochrome P450 in a mouse model of ulcerative colitis changes with pathological conditions. J Gastroenterol Hepatol 30(11):1618–1626. https://doi.org/10.1111/jgh.12966

114. Liu J, Zhou F, Chen Q, Kang A, Lu M, Liu W, Zang X, Wang G, Zhang J (2015) Chronic inflammation up-regulates P-gp in peripheral mononuclear blood cells via the STAT3/Nf-kappab pathway in 2,4,6-trinitrobenzene sulfonic acid-induced colitis mice. Sci Rep 5:13558. https://doi.org/10.1038/srep13558

115. Nyagode BA, Jahangardi R, Merrell MD, Tansey MG, Morgan ET (2014) Selective effects of a therapeutic protein targeting tumor necrosis factor-alpha on cytochrome P450 regulation during infectious colitis: implications for disease-dependent drug-drug interactions. Pharmacol Res Perspect 2(1):e00027. https://doi.org/10.1002/prp2.27

116. Nyagode BA, Lee CM, Morgan ET (2010) Modulation of hepatic cytochrome P450s by Citrobacter rodentium infection in interleukin-6- and interferon-{gamma}-null mice. J Pharmacol Exp Ther 335(2):480–488. https://doi.org/10.1124/jpet.110.171488

117. Sugimoto M, Furuta T, Shirai N, Ikuma M, Hishida A, Ishizaki T (2006) Influences of proinflammatory and anti-inflammatory cytokine polymorphisms on eradication rates of clarithromycin-sensitive strains of Helicobacter pylori by triple therapy. Clin Pharmacol Ther 80(1):41–50. https://doi.org/10.1016/j.clpt.2006.03.007. S0009-9236(06)00120-2 [pii]

118. Miftahussurur M, Yamaoka Y (2015) Helicobacter pylori virulence genes and host genetic polymorphisms as risk factors for peptic ulcer disease. Expert Rev Gastroenterol Hepatol 9(12):1535–1547. https://doi.org/10.1586/17474124.2015.1095089

119. Zambon CF, Fasolo M, Basso D, D'Odorico A, Stranges A, Navaglia F, Fogar P, Greco E, Schiavon S, Padoan A, Fadi E, Sturniolo GC, Plebani M, Pedrazzoli S (2007) Clarithromycin resistance, tumor necrosis factor alpha gene polymorphism and mucosal inflammation affect H. pylori eradication success. J Gastrointest Surg 11(11):1506–1514.; discussion 1514. https://doi.org/10.1007/s11605-007-0246-4

120. Furuta T, Shirai N, Takashima M, Xiao F, Sugimura H (2002) Effect of genotypic differences in interleukin-1 beta on gastric acid secretion in Japanese patients infected with Helicobacter pylori. Am J Med 112(2):141–143. doi:S0002934301010361 [pii]

121. Furuta T, Shirai N, Xiao F, El-Omar EM, Rabkin CS, Sugimura H, Ishizaki T, Ohashi K (2004) Polymorphism of interleukin-1beta affects the eradication rates of Helicobacter pylori by triple therapy. Clin Gastroenterol Hepatol 2(1):22–30. doi:S154235650300288X [pii]

122. Schmidt C, Hocherl K, Bucher M (2007) Cytokine-mediated regulation of urea transporters during experimental endotoxemia. Am J Physiol Renal Physiol 292(5):F1479–F1489. https://doi.org/10.1152/ajprenal.00460.2006. 00460.2006 [pii]

123. Schmidt C, Hocherl K, Bucher M (2007) Regulation of renal glucose transporters during severe inflammation. Am J Physiol Renal Physiol 292(2):F804–F811. https://doi.org/10.1152/ajprenal.00258.2006. 00258.2006 [pii]

124. Schmidt C, Hocherl K, Schweda F, Kurtz A, Bucher M (2007) Regulation of renal sodium transporters during severe inflammation. J Am Soc Nephrol 18(4):1072–1083. https://doi.org/10.1681/ASN.2006050454. ASN.2006050454 [pii]

125. Heemskerk S, Wouterse AC, Russel FG, Masereeuw R (2008) Nitric oxide down-regulates the expression of organic cation transporters (OCT) 1 and 2 in rat kidney during endotoxemia. Eur J Pharmacol 584(2–3):390–397. https://doi.org/10.1016/j.ejphar.2008.02.006. S0014-2999(08)00153-2 [pii]

126. Heemskerk S, van Koppen A, van den Broek L, Poelen GJ, Wouterse AC, Dijkman HB, Russel FG, Masereeuw R (2007) Nitric oxide differentially regulates renal ATP-binding cassette transporters during endotoxemia. Pflugers Arch 454(2):321–334. https://doi.org/10.1007/s00424-007-0210-x

127. Huls M, van den Heuvel JJ, Dijkman HB, Russel FG, Masereeuw R (2006) ABC transporter expression profiling after ischemic reperfusion injury in mouse kidney. Kidney Int 69(12):2186–2193. https://doi.org/10.1038/sj.ki.5000407. 5000407 [pii]

128. Masereeuw R, Moons MM, Russel FG (1997) Rhodamine 123 accumulates extensively in the isolated perfused rat kidney and is secreted by the organic cation system. Eur J Pharmacol 321(3):315–323. doi:S0014-2999(96)00957-0 [pii]

129. Graff CL, Pollack GM (2004) Drug transport at the blood-brain barrier and the choroid plexus. Curr Drug Metab 5(1):95–108

130. Kusuhara H, Sugiyama Y (2005) Active efflux across the blood-brain barrier: role of the solute carrier family. NeuroRx 2(1):73–85

131. de Vries HE, Kuiper J, de Boer AG, Van Berkel TJ, Breimer DD (1997) The blood-brain barrier in neuroinflammatory diseases. Pharmacol Rev 49(2):143–155

132. Eikelenboom P, Bate C, Van Gool WA, Hoozemans JJ, Rozemuller JM, Veerhuis R, Williams A (2002) Neuroinflammation in Alzheimer's disease and prion disease. Glia 40(2):232–239

133. Ghafouri M, Amini S, Khalili K, Sawaya BE (2006) HIV-1 associated dementia: symptoms and causes. Retrovirology 3:28
134. Whitton PS (2007) Inflammation as a causative factor in the aetiology of Parkinson's disease. Br J Pharmacol 150(8):963–976
135. Bauer B, Hartz AM, Pekcec A, Toellner K, Miller DS, Potschka H (2008) Seizure-induced up-regulation of P-glycoprotein at the blood-brain barrier through glutamate and cyclooxygenase-2 signaling. Mol Pharmacol 73(5):1444–1453
136. Kortekaas R, Leenders KL, van Oostrom JC, Vaalburg W, Bart J, Willemsen AT, Hendrikse NH (2005) Blood-brain barrier dysfunction in parkinsonian midbrain in vivo. Ann Neurol 57(2):176–179
137. Langford D, Grigorian A, Hurford R, Adame A, Ellis RJ, Hansen L, Masliah E (2004) Altered P-glycoprotein expression in AIDS patients with HIV encephalitis. J Neuropathol Exp Neurol 63(10):1038–1047
138. Loscher W, Potschka H (2005) Drug resistance in brain diseases and the role of drug efflux transporters. Nat Rev Neurosci 6(8):591–602
139. Roberts DJ, Goralski KB, Renton KW, Julien LC, Webber AM, Sleno L, Volmer DA, Hall RI (2009) Effect of acute inflammatory brain injury on accumulation of morphine and morphine 3- and 6-glucuronide in the human brain. Crit Care Med 37(10):2767–2774. https://doi.org/10.1097/CCM.0b013e3181b755d5. 00003246-200910000-00014 [pii]
140. Vogelgesang S, Warzok RW, Cascorbi I, Kunert-Keil C, Schroeder E, Kroemer HK, Siegmund W, Walker LC, Pahnke J (2004) The role of P-glycoprotein in cerebral amyloid angiopathy; implications for the early pathogenesis of Alzheimer's disease. Curr Alzheimer Res 1(2):121–125
141. Beaulieu E, Demeule M, Ghitescu L, Beliveau R (1997) P-glycoprotein is strongly expressed in the luminal membranes of the endothelium of blood vessels in the brain. Biochem J 326(Pt 2):539–544
142. Virgintino D, Robertson D, Errede M, Benagiano V, Girolamo F, Maiorano E, Roncali L, Bertossi M (2002) Expression of P-glycoprotein in human cerebral cortex microvessels. J Histochem Cytochem 50(12):1671–1676
143. Choo EF, Leake B, Wandel C, Imamura H, Wood AJ, Wilkinson GR, Kim RB (2000) Pharmacological inhibition of P-glycoprotein transport enhances the distribution of HIV-1 protease inhibitors into brain and testes. Drug Metab Dispos 28(6):655–660
144. King M, Su W, Chang A, Zuckerman A, Pasternak GW (2001) Transport of opioids from the brain to the periphery by P-glycoprotein: peripheral actions of central drugs. Nat Neurosci 4(3):268–274
145. Lankas GR, Cartwright ME, Umbenhauer D (1997) P-glycoprotein deficiency in a subpopulation of CF-1 mice enhances avermectin-induced neurotoxicity. Toxicol Appl Pharmacol 143(2):357–365
146. Luurtsema G, Molthoff CF, Windhorst AD, Smit JW, Keizer H, Boellaard R, Lammertsma AA, Franssen EJ (2003) (R)- and (S)-[11C]verapamil as PET-tracers for measuring P-glycoprotein function: in vitro and in vivo evaluation. Nucl Med Biol 30(7):747–751
147. Thuerauf N, Fromm MF (2006) The role of the transporter P-glycoprotein for disposition and effects of centrally acting drugs and for the pathogenesis of CNS diseases. Eur Arch Psychiatry Clin Neurosci 256(5):281–286
148. Uhr M, Tontsch A, Namendorf C, Ripke S, Lucae S, Ising M, Dose T, Ebinger M, Rosenhagen M, Kohli M, Kloiber S, Salyakina D, Bettecken T, Specht M, Putz B, Binder EB, Muller-Myhsok B, Holsboer F (2008) Polymorphisms in the drug transporter gene ABCB1 predict antidepressant treatment response in depression. Neuron 57(2):203–209
149. Greenberg ML, Fisher PG, Freeman C, Korones DN, Bernstein M, Friedman H, Blaney S, Hershon L, Zhou T, Chen Z, Kretschmar C (2005) Etoposide, vincristine, and cyclosporin A with standard-dose radiation therapy in newly diagnosed diffuse intrinsic brainstem gliomas: a pediatric oncology group phase I study. Pediatr Blood Cancer 45(5):644–648
150. Sadeque AJ, Wandel C, He H, Shah S, Wood AJ (2000) Increased drug delivery to the brain by P-glycoprotein inhibition. Clin Pharmacol Ther 68(3):231–237

151. Chen X, Zhou ZW, Xue CC, Li XX, Zhou SF (2007) Role of P-glycoprotein in restricting the brain penetration of tanshinone IIA, a major active constituent from the root of Salvia miltiorrhiza Bunge, across the blood-brain barrier. Xenobiotica 37(6):635–678

152. Imbert F, Jardin M, Fernandez C, Gantier JC, Dromer F, Baron G, Mentre F, Van Beijsterveldt L, Singlas E, Gimenez F (2003) Effect of efflux inhibition on brain uptake of itraconazole in mice infected with Cryptococcus neoformans. Drug Metab Dispos 31(3):319–325

153. XY Y, Lin SG, Chen X, Zhou ZW, Liang J, Duan W, Chowbay B, Wen JY, Chan E, Cao J, Li CG, Zhou SF (2007) Transport of cryptotanshinone, a major active triterpenoid in Salvia miltiorrhiza Bunge widely used in the treatment of stroke and Alzheimer's disease, across the blood-brain barrier. Curr Drug Metab 8(4):365–378

154. Zhao YL, Du J, Kanazawa H, Cen XB, Takagi K, Kitaichi K, Tatsumi Y, Takagi K, Ohta M, Hasegawa T (2002) Shiga-like toxin II modifies brain distribution of a P-glycoprotein substrate, doxorubicin, and P-glycoprotein expression in mice. Brain Res 956(2):246–253

155. Zhao YL, Du J, Kanazawa H, Sugawara A, Takagi K, Kitaichi K, Tatsumi Y, Takagi K, Hasegawa T (2002) Effect of endotoxin on doxorubicin transport across blood-brain barrier and P-glycoprotein function in mice. Eur J Pharmacol 445(1–2):115–123

156. Hartz AM, Bauer B, Block ML, Hong JS, Miller DS (2008) Diesel exhaust particles induce oxidative stress, proinflammatory signaling, and P-glycoprotein up-regulation at the blood-brain barrier. FASEB J 22(8):2723–2733

157. Seelbach MJ, Brooks TA, Egleton RD, Davis TP (2007) Peripheral inflammatory hyperalgesia modulates morphine delivery to the brain: a role for P-glycoprotein. J Neurochem 102(5):1677–1690

158. McCaffrey G, Staatz WD, Sanchez-Covarrubias L, Finch JD, Demarco K, Laracuente ML, Ronaldson PT, Davis TP (2012) P-glycoprotein trafficking at the blood-brain barrier altered by peripheral inflammatory hyperalgesia. J Neurochem 122(5):962–975. https://doi.org/10.1111/j.1471-4159.2012.07831.x

159. Bauer B, Hartz AM, Fricker G, Miller DS (2005) Modulation of p-glycoprotein transport function at the blood-brain barrier. Exp Biol Med (Maywood, NJ) 230(2):118–127

160. Hartz AM, Bauer B, Fricker G, Miller DS (2004) Rapid regulation of P-glycoprotein at the blood-brain barrier by endothelin-1. Mol Pharmacol 66(3):387–394

161. Poller B, Drewe J, Krahenbuhl S, Huwyler J, Gutmann H (2010) Regulation of BCRP (ABCG2) and P-glycoprotein (ABCB1) by cytokines in a model of the human blood-brain barrier. Cell Mol Neurobiol 30(1):63–70. https://doi.org/10.1007/s10571-009-9431-1

162. Ronaldson PT, Ashraf T, Bendayan R (2010) Regulation of multidrug resistance protein 1 by tumor necrosis factor alpha in cultured glial cells: involvement of nuclear factor-kappaB and c-Jun N-terminal kinase signaling pathways. Mol Pharmacol 77(4):644–659. https://doi.org/10.1124/mol.109.059410. mol.109.059410 [pii]

163. Ronaldson PT, Bendayan R (2006) HIV-1 viral envelope glycoprotein gp120 triggers an inflammatory response in cultured rat astrocytes and regulates the functional expression of P-glycoprotein. Mol Pharmacol 70(3):1087–1098

164. Lazarowski A, Czornyj L, Lubienieki F, Girardi E, Vazquez S, D'Giano C (2007) ABC transporters during epilepsy and mechanisms underlying multidrug resistance in refractory epilepsy. Epilepsia 48(Suppl 5):140–149

165. Spudich A, Kilic E, Xing H, Kilic U, Rentsch KM, Wunderli-Allenspach H, Bassetti CL, Hermann DM (2006) Inhibition of multidrug resistance transporter-1 facilitates neuroprotective therapies after focal cerebral ischemia. Nat Neurosci 9(4):487–488

166. Jin L, Li J, Nation RL, Nicolazzo JA (2011) Impact of p-glycoprotein inhibition and lipopolysaccharide administration on blood-brain barrier transport of colistin in mice. Antimicrob Agents Chemother 55(2):502–507. https://doi.org/10.1128/AAC.01273-10

167. Daniels BP, Holman DW, Cruz-Orengo L, Jujjavarapu H, Durrant DM, Klein RS (2014) Viral pathogen-associated molecular patterns regulate blood-brain barrier integrity via competing innate cytokine signals. MBio 5(5):e01476–e01414. https://doi.org/10.1128/mBio.01476-14

168. Wong D, Dorovini-Zis K, Vincent SR (2004) Cytokines, nitric oxide, and cGMP modulate the permeability of an in vitro model of the human blood-brain barrier. Exp Neurol 190(2):446–455. https://doi.org/10.1016/j.expneurol.2004.08.008

169. Tsao N, Hsu HP, CM W, Liu CC, Lei HY (2001) Tumour necrosis factor-alpha causes an increase in blood-brain barrier permeability during sepsis. J Med Microbiol 50(9):812–821. https://doi.org/10.1099/0022-1317-50-9-812

170. Vahakangas K, Myllynen P (2009) Drug transporters in the human blood-placental barrier. Br J Pharmacol 158(3):665–678. https://doi.org/10.1111/j.1476-5381.2009.00336.x. BPH336 [pii]

171. Evseenko DA, Paxton JW, Keelan JA (2006) ABC drug transporter expression and functional activity in trophoblast-like cell lines and differentiating primary trophoblast. Am J Physiol Regul Integr Comp Physiol 290(5):R1357–R1365. https://doi.org/10.1152/ajpregu.00630.2005. 00630.2005 [pii]

172. Jonker JW, Smit JW, Brinkhuis RF, Maliepaard M, Beijnen JH, Schellens JH, Schinkel AH (2000) Role of breast cancer resistance protein in the bioavailability and fetal penetration of topotecan. J Natl Cancer Inst 92(20):1651–1656

173. Lankas GR, Wise LD, Cartwright ME, Pippert T, Umbenhauer DR (1998) Placental P-glycoprotein deficiency enhances susceptibility to chemically induced birth defects in mice. Reprod Toxicol 12(4):457–463. doi:S0890623898000276 [pii]

174. Evseenko DA, Paxton JW, Keelan JA (2007) Independent regulation of apical and basolateral drug transporter expression and function in placental trophoblasts by cytokines, steroids, and growth factors. Drug Metab Dispos 35(4):595–601. https://doi.org/10.1124/dmd.106.011478. dmd.106.011478 [pii]

175. Smit JW, Huisman MT, van Tellingen O, Wiltshire HR, Schinkel AH (1999) Absence or pharmacological blocking of placental P-glycoprotein profoundly increases fetal drug exposure. J Clin Invest 104(10):1441–1447. https://doi.org/10.1172/JCI7963

176. Zhou L, Naraharisetti SB, Wang H, Unadkat JD, Hebert MF, Mao Q (2008) The breast cancer resistance protein (Bcrp1/Abcg2) limits fetal distribution of glyburide in the pregnant mouse: an Obstetric-Fetal Pharmacology Research Unit Network and University of Washington Specialized Center of Research Study. Mol Pharmacol 73(3):949–959. https://doi.org/10.1124/mol.107.041616. mol.107.041616 [pii]

177. Molsa M, Heikkinen T, Hakkola J, Hakala K, Wallerman O, Wadelius M, Wadelius C, Laine K (2005) Functional role of P-glycoprotein in the human blood-placental barrier. Clin Pharmacol Ther 78(2):123–131. https://doi.org/10.1016/j.clpt.2005.04.014. S0009-9236(05)00188-8 [pii]

178. Sudhakaran S, Rayner CR, Li J, Kong DC, Gude NM, Nation RL (2008) Inhibition of placental P-glycoprotein: impact on indinavir transfer to the foetus. Br J Clin Pharmacol 65(5):667–673. https://doi.org/10.1111/j.1365-2125.2007.03067.x. BCP3067 [pii]

179. Hemauer SJ, Patrikeeva SL, Nanovskaya TN, Hankins GD, Ahmed MS (2010) Role of human placental apical membrane transporters in the efflux of glyburide, rosiglitazone, and metformin. Am J Obstet Gynecol 202(4):383 e381–383 e387. https://doi.org/10.1016/j.ajog.2010.01.035. S0002-9378(10)00065-7 [pii]

180. Pollex E, Lubetsky A, Koren G (2008) The role of placental breast cancer resistance protein in the efflux of glyburide across the human placenta. Placenta 29(8):743–747. https://doi.org/10.1016/j.placenta.2008.05.001. S0143-4004(08)00148-3 [pii]

181. Lye P, Bloise E, Javam M, Gibb W, Lye SJ, Matthews SG (2015) Impact of bacterial and viral challenge on multidrug resistance in first- and third-trimester human placenta. Am J Pathol 185(6):1666–1675. https://doi.org/10.1016/j.ajpath.2015.02.013

182. Hamai Y, Fujii T, Yamashita T, Nishina H, Kozuma S, Mikami Y, Taketani Y (1997) Evidence for an elevation in serum interleukin-2 and tumor necrosis factor-alpha levels before the clinical manifestations of preeclampsia. Am J Reprod Immunol 38(2):89–93

183. Saji F, Samejima Y, Kamiura S, Sawai K, Shimoya K, Kimura T (2000) Cytokine production in chorioamnionitis. J Reprod Immunol 47(2):185–196. doi:S0165-0378(00)00064-4 [pii]

184. Steinborn A, Niederhut A, Solbach C, Hildenbrand R, Sohn C, Kaufmann M (1999) Cytokine release from placental endothelial cells, a process associated with preterm labour in the absence of intrauterine infection. Cytokine 11(1):66–73. https://doi.org/10.1006/cyto.1998.0399. S1043-4666(98)90399-4 [pii]

185. Chen YH, Wang JP, Wang H, Sun MF, Wei LZ, Wei W, DX X (2005) Lipopolysaccharide treatment downregulates the expression of the pregnane X receptor, cyp3a11 and mdr1a genes in mouse placenta. Toxicology 211(3):242–252. https://doi.org/10.1016/j.tox.2005.03.011. S0300-483X(05)00168-X [pii]

186. Petrovic V, Wang JH, Piquette-Miller M (2008) Effect of endotoxin on the expression of placental drug transporters and glyburide disposition in pregnant rats. Drug Metab Dispos 36(9):1944–1950. https://doi.org/10.1124/dmd.107.019851. dmd.107.019851 [pii]

187. Gedeon C, Behravan J, Koren G, Piquette-Miller M (2006) Transport of glyburide by placental ABC transporters: implications in fetal drug exposure. Placenta 27(11–12):1096–1102. https://doi.org/10.1016/j.placenta.2005.11.012. S0143-4004(05)00312-7 [pii]

188. Mason CW, Buhimschi IA, Buhimschi CS, Dong Y, Weiner CP, Swaan PW (2011) ATP-binding cassette transporter expression in human placenta as a function of pregnancy condition. Drug Metab Dispos 39(6):1000–1007. https://doi.org/10.1124/dmd.111.038166

189. Mantovani A, Allavena P, Sica A, Balkwill F (2008) Cancer-related inflammation. Nature 454(7203):436–444. https://doi.org/10.1038/nature07205. nature07205 [pii]

190. Conze D, Weiss L, Regen PS, Bhushan A, Weaver D, Johnson P, Rincon M (2001) Autocrine production of interleukin 6 causes multidrug resistance in breast cancer cells. Cancer Res 61(24):8851–8858

191. Wang Y, Niu XL, Qu Y, Wu J, Zhu YQ, Sun WJ, Li LZ (2010) Autocrine production of interleukin-6 confers cisplatin and paclitaxel resistance in ovarian cancer cells. Cancer Lett 295(1):110–123. https://doi.org/10.1016/j.canlet.2010.02.019. S0304-3835(10)00120-5 [pii]

192. Kacevska M, Robertson GR, Clarke SJ, Liddle C (2008) Inflammation and CYP3A4-mediated drug metabolism in advanced cancer: impact and implications for chemotherapeutic drug dosing. Expert Opin Drug Metab Toxicol 4(2):137–149. https://doi.org/10.1517/17425255.4.2.137

193. Gottesman MM, Fojo T, Bates SE (2002) Multidrug resistance in cancer: role of ATP-dependent transporters. Nat Rev Cancer 2(1):48–58

194. Chen Z, Shi T, Zhang L, Zhu P, Deng M, Huang C, Hu T, Jiang L, Li J (2016) Mammalian drug efflux transporters of the ATP binding cassette (ABC) family in multidrug resistance: a review of the past decade. Cancer Lett 370(1):153–164. https://doi.org/10.1016/j.canlet.2015.10.010

195. Longley DB, Johnston PG (2005) Molecular mechanisms of drug resistance. J Pathol 205(2):275–292

196. Szakacs G, Paterson JK, Ludwig JA, Booth-Genthe C, Gottesman MM (2006) Targeting multidrug resistance in cancer. Nat Rev 5(3):219–234

197. Duan Z, Lamendola DE, Penson RT, Kronish KM, Seiden MV (2002) Overexpression of IL-6 but not IL-8 increases paclitaxel resistance of U-2OS human osteosarcoma cells. Cytokine 17(5):234–242. https://doi.org/10.1006/cyto.2001.1008. S1043466601910087 [pii]

198. Mosaffa F, Lage H, Afshari JT, Behravan J (2009) Interleukin-1 beta and tumor necrosis factor-alpha increase ABCG2 expression in MCF-7 breast carcinoma cell line and its mitoxantrone-resistant derivative, MCF-7/MX. Inflamm Res 58(10):669–676. https://doi.org/10.1007/s00011-009-0034-6

199. Mosaffa F, Kalalinia F, Lage H, Afshari JT, Behravan J (2012) Pro-inflammatory cytokines interleukin-1 beta, interleukin 6, and tumor necrosis factor-alpha alter the expression and function of ABCG2 in cervix and gastric cancer cells. Mol Cell Biochem 363(1–2):385–393. https://doi.org/10.1007/s11010-011-1191-9

200. Charles KA, Rivory LP, Brown SL, Liddle C, Clarke SJ, Robertson GR (2006) Transcriptional repression of hepatic cytochrome P450 3A4 gene in the presence of cancer. Clin Cancer Res 12(24):7492–7497. https://doi.org/10.1158/1078-0432.CCR-06-0023. 12/24/7492 [pii]

201. Helsby NA, Lo WY, Sharples K, Riley G, Murray M, Spells K, Dzhelai M, Simpson A, Findlay M (2008) CYP2C19 pharmacogenetics in advanced cancer: compromised function independent of genotype. Br J Cancer 99(8):1251–1255. https://doi.org/10.1038/sj.bjc.6604699. 6604699 [pii]

202. Okuno H, Kitao Y, Takasu M, Kano H, Kunieda K, Seki T, Shiozaki Y, Sameshima Y (1990) Depression of drug metabolizing activity in the human liver by interferon-alpha. Eur J Clin Pharmacol 39(4):365–367

203. Williams SJ, Farrell GC (1986) Inhibition of antipyrine metabolism by interferon. Br J Clin Pharmacol 22(5):610–612

204. Pageaux GP, le Bricquir Y, Berthou F, Bressot N, Picot MC, Blanc F, Michel H, Larrey D (1998) Effects of interferon-alpha on cytochrome P-450 isoforms 1A2 and 3A activities in patients with chronic hepatitis C. Eur J Gastroenterol Hepatol 10(6):491–495

205. Becquemont L, Chazouilleres O, Serfaty L, Poirier JM, Broly F, Jaillon P, Poupon R, Funck-Brentano C (2002) Effect of interferon alpha-ribavirin bitherapy on cytochrome P450 1A2 and 2D6 and N-acetyltransferase-2 activities in patients with chronic active hepatitis C. Clin Pharmacol Ther 71(6):488–495. https://doi.org/10.1067/mcp.2002.124468. S0009923602000097 [pii]

206. Islam M, Frye RF, Richards TJ, Sbeitan I, Donnelly SS, Glue P, Agarwala SS, Kirkwood JM (2002) Differential effect of IFNalpha-2b on the cytochrome P450 enzyme system: a potential basis of IFN toxicity and its modulation by other drugs. Clin Cancer Res 8(8):2480–2487

207. Ghany MG, Strader DB, Thomas DL, Seeff LB, American Association for the Study of Liver D (2009) Diagnosis, management, and treatment of hepatitis C: an update. Hepatology 49(4):1335–1374. https://doi.org/10.1002/hep.22759

208. Brennan BJ, ZX X, Grippo JF (2013) Effect of peginterferon alfa-2a (40KD) on cytochrome P450 isoenzyme activity. Br J Clin Pharmacol 75(2):497–506. https://doi.org/10.1111/j.1365-2125.2012.04373.x

209. Gupta SK, Kolz K, Cutler DL (2011) Effects of multiple-dose pegylated interferon alfa-2b on the activity of drug-metabolizing enzymes in persons with chronic hepatitis C. Eur J Clin Pharmacol 67(6):591–599. https://doi.org/10.1007/s00228-010-0972-5

210. Scavone C, Sportiello L, Rafaniello C, Mascolo A, Sessa M, Rossi F, Capuano A (2016) New era in treatment options of chronic hepatitis C: focus on safety of new direct-acting antivirals (DAAs). Expert Opin Drug Saf 15(sup2):85–100. https://doi.org/10.1080/14740338.2016.1221396

211. AASLD-IDSA (2017) Recommendations for testing, managing and treating hepatitis C. http://www.hcvguidelines.org. Accessed 2 Feb 2017

212. Myers RP, Shah H, Burak KW, Cooper C, Feld JJ (2015) An update on the management of chronic hepatitis C: 2015 Consensus guidelines from the Canadian Association for the Study of the Liver. Can J Gastroenterol Hepatol 29(1):19–34

213. Burgess S, Partovi N, Yoshida EM, Erb SR, Azalgara VM, Hussaini T (2015) Drug interactions with direct-acting antivirals for hepatitis C: implications for HIV and transplant patients. Ann Pharmacother 49(6):674–687. https://doi.org/10.1177/1060028015576180

214. Micromedex® 2.0 (electronic version). Truven Health Analytics, Greenwood Village. http://www.micromedexsolutions.com/. Accessed 2 Feb 2017

215. Reesink HW, Fanning GC, Farha KA, Weegink C, Van Vliet A, Van 't Klooster G, Lenz O, Aharchi F, Marien K, Van Remoortere P, de Kock H, Broeckaert F, Meyvisch P, Van Beirendonck E, Simmen K, Verloes R (2010) Rapid HCV-RNA decline with once daily TMC435; a phase I study in healthy volunteers and hepatitis C patients. Gastroenterology 138(3):913–921. https://doi.org/10.1053/j.gastro.2009.10.033

216. Kumar D (2010) Emerging viruses in transplantation. Curr Opin Infect Dis 23(4):374–378. https://doi.org/10.1097/QCO.0b013e32833bc19d. 00001432-201008000-00013 [pii]

217. Sayegh MH, Carpenter CB (2004) Transplantation 50 years later--progress, challenges, and promises. N Engl J Med 351(26):2761–2766. https://doi.org/10.1056/NEJMon043418. 351/26/2761 [pii]

218. Dharnidharka VR, Stablein DM, Harmon WE (2004) Post-transplant infections now exceed acute rejection as cause for hospitalization: a report of the NAPRTCS. Am J Transplant 4(3):384–389
219. Monforte V, Bullich S, Pou L, Bravo C, Lopez R, Gavalda J, Roman A (2003) Blood cyclosporine C0 and C2 concentrations and cytomegalovirus infections following lung transplantation. Transplant Proc 35(5):1992–1993. doi:S0041134503006894 [pii]
220. Kuypers DR, Claes K, Evenepoel P, Maes B, Vanrenterghem Y (2004) Clinical efficacy and toxicity profile of tacrolimus and mycophenolic acid in relation to combined long-term pharmacokinetics in de novo renal allograft recipients. Clin Pharmacol Ther 75(5):434–447. https://doi.org/10.1016/j.clpt.2003.12.009. S0009923603007707 [pii]
221. Latorre A, Morales E, Gonzalez E, Herrero JC, Ortiz M, Sierra P, Dominguez-Gil B, Torres A, Munoz MA, Andres A, Manzanares C, Morales JM (2002) Clinical management of renal transplant patients with hepatitis C virus infection treated with cyclosporine or tacrolimus. Transplant Proc 34(1):63–64. doi:S0041134501026781 [pii]
222. Mignogna MD, Fedele S, Lo Russo L, Bonadies G, Nappa S, Lo Muzio L (2005) Acute cyclosporine nephrotoxicity in a patient with oral pemphigus vulgaris and HIV infection on antiretroviral therapy. J Am Acad Dermatol 53(6):1089–1090. https://doi.org/10.1016/j.jaad.2005.07.054. S0190-9622(05)02326-1 [pii]
223. Vercauteren SB, Bosmans JL, Elseviers MM, Verpooten GA, De Broe ME (1998) A meta-analysis and morphological review of cyclosporine-induced nephrotoxicity in auto-immune diseases. Kidney Int 54(2):536–545. https://doi.org/10.1046/j.1523-1755.1998.00017.x
224. Strehlau J, Pape L, Offner G, Nashan B, Ehrich JH (2000) Interleukin-2 receptor antibody-induced alterations of ciclosporin dose requirements in paediatric transplant recipients. Lancet 356(9238):1327–1328
225. Elkahwaji J, Robin MA, Berson A, Tinel M, Letteron P, Labbe G, Beaune P, Elias D, Rougier P, Escudier B, Duvillard P, Pessayre D (1999) Decrease in hepatic cytochrome P450 after interleukin-2 immunotherapy. Biochem Pharmacol 57(8):951–954
226. Kuek A, Hazleman BL, Ostor AJ (2007) Immune-mediated inflammatory diseases (IMIDs) and biologic therapy: a medical revolution. Postgrad Med J 83(978):251–260. https://doi.org/10.1136/pgmj.2006.052688. 83/978/251 [pii]
227. Keizer RJ, Huitema AD, Schellens JH, Beijnen JH (2010) Clinical pharmacokinetics of therapeutic monoclonal antibodies. Clin Pharmacokinet 49(8):493–507. https://doi.org/10.2165/11531280-000000000-00000
228. Reichert JM (2016) Antibodies to watch in 2016. MAbs 8(2):197–204. https://doi.org/10.1080/19420862.2015.1125583
229. Morgan ET (2009) Impact of infectious and inflammatory disease on cytochrome P450-mediated drug metabolism and pharmacokinetics. Clin Pharmacol Ther 85(4):434–438. https://doi.org/10.1038/clpt.2008.302. clpt2008302 [pii]
230. Gupta R, JJ W, Levin E, Koo JY, Liao W (2013) Possible drug-drug interaction between adalimumab and duloxetine and/or pregabalin in a psoriasis patient. J Drugs Dermatol 12(10):1089
231. Zhou H, Parks V, Patat A, Le Coz F, Simcoe D, Korth-Bradley J (2004) Absence of a clinically relevant interaction between etanercept and digoxin. J Clin Pharmacol 44(11):1244–1251. https://doi.org/10.1177/0091270004268050. 44/11/1244 [pii]
232. Zhou H, Patat A, Parks V, Buckwalter M, Metzger D, Korth-Bradley J (2004) Absence of a pharmacokinetic interaction between etanercept and warfarin. J Clin Pharmacol 44(5):543–550. https://doi.org/10.1177/0091270004264164. 44/5/543 [pii]
233. Lee EB, Daskalakis N, Xu C, Paccaly A, Miller B, Fleischmann R, Bodrug I, Kivitz A (2016) Disease-drug interaction of sarilumab and simvastatin in patients with rheumatoid arthritis. Clin Pharmacokinet. https://doi.org/10.1007/s40262-016-0462-8
234. Zhuang Y, de Vries DE, Xu Z, Marciniak SJ Jr, Chen D, Leon F, Davis HM, Zhou H (2015) Evaluation of disease-mediated therapeutic protein-drug interactions between an anti-interleukin-6 monoclonal antibody (sirukumab) and cytochrome P450 activities in a phase

1 study in patients with rheumatoid arthritis using a cocktail approach. J Clin Pharmacol 55(12):1386–1394. https://doi.org/10.1002/jcph.561

235. Tran JQ, Othman AA, Wolstencroft P, Elkins J (2016) Therapeutic protein-drug interaction assessment for daclizumab high-yield process in patients with multiple sclerosis using a cocktail approach. Br J Clin Pharmacol 82(1):160–167. https://doi.org/10.1111/bcp.12936

236. Enioutina EY, Bareyan D, Daynes RA (2009) TLR-induced local metabolism of vitamin D3 plays an important role in the diversification of adaptive immune responses. J Immunol 182(7):4296–4305

237. Pellegrino P, Perrotta C, Clementi E, Radice S (2015) Vaccine-drug interactions: cytokines, cytochromes, and molecular mechanisms. Drug Saf 38(9):781–787. https://doi.org/10.1007/s40264-015-0330-8

238. Kuo AM, Brown JN, Clinard V (2012) Effect of influenza vaccination on international normalized ratio during chronic warfarin therapy. J Clin Pharm Ther 37(5):505–509. https://doi.org/10.1111/j.1365-2710.2012.01341.x

239. Raaska K, Neuvonen PJ (2014) Infections and possible vaccine-drug interactions. Eur J Clin Pharmacol 70(7):889–890. https://doi.org/10.1007/s00228-014-1688-8

240. Scavone JM, Blyden GT, Greenblatt DJ (1989) Lack of effect of influenza vaccine on the pharmacokinetics of antipyrine, alprazolam, paracetamol (acetaminophen) and lorazepam. Clin Pharmacokinet 16(3):180–185. https://doi.org/10.2165/00003088-198916030-00004

241. Pasanen M, Rannala Z, Tooming A, Sotaniemi EA, Pelkonen O, Rautio A (1997) Hepatitis A impairs the function of human hepatic CYP2A6 in vivo. Toxicology 123(3):177–184. doi:S0300483X97001194 [pii]

242. Anolik R, Kolski GB, Schaible DH, Ratner J (1982) Transient alteration of theophylline half-life: possible association with Herpes simplex infection. Ann Allergy 49(2):109–111

243. Trenholme GM, Williams RL, Rieckmann KH, Frischer H, Carson PE (1976) Quinine disposition during malaria and during induced fever. Clin Pharmacol Ther 19(4):459–467

244. Masimirembwa CM, Beke M, Hasler JA, Tang BK, Kalow W (1995) Low CYP1A2 activity in rural Shona children of Zimbabwe. Clin Pharmacol Ther 57(1):25–31. https://doi.org/10.1016/0009-9236(95)90262-7. 0009-9236(95)90262-7 [pii]

245. Satarug S, Lang MA, Yongvanit P, Sithithaworn P, Mairiang E, Mairiang P, Pelkonen P, Bartsch H, Haswell-Elkins MR (1996) Induction of cytochrome P450 2A6 expression in humans by the carcinogenic parasite infection, opisthorchiasis viverrini. Cancer Epidemiol Biomark Prev 5(10):795–800

246. Shedlofsky SI, Israel BC, McClain CJ, Hill DB, Blouin RA (1994) Endotoxin administration to humans inhibits hepatic cytochrome P450-mediated drug metabolism. J Clin Invest 94(6):2209–2214. https://doi.org/10.1172/JCI117582

247. Shedlofsky SI, Israel BC, Tosheva R, Blouin RA (1997) Endotoxin depresses hepatic cytochrome P450-mediated drug metabolism in women. Br J Clin Pharmacol 43(6):627–632

248. Haas CE, Kaufman DC, Jones CE, Burstein AH, Reiss W (2003) Cytochrome P450 3A4 activity after surgical stress. Crit Care Med 31(5):1338–1346. https://doi.org/10.1097/01.CCM.0000063040.24541.49

249. Carcillo JA, Doughty L, Kofos D, Frye RF, Kaplan SS, Sasser H, Burckart GJ (2003) Cytochrome P450 mediated-drug metabolism is reduced in children with sepsis-induced multiple organ failure. Intensive Care Med 29(6):980–984. https://doi.org/10.1007/s00134-003-1758-3

250. Novotny AR, Emmanuel K, Maier S, Westerholt A, Weighardt H, Stadler J, Bartels H, Schwaiger M, Siewert JR, Holzmann B, Heidecke CD (2007) Cytochrome P450 activity mirrors nitric oxide levels in postoperative sepsis: predictive indicators of lethal outcome. Surgery 141(3):376–384. https://doi.org/10.1016/j.surg.2006.08.011. S0039-6060(06)00568-X [pii]

Chapter 6
Interactions Between Herbs and Anti-infective Medications

Surulivelrajan Mallayasamy and Scott R. Penzak

6.1 Introduction

Approximately 59 million people in the Unit predicting such herb-drug interactions ed States have used at least one complementary and alternative medicine (CAM) in a year, resulting in an out-of-pocket expenditure of 30.2 billion dollars. Patients with HIV infection represent an important segment of this population. Because of their ability to modulate a variety of cytochrome P450 (CYP) enzymes and drug transport proteins such as P-glycoprotein (P-gp), a number of herbs have been shown to interact with coadministered medications. Unfortunately, in vitro microsomal studies often fail to predict results obtained in humans. The herb associated with the greatest number of drug interactions in humans is St. John's wort *(Hypericum perforatum)*. As a potent inducer of CYP and P-gp, St. John's wort has been shown to reduce the plasma concentrations of certain coadministered medications by >50%. Other herbs have been shown to induce the metabolism of coadministered medications as well. However, the magnitude of these interactions is markedly less than that produced by St. John's wort. Nonetheless, even mild herb-drug interactions may be clinically relevant for coadministered medications with narrow therapeutic indices. To this end, the need for rigorous studies to identify potentially significant herb-drug interactions continues. Clinicians caring for patients taking CAM therapy should maintain a high degree of suspicion for

S. Mallayasamy
Department of Pharmacy Practice, MCOPS, Manipal University, Manipal, India

Department of Pharmacotherapy, University of North Texas System College of Pharmacy, Fort Worth, TX, USA

S.R. Penzak (✉)
Department of Pharmacotherapy, University of North Texas System College of Pharmacy, Fort Worth, TX, USA
e-mail: scott.penzak@unthsc.edu

© Springer International Publishing AG 2018
M.P. Pai et al. (eds.), *Drug Interactions in Infectious Diseases: Mechanisms and Models of Drug Interactions*, Infectious Disease,
https://doi.org/10.1007/978-3-319-72422-5_6

herb-drug interactions in the face of unexplained toxicity or loss of efficacy and be familiar with resources that can help manage or avoid herb-drug interactions.

Herbal supplements have been widely used in the East for centuries; more recently their use has expanded to include areas of the Western world such as the United States and Canada. The increased use in CAM is multifactorial and includes a general desire for good health and wellness and disease prevention and treatment. Thus, many consumers believe CAM is safer than prescription drugs because they contain "natural" ingredients [2].The majority of CAM includes herbal supplements, which are generally defined as any form of a plant or plant product, including stems, flowers, leaves, roots, and seeds [2]. Herbal supplements can contain a single herb or combinations of herbs that possess complementary effects. In the United States, herbs are regarded as dietary supplements (i.e., food products) and are not subject to intense regulatory oversight by the US Food and Drug Administration (FDA) [3]. However, herbs and other dietary supplements are subject to regulation as specified in the Dietary Supplement Health and Education Act of 1994 [4]. As a result, herbal supplements may not claim to "treat, prevent, cure, or diagnose a specific disease," as such claims are limited to medications that have been proven to be safe and effective by the FDA.

The USP Dietary Supplement Verification Program was developed to assess the integrity of dietary supplements. This program performs comprehensive laboratory testing of dietary supplements and their ingredients against standards found in the US Pharmacopeia and the National Formulary (USP-NF). Products that meet the program's criteria are labeled with a *USP Verified* logo that can be placed on labels, packaging, and promotional materials. This logo allows customers and health-care practitioners to identify herbal products that are *USP Verified*. This logo has appeared on more than 400 million supplementary labels since its introduction [5]. The dietary supplement manufacturers who participate in this program do so voluntarily [6].

Despite a relative paucity of scientific data regarding the safety and efficacy of herbal products, a significant number of patient populations report using these supplements on a regular basis. These patients typically include those with chronic medical conditions such as breast cancer (12%), liver disease (7%), asthma (93%), rheumatological disorders (26%), and gastrointestinal conditions (42%) [7–9]. An additional group of individuals who commonly report using herbal products are those suffering from, or desiring to prevent, an infectious process. A variety of herbs have been touted for the treatment and/or prevention of the common cold, urinary tract infections, upper respiratory tract infections, prostatitis, hepatitis, and the human immunodeficiency virus (HIV) [10–14]. Patients with HIV infection undoubtedly represent the largest group of CAM users in the infectious disease arena. Surveys around the world have shown that approximately 50–70% of patients with HIV infection are using CAMs along with their antiretroviral medications [15–19]. Patients with HIV infection take herbal supplements for purported antiviral activity, "boosting" of the immune system, the treatment or prevention of opportunistic infections, and treatment of medication-related side effects such as gastrointestinal disturbances, peripheral neuropathy, weight loss, and fatigue [14].

Due to frequent CAM use in patients taking prescription medications, there is a strong possibility of clinically relevant drug interactions between these classes of compounds. Although a number of studies have reported interactions between prescription drugs and herbal preparations, countless herb-drug combinations remain unstudied. Most patients who use CAM also do not readily report this information to their health-care provider [1]. As a result, potentially dangerous herb-drug interactions are likely to go unnoticed in many patients.

6.2 Potential Mechanisms of Herb-Drug Interactions

A growing number of preclinical and clinical studies have shown a variety of herbal preparations are capable of modulating drug metabolism and transport at various anatomical sites, most notably in the liver and intestines. Greater than 90% of oxidative metabolism in the liver can be attributed to six cytochrome P450 enzymes. These include CYP1A2, CYP2C9, CYP2C19, CYP2D6, CYP2E1, and CYP3A4 [20]. A number of anti-infective medications including antibacterials, antifungals, and antiretrovirals such as HIV protease inhibitors and non-nucleoside reverse transcriptase inhibitors are metabolized through one or more of these enzymatic pathways [21]. As will be discussed in detail below, a number of herbal constituents have been shown to inhibit and/or induce CYP enzymes, thereby increasing or decreasing the plasma concentrations of coadministered medications. This may result in untoward toxicity or reduced efficacy (i.e., antimicrobial failure) depending on the nature of the interaction. Similarly, a number of herbs have been shown to modulate the activity of uridine diphosphate (UDP) glucuronosyltransferases (UGT), and drug transport proteins such as P-glycoprotein (P-gp), multidrug resistance proteins (MRPs), organic anion-transporting polypeptides (OATP), and organic anion transporters (OATs) [22–24]. Modulation of these metabolic and transport proteins by herbal products may also alter the distribution and/or systemic exposure of concurrently administered medications and potentially result in adverse events or poor efficacy [2]. Common herbal preparations that have been shown to modulate CYP and/or P-gp activity in humans are presented in Table 6.1 [25–52]. Known pharmacokinetic interactions between herbal supplements and anti-infective agents are described below; of the studies discussed, those conducted in humans are highlighted in Table 6.2 [30, 34, 39, 44, 53–61].

In addition to inhibition and/or induction of various metabolic enzymes and transport proteins, herbal preparations may alter drug exposure secondary to other mechanisms such as changes in drug absorption. Soluble and insoluble fibers such as psyllium, plantago ovate husk, guar gum, and alginate fiber act similar to bile-sequestering agents and can hinder the absorption of coadministered drugs [62].

Table 6.1 Influence of herbal compounds of cytochrome P450 (CYP) and P-glycoprotein (P-gp) activity in humansa

	CYP1A2	CYP2C9	CYP2C19	CYP2D6	CYP2E1	CYP3A4/5	P-gp
St. John's wort (*Hypericum perforatum*) [25–30]	↑	↑[b]	↑[c]	↑	↑	↑	↑
Garlic (*Allium sativum*) [25, 31]	ND	ND	ND	↔	↓	↔	ND
Milk thistle (*Silybum marianum*) [32, 33]	↔	ND	ND	↔	↔	↔	↔
Ginkgo biloba [25, 34–37]	↔	↔	↑	↔	↔	↑, ↔	↔
Echinacea (*Echinacea purpurea, Echinacea angustifolia, Echinacea pallida*) [26, 32, 39–41]	↓, ↔	↔	ND	↔	↔	↑[e], ↓[e], ↔[e]	↔
Panax ginseng [25, 28, 42, 107, 108]	↔	↔[f]	ND	↔	↔	↑ ↔	↔[g]
Black cohosh (*Actaea racemosa*) [26, 33, 43]	↔	ND	ND	↔	↔	↔	↔
Goldenseal (*Hydrastis canadensis*) [43–45]	↔	ND	ND	↓	↔	↔, ↓	↔
Kava kava (*Piper methysticum*) [43, 45]	↔	ND	ND	↔	↑	↔	↔
Valerian (*Valeriana officinalis*) [43, 46]	↔	ND	ND	↔	↔	↔	ND
Grape seed (*Vitis vinifera*) [47]	↑	ND	ND	ND	ND	ND	ND
Green tea (*Camellia sinensis*) [48]	ND	ND	ND	↔	ND	↔	ND
Ginger (*Zingiber officinale*) [35]	ND	ND	ND	ND	ND	↔	ND
Hawthorn (*Crataegus monogyna, Crataegus laevigata, Crataegus oxyacantha*) [49]	ND	ND	ND	ND	ND	ND	↔
Saw palmetto (*Serenoa repens*) [32, 50]	↔	ND	ND	↔	↔	↔	ND
Soy (*Glycine max*) [51, 52]	ND	↔[g]	ND	ND	ND	↔[h]	ND
Fermented red ginseng [102]	↔	↔	↔	↔	ND	↔	↓

Arrows pointing upward indicate increased enzymatic activity
Arrows pointing downward indicate decreased enzymatic activity
Arrows pointing side-to-side indicate no significant change in enzymatic activity
"ND" indicates that no human data were located

(continued)

Table 6.1 (continued)

[a]No human data were found for the following herbs with regard to their ability to modulate specific CYP pathways: African potato (*Hypoxis hemerocallidea*), sutherlandia (*Sutherlandia frutescens*), devil's claw (*Harpagophytum procumbens*), and evening primrose (*Oenothera biennis*)
[b]Increased enzymatic activity was observed regardless of CYP2C9 genotype [29]
[c]Increased enzymatic activity was observed in CYP2C19 wild-type subjects, not in CYP2C19 poor metabolizers [27]
[d]Increased enzymatic activity was observed in CYP2C19 homozygous-extensive metabolizers, heterozygous-extensive metabolizers, and poor metabolizers [37]
[e]Decreased activity of intestinal CYP3A and increased activity of hepatic CYP3A4 were observed in one study [40]. A second study observed an increase in CYP3A activity [39] while a third showed no effect [32]
[f]Study examined the effects of Asian ginseng on the CYP2C9 substrate S-warfarin [42]
[g]Study assessed the influence of soy extract on the pharmacokinetics of losartan, which is metabolized by both CYP2C9 and CYP3A4; the study was conducted in an all-female population (18 healthy volunteers) [51]
[h]Study assessed CYP3A4 activity, using 6-β-hydroxycortisol/cortisol ratios, before and after soy extract for 14 days in 20 healthy females. CYP3A4 activity was not significantly changed by soy extract administration [52]

6.3 Interactions Between Herbs and Anti-infective Medications

6.3.1 *St. John's Wort* (Hypericum perforatum)

St. John's wort is used for a variety of ailments including depression, anxiety, dysthymia, attention deficit hyperactivity disorder (ADHD), chronic fatigue syndrome, insomnia, HIV/AIDS, hepatitis C, and numerous others [63]. St. John's wort is a potent inducer of various CYP isoforms as well as P-gp [25–30, 64]. As a result, St. John's wort interacts with numerous medications, in some cases drastically reducing their systemic exposure [65]. Among anti-infectives whose plasma concentrations are significantly reduced by St. John's wort are the CYP3A4 substrate indinavir (57% $\downarrow$ in indinavir area under the concentration-time curve from zero to 8 hrs [AUC_{0-8}]) and the CYP2B6 and CYP3A4 substrate nevirapine (35%$\uparrow$ in nevirapine apparent oral clearance [Cl/F]) [53, 66]. To this end, St. John's wort should be avoided in combination with all unboosted HIV protease inhibitors and non-nucleoside reverse transcriptase inhibitors since all are metabolized to varying degrees by CYP3A4 as well as other CYP isoforms. Of note, a study showed that ritonavir 300 mg twice daily, given as a boosting agent for concurrent protease inhibitor therapy, masks CYP3A4 induction by St. John's wort [30]. Whether this occurs with lower boosting doses of ritonavir (i.e. 100 mg twice daily) is unknown.

St. John's wort was also found to interact with the azole antifungal voriconazole in a slightly more complex manner. Fifteen days of St. John's wort administration reduced the area under the plasma concentration-versus-time curve from zero to infinity ($AUC_{0-\infty}$) of voriconazole by 59% ($P = 0.0004$) [54]. Voriconazole is metabolized by CYP2C19, CYP3A4, and, to a lesser extent, CYP2C9 [67]. During the

Table 6.2 Summary of drug interaction studies involving herbal supplements and anti-infective agents conducted in healthy human volunteers

References	Herbal preparation (regimen)	N^a	Standardization	Coadministered drug regimen	Primary outcome, suggested mechanism	Conclusion
[53]	St. John's wort (300 mg TID × 14 days)	8	0.3% hypericin	Indinavir (800 mg q8h × 4 doses)[b]	Indinavir $AUC_{0-8} \downarrow 57\%$ ($p = 0.0008$); CYP3A4 induction	Avoid St. John's wort with unboosted indinavir and other CYP3A4 substrates
[54]	St. John's wort (300 mg TID × 15 days)	16	Extract L1 160	Voriconazole (400 mg single doses)[b]	Voriconazole $AUC_{0-\infty} \downarrow$ 59% ($p = 0.0004$); intestinal and hepatic CYP2C19, CYP3A4, and P-gp induction	Avoid St. John's wort with voriconazole and other CYP2C19, CYP3A4, and P-gp substrates
[30]	St. John's wort (300 mg TID × 14 days)	12	Extract L1 160	Midazolam (4 mg PO × 1, and 2 mg IV × 1)[c], ritonavir (300 mg BID × 14 days)	IV and PO midazolam (AUC_{0-6} and AUC_{0-8}, respectively) $\uparrow$ 180% and 412%, respectively, vs. baseline after ritonavir + St. John's wort ($p < 0.05$ for each). Two days after stopping St. John's wort and ritonavir, IV and PO midazolam AUCs were reduced below baseline values ($p < 0.001$ for each)	The CYP3A inhibitory effects of ritonavir at 300 mg BID superseded CYP3A induction by St. John's wort. CYP3A induction by St. John's wort was unmasked 2 days after stopping its coadministration × 14 days with ritonavir. Note that the ritonavir dose used in this study was threefold higher than the typical ritonavir boosting dose
[55]	Garlic caplets; maximum allicin formula (BID × 20 days)	10	Allicin (4.64 mg/ caplet) and alliin (11.2 mg/caplet)	Saquinavir (soft gel capsule) (1200 mg TID × 10 doses)[b]	Saquinavir $AUC_{0-8} \downarrow$ 51% ($p < 0.028$); $\downarrow$ saquinavir absorption via unknown mechanism; possible induction of intestinal CYP3A4 and/or P-gp	Avoid garlic supplements with unboosted saquinavir and possibly other CYP3A4 substrates

[56]	Garlic soft gel capsules (10 mg BID × 4 days)	10	Allicin content < 50 μg/g extract	Ritonavir capsule (400 mg × 1 dose)[b]	Ritonavir $AUC_{0-\infty}$ ↓ 17% ($p = 0.094$)	Short-term garlic supplementation does not alter ritonavir exposure but may exacerbate GI side effects associated with ritonavir
[57]	Milk thistle (175 mg TID × 21 days)	10	153 mg Silymarin confirmed	Indinavir (800 mg q8h × 4 doses)[b]	Indinavir AUC_{0-8} ↓ 9% ($p = 0.20$)	Milk thistle, in commonly administered doses, should not alter the PK of indinavir
[58]	Milk thistle (160 mg TID × 14 days)	10	173 mg Silymarin confirmed	Indinavir (800 mg q8h × 4 doses)[b]	Indinavir AUC_{0-8} ↓ 6% ($p = 0.64$)	Milk thistle, in commonly administered doses, should not alter the PK of indinavir
[59]	Milk thistle (450 mg TID × 28 days)	16	456 mg Silymarin confirmed	Indinavir (800 mg q8h × 4 doses)[b]	Indinavir AUC_{0-8} ↓ 4% ($p = 0.78$)	Milk thistle, in commonly administered doses, should not alter the PK of indinavir
[60]	Milk thistle (140 mg/day × 9 days)	12	140 mg Silymarin	Metronidazole (400 mg Q8h × 3 days and 9 days with a 7 day washout period)[b]	Metronidazole AUC_{0-8} ↓ 29% ($p < 0.001$); intestinal P-gp and CYP3A4 induction suggested by authors (unlikely based on data from additional studies)	Avoid multiple dose administration of milk thistle with metronidazole
[62]	Milk thistle (450 mg/day × 14 days)	15	150 mg silymarin	Darunavir-ritonavir 600/100 mg BID × 4 weeks	Darunavir AUC_{0-12} reduction was not significant	Milk thistle can be safely coadministered with darunavir-ritonavir combination
[34]	Ginkgo biloba extract (120 mg BID × 14 days in combination with lopinavir-ritonavir)	14	Flavonol glycoside and terpene lactone content consistent with product label	Lopinavir (400 mg) + ritonavir (100 mg) (BID × 29.5 days)[b]	Lopinavir AUC_{0-12} ↓ 1.6% ($p = 0.42$), ritonavir AUC_{0-12} ↓ 6.5% ($p = 0.28$); concurrent administration of the CYP3A4 inhibitor ritonavir, prevented CYP3A4 induction by Ginkgo biloba	Ginkgo biloba, in commonly administered doses, should not alter the PK of lopinavir-ritonavir or other ritonavir-boosted PI combinations

(continued)

Table 6.2 (continued)

References	Herbal preparation (regimen)	N^a	Standardization	Coadministered drug regimen	Primary outcome, suggested mechanism	Conclusion
[96]	120 mg *Ginkgo biloba* BID × 14 days. Day 15, 120 mg of *Ginkgo biloba* and 400 mg of raltegravir administered together	8	Flavonoids and terpene lactone as per product label	Raltegravir (400 mg) single dose	Concurrent administration of *Ginkgo biloba* did not alter $AUC_{0-\infty}$ of raltegravir	*Ginkgo biloba* can be safely administered with raltegravir
[39]	*Echinacea Purpurea* (500 mg TID × 28 days)	13	Standardized amounts of alkylamides, polysaccharides, and cichoric acid	Lopinavir (400 mg) + ritonavir (100 mg) (BID × 29.5 days)^b	Lopinavir AUC_{0-12} ↓ 3.7% ($p = 0.82$), ritonavir AUC_{0-12} ↓ 8.1% ($p = 0.76$); concurrent administration of the CYP3A4 inhibitor ritonavir, prevented CYP3A4 induction by *Ginkgo biloba*	*Echinacea purpurea*, in commonly administered doses, should not alter the PK of lopinavir-ritonavir or other ritonavir- boosted PI combinations
[61]	African potato (*Hypoxis hemerocallidea*) (15 mg/kg/day of hypoxoside, prepared by traditional decoction, × 14 days)	10	Samples analyzed a priori for hypoxoside content	Efavirenz 600 mg (2 single doses)^b	Efavirenz AUC_{0-48} ↓ 2.1% (90% CIs for AUC_{0-48} and C_{max} were within 80–125%; hence no interaction was observed)	Coadministration of African potato is unlikely to alter efavirenz PK
[44]	Goldenseal root (1140 mg BID × 14 days)	10	Total alkaloid content consistent with product label	Indinavir 800 mg (two single doses)^b	Indinavir CL/F ↓ 4.6% (p = NS)	Coadministration of goldenseal root is unlikely to alter the PK of indinavir; interactions with other CYP3A substrates may depend on the relative degree of hepatic vs. intestinal metabolism of the coadministered drug

| [113] | *Sutherlandia frutescens* Tablets (300 mg BID × 14 days) | 12 | Triterpenoid glycosides and flavonol glycosides were estimated | Atazanavir sulfate 200 mg (two single doses) | Atazanavir C_{max} and AUC_{0-24} ↓ | It is advisable to avoid coadministration of *Sutherlandia frutescens* preparations with atazanavir |

[a]*N* number of healthy volunteers
[b]Pharmacokinetics determined before and after herbal administration
[c]Midazolam AUC_{0-6} was determined at baseline, after 14 days for St

Abbreviations: AUC area under the concentration vs. time curve, *CL/F* apparent oral clearance, *IV* intravenous administration, *NS* not statistically significant John's wort + ritonavir and 2 days after cessation of both agents, *PI* protease inhibitor, *PO* by mouth, *Q8h* every 8 h, *BID* two times daily, *TID* three times daily

first day of St. John's wort administration, the voriconazole AUC_{0-10} actually increased by 22% ($P = 0.02$) suggesting that St. John's wort caused a short-term clinically insignificant increase in voriconazole exposure followed by a prolonged excessive reduction in voriconazole concentrations. The mechanism by which St. John's wort (hyperforin) induces a variety of metabolic and transport proteins is directly related to its potent ability to bind to and subsequently activate the pregnane X receptor (PXR) [68]. PXR is a key regulator of xenobiotic-inducible CYP3A gene expression; PXR also regulates the CYP2B and CYP2C subfamilies in addition to UDP-glucuronosyltransferases, OATP2, the multidrug-resistant protein (MDR1, which encodes for P-glycoprotein), and MRPs 2 and 3 [65]. To this end, long-term exposure to St. John's wort (>12 days) has the ability to significantly reduce the systemic exposure of coadministered medications that are metabolized by CYP enzymes and transport proteins that are regulated by PXR. St. John's wort should be avoided by individuals taking such medications.

6.3.2   Garlic (Allium sativum)

Garlic is used for the treatment of hypertension, hyperlipidemia (including drug-induced hyperlipidemia in patients with HIV infection), coronary heart disease, age-related vascular changes, chronic fatigue syndrome, and menstrual disorders [63]. In addition, garlic has been used for its antibacterial, anthelmintic, antiviral, immunostimulant, and antithrombotic effects. The major active components of garlic are organosulfur compounds [69]. Alliin (S-allylcysteine sulfoxide) – a major constituent of garlic – is converted by alliinase to allicin. Allicin is then further transformed to additional garlic compounds including diallyl sulfide. These organosulfur compounds have been shown to modulate CYP isoforms in vitro and in vivo [70]. Indeed, various garlic preparations were shown to inhibit human CYP2C9, 2C19, 3A4, 3A5, and 3A7 activity in vitro, whereas CYP2D6 activity was unaltered [71]. CYP2E1 activity, measured using chlorzoxazone as a probe compound, was reduced by 39% ($P = 0.30$) in healthy volunteers receiving garlic oil for 4 weeks [25].

The impact of garlic supplementation on CYP3A4 activity in humans has yielded inconsistent results. Piscitelli et al. found that 3 weeks of twice daily garlic administration (containing 4.64 mg and 11.2 mg of allicin and alliin per caplet, respectively) resulted in a mean 51% decrease in steady-state saquinavir (Fortovase®) AUC_{0-8} and a 54% decrease in maximum concentration (C_{max}) in 10 healthy volunteers [55]. After a 10-day washout period, saquinavir AUC_{0-8} and C_{max} values only returned to 60–70% of baseline (control) values. Of note, saquinavir is a CYP3A4 and a P-gp substrate, leading to speculation that the garlic-saquinavir interaction occurred primarily due to induction of P-gp rather than CYP3A4. This is consistent with results from other investigations of long-term garlic administration (14–28 days) that did not observe changes in CYP3A4 activity [25, 31].

Markowicz et al. administered garlic extract (1800 μg allicin, twice daily) to healthy volunteers for 2 weeks to determine the influence of garlic on CYP3A4 activity using alprazolam as a metabolic probe [31]. There were no differences in alprazolam pharmacokinetics following garlic administration. Similarly, administration of garlic oil (500 mg three times daily, allicin content unspecified) to healthy volunteers for 28 days did not alter CYP3A activity using midazolam as a probe [25].

Inconsistencies in the literature regarding the ability of garlic supplements to modulate CYP activity (CYP3A4 in particular) may be due to several reasons. Commercially available garlic supplements have been noted to contain varying amounts of organosulfur compounds (i.e., alliin and allicin), which have been implicated in modulating several CYP isoforms [72, 73]. Garlic also contains numerous flavonoids and isoflavonoids that may alter CYP activity leading to differences among various garlic supplements with regard to their ability to modulate CYP activity [69]. As a result, it is difficult to predict drug interactions with garlic *a priori*.

Erring on the side of caution, HIV protease inhibitors and NNRTIs should not be coadministered with garlic supplements. Other anti-infective agents that are metabolized by CYP3A4 should be used with caution in patients taking long-term garlic supplementation.

A potential pharmacodynamic interaction between garlic and anti-infectives involves garlic's penchant for causing gastrointestinal toxicities such as mouth and gastrointestinal burning or irritation, heartburn, flatulence, nausea, vomiting, and diarrhea [63]. When taken in combination with other medications that commonly cause gastrointestinal distress (i.e., numerous antibiotics, certain antifungals, and ritonavir), patients may experience additive gastrointestinal toxicity. Indeed, Laroche et al. reported two HIV-infected patients taking garlic supplements for >2 weeks who developed severe gastrointestinal toxicity after commencing therapy with a ritonavir-containing antiretroviral regimen [74]. Separating doses of garlic supplements and prescription medications by several hours may help to alleviate gastrointestinal side effects caused by the coadministration of garlic with anti-infectives known to cause G.I. distress.

6.3.3 Milk Thistle (Silybum marianum)

Orally, milk thistle is used for liver disorders including hepatotoxicity due to highly active antiretroviral therapy (HAART) in patients with HIV infection, jaundice, chronic inflammatory liver disease, hepatic cirrhosis, and chronic hepatitis. Milk thistle has also been used for other diverse conditions such as loss of appetite, dyspepsia, diabetes, hangover, malaria, and depression [63]. Based on in vitro data that showed milk thistle inhibited CYP3A4 and CYP2C9, several studies examined the influence of milk thistle on the HIV protease inhibitor and CYP3A4 substrate indinavir [75, 76, 82]. Piscitelli et al. observed a nonsignificant 9% decrease in indinavir AUC_{0-8} at steady state following 3 weeks of dosing with milk thistle (175 mg [153 mg silymarin, which is the standardized extract of milk thistle seeds] three times daily) in ten healthy volunteers [57]. Similarly, DiCenzo et al. observed a nonsignificant 6% reduction in steady-state indinavir exposure after 2 weeks of silymarin administration (160 mg three times daily) to ten healthy volunteers [58]. In a third study, Mills et al. reported a 4.4% decrease in steady-state indinavir AUC_{0-8} ($P = 0.78$) after 28 days of milk thistle dosing (450 mg capsules 3 times daily) in 16 healthy subjects [59]. Mills and colleagues also conducted a meta-analysis of these three drug interaction studies between milk thistle and indinavir; their analysis revealed a nonsignificant mean difference of 1% in indinavir steady-state AUC_{0-8} ($P = 0.97$). In an open-label study in 15 HIV-infected patients receiving treatment with darunavir/ritonavir, 2 weeks of milk thistle coadministration (150 mg thrice daily) did not result in significant changes in darunavir AUC_{0-12} or C_{trough} (90% confidence intervals around both pharmacokinetic parameters included the value 1.0) [77]. Consistent with these data, milk thistle did not alter the pharmacokinetics of the CYP3A4 and UGT1A1 substrate irinotecan or the CYP3A substrate midazolam [32, 78].

In contrast, administration of silymarin (140 mg/day for 9 days) reduced the steady-state AUC_{0-8} of metronidazole (400 mg orally every 8 h) by 29% in 12 healthy volunteers [60]. Metronidazole is a putative substrate for CYP3A4, CYP2C9, and P-gp [60, 79]. Given the lack of an interaction between milk thistle and other CYP3A4 substrates (indinavir, midazolam, and irinotecan), it is unlikely that the interaction between milk thistle and metronidazole occurred via CYP3A4 induction by the former. The authors of this study suggest that induction of intestinal P-gp by milk thistle may have contributed to the interaction. Gurley et al. noted an approximate 9% decrease ($P = 0.06$) in the AUC_{0-24} of the P-gp substrate digoxin (which is not metabolized by CYP enzymes) after milk thistle administration (300 mg three times daily) for 14 days [33]. Although this difference trended toward statistical significance, the magnitude of the interaction is not likely to be clinically relevant. In addition, it is unlikely that milk thistle induced the metabolism of metronidazole through CYP2C9 since preliminary data suggest that metronidazole would be more apt to reduce – as opposed to enhance – the catalytic activity of this isoform [76, 80]. To this end, with the possible exception of metronidazole, milk thistle appears to have limited clinical impact on anti-infectives metabolized via CYP and/or transported by P-gp (Tables 6.1 and 6.2).

Preliminary data suggest that milk thistle may decrease the activity of organic anion transporting polypeptide 1B1 (OATP1B1) and increase or decrease the plasma concentrations of medications that undergo glucuronidation [81, 83]. Further study is necessary to determine whether these putative interactions are clinically relevant.

6.3.4   Ginkgo biloba (Ginkgo biloba)

Ginkgo biloba extract (GBE), one of the most popular herbal medicines in the world, is used for dementia, including Alzheimer's disease. Ginkgo is also used for conditions associated with cerebral vascular insufficiency including memory loss, headache, vertigo, difficulty concentrating, mood disturbances, and hearing disorders [63]. Patients with HIV infection take GBE for a variety of conditions including AIDS-related dementia, depressive disorders, and CNS side effects associated with antiretroviral use [34]. GBE is characterized by 22–27% flavone glycosides, consisting primarily of quercetin and kaempferol, and 5–7% terpene lactones, which include ginkgolides and bilobalide [63, 84].

Several studies utilizing rat models were conducted to evaluate the effect of standardized ginkgo extracts on CYP3A activity using various probe drugs. In general, results from these animal studies showed induction of 3A activity, though at significantly higher doses of GBE than would normally be administered to humans (as high as 100 times the normal human doses) [85, 88]. One investigation in rats noted a decrease in the hypotensive effect of the CYP3A substrate nicardipine after GBE administration, suggesting possible CYP3A4 induction by GBE [86]. In contrast, liver microsomal studies and fluorometric microtitre plate assays have shown inhibition of CYP3A4 using a wide variety of GBE concentrations [73, 87, 89, 90].

Similar to results from the preclinical investigations discussed above, several drug interaction studies conducted with GBE have also shown inconsistent findings. Gurley et al. found that 28 days of GBE (60 mg, 4 times daily) had no apparent effect on CYP3A activity using midazolam as a probe drug in 12 healthy subjects [25]. Another study assessed the influence of an 18-day course of GBE (120 mg/day) on the pharmacokinetics of the CYP3A4 substrate nifedipine [91]. GBE did not significantly alter the mean AUC or C_{max} of nifedipine in eight healthy volunteers; however, two subjects did experience a doubling in C_{max}, which the investigators attributed to GBE. Due to the small sample size and lack of statistically significant findings, results from this study can best be described as inconclusive.

Due to the discordance in results among studies assessing the influence of GBE on CYP3A activity, we conducted a study in 14 healthy volunteers to determine the influence of GBE on the pharmacokinetics of the protease inhibitor combination lopinavir-ritonavir and the respective CYP3A and P-gp probes midazolam and fexofenadine [34]. Single-dose fexofenadine pharmacokinetics were unaltered by GBE (120 mg twice daily for 28 days), suggesting that the herb does not significantly modulate P-gp activity. Conversely the geometric mean midazolam $AUC_{0-\infty}$ following single doses was reduced by 34% ($P = 0.03$) after 28 days of GBE administration, thus suggesting mild induction of CYP3A by GBE. Lastly, volunteers received 2 weeks of lopinavir-ritonavir (400/100 mg twice daily) alone and then in combination with GBE 120 mg twice daily. Geometric mean ratios (GMRs) of lopinavir and ritonavir AUC_{0-12} (post-GBE/pre-GBE) were 1.02 ($P = 0.42$) and 0.93 ($P = 0.28$), respectively, indicating that GBE had no effect on either lopinavir or ritonavir exposure despite the fact that both of these agents are metabolized by CYP3A4 [34].

The reason lopinavir exposure was not affected by GBE is likely due to the coadministration of ritonavir, a potent CYP3A inhibitor. Ritonavir is capable of abating CYP3A induction associated with other enzyme inducers, such as efavirenz and rifabutin [92, 93]. Based on these results, it appears unlikely that GBE would reduce the systemic exposure of protease inhibitors that are boosted with low-dose ritonavir. However, it is possible that GBE may reduce the plasma concentrations of protease inhibitors not boosted by ritonavir. In addition, GBE may reduce the systemic exposure of other anti-infective agents metabolized by CYP3A including the CCR5 co-receptor antagonist, maraviroc, clarithromycin, erythromycin atovaquone, and the non-nucleoside reverse transcriptase inhibitors (NNRTIs) nevirapine, delavirdine, and efavirenz [21]. In support of this hypothesis, there is a single case report of reduced efavirenz plasma concentrations and virologic failure in an HIV-infected patient taking *Ginkgo biloba*. After developing a K103 N mutation and an HIV-1 RNA increase from <50 to 1780 copies/mL, plasma efavirenz concentrations were determined from stored samples dating back 2 years. Over a 14-month period when the patient was taking *Ginkgo biloba* along with efavirenz, he experienced a 62% decrease in efavirenz concentrations (from 1.26 to 0.48 mg/L; therapeutic range, 1.0–4.0 mg/L). The authors of this report surmised that terpenoids from the ginkgo extract reduced plasma efavirenz concentrations by inducing CYP3A4 or P-gp [94].

The influence of *Ginkgo biloba* extract (120 mg twice daily for 15 days) on the pharmacokinetics of a single 400 mg dose of raltegravir was studied in an open-label, randomized, two-period, crossover trial in 18 healthy volunteers. Raltegravir, an HIV integrase inhibitor, is primarily metabolized by uridine diphosphate glucuronosyl-transferase (UGT) 1A1, with UGT1A3 and UGT1A9 playing lesser roles [95]. The GMRs (90% confidence intervals) of $AUC_{0-\infty}$ and the C_{max} of raltegravir with *Ginkgo biloba* versus raltegravir alone were 1.21 (0.93–1.58) and 1.44 (1.03–2.02). These data indicate that *Ginkgo biloba* did not alter the systemic exposure of raltegravir and the two compounds can likely be safely coadministered, The increase in raltegravir C_{max} with *Ginkgo biloba* is not expected to be clinically meaningful [96].

Beyond CYP3A, human data suggest that GBE does not modulate the activity of CYP1A2, CYP2C9, and CYP2D6 (Table 6.1) and is therefore unlikely to interact with anti-infective medications metabolized through these pathways.

6.3.5 *Echinacea* (**Echinacea purpurea, Echinacea angustifolia, Echinacea pallida**)

Echinacea is used for treating and preventing upper respiratory infections including the common cold. Echinacea is also used as an immunostimulant to help counter a variety of other infections, including vaginal candidiasis, urinary tract infections, and genital herpes [63]. Of the three common *Echinacea* species listed above, the majority of research has been conducted with *Echinacea purpurea*. However, the potential for drug interactions among the three *Echinacea* species may differ due to varying amounts of alkylamide content within the different species [97].

At least two studies have characterized the effect of *E. purpurea* root on CYP3A activity in healthy volunteers [32, 40]. Using single doses of oral and intravenous midazolam as a probe compound for intestinal and hepatic CYP3A activity, respectively, Gorski et al. reported an 85% increase in the intestinal availability of midazolam ($P = 0.015$) and a 15% decrease in the hepatic availability of midazolam ($P = 0.006$) after 12 subjects received a total daily dose of 1600 mg of *E. purpurea* for 8 days [40]. These data suggest that *E. purpurea* selectively modulates CYP3A activity in the liver and intestine. Conversely, Gurley et al. found that 28 days of *E. purpurea* whole plant extract administration (800 mg twice daily) did not significantly alter CYP3A activity in 12 healthy volunteers as measured by serum ratios of 1-hydroxymidazolam/midazolam collected 1 h post-dose [32]. Due to the conflicting nature of the data presented by Gorski et al. and Gurley et al., we conducted a study to assess the influence of *E. purpurea* on the pharmacokinetics of lopinavir-ritonavir and the CYP3A and P-gp probe drugs oral midazolam and fexofenadine, respectively [39].

Healthy volunteers received lopinavir-ritonavir (400/100 mg) alone for 2 weeks and in combination with *Echinacea purpurea* 500 mg three times daily for 2 weeks. Lopinavir and ritonavir pharmacokinetics were determined pre- and post-*E. purpurea* administration. Study subjects also received single doses of midazolam (8 mg orally) and fexofenadine (120 mg orally) before and after 28 days of *Echinacea purpurea* to characterize CYP3A and P-gp activity, respectively. Neither lopinavir nor ritonavir pharmacokinetics were significantly altered by 2 weeks of *Echinacea* administration. The GMRs for lopinavir AUC_{0-12} and maximum concentration (post-*Echinacea*/pre-*Echinacea*) were 0.96 and 1.00, respectively ($P > 0.05$ for both comparisons). Similarly, fexofenadine pharmacokinetics did not significantly differ pre- and post-*Echinacea* administration ($P > 0.05$). However, the GMR (post-*Echinacea*/pre-*Echinacea*) for midazolam $AUC_{0-\infty}$ was 0.73 ($P = 0.008$), which is suggestive of a mild induction effect on CYP3A by *E. purpurea* [39]. Despite this mild induction, it is not surprising that lopinavir pharmacokinetics were unaltered by *E. purpurea* given the concurrent administration of the potent CYP3A inhibitor ritonavir [34, 92, 93].

Results from this study suggest that *E. purpurea* may cause mild reductions ($\cong 25$–30%) in plasma concentrations of CYP3A substrates that are not routinely coadministered with potent CYP3A inhibitors; the clinical relevance of such interactions is apt to be greater in patients receiving medications whose plasma concentrations must be maintained above threshold values for optimal pharmacologic efficacy. Such medications may include unboosted HIV protease inhibitors, non-nucleoside reverse transcriptase inhibitors, and certain azole antifungals and macrolides antibiotics.

Due to the selective effects of *E. purpurea* on intestinal versus hepatic CYP3A activity as shown by Gorski et al., the influence of *E. purpurea* on the net exposure of a coadministered CYP3A substrate will likely depend on the extraction ratio of the concurrent medication [40]. Drugs that are poorly absorbed due to significant intestinal metabolism via CYP3A may experience an increase in oral bioavailability secondary to intestinal CYP3A inhibition by *E. purpurea*. Conversely, CYP3A substrates with good oral bioavailability and a low clearance may undergo increased oral clearance secondary to induction of hepatic CYP3A by *E. purpurea* [40].

To this end, it is difficult to predict interactions between *E. purpurea* and CYP3A substrates, as the presence or absence of such interactions likely depends on the relative extraction of the coadministered drug by hepatic versus intestinal CYP3A.

In addition to its effect on CYP3A, *Echinacea* was found to inhibit CYP1A2 as evidenced by a 30% increase in plasma concentrations of the CYP1A2 substrate caffeine, when it was coadministered with *Echinacea* for 8 days [40]. To date, there are no anti-infective agents that are primarily metabolized by CYP1A2, thus making CYP1A2-mediated drug interactions between *Echinacea* spp. and anti-infectives unlikely.

Lastly, there are theoretical concerns regarding the use of *Echinacea* spp. in patients with HIV infection. Patients with HIV may take *Echinacea* for its immuno-stimulatory effects or for the short-term treatment/prevention of upper respiratory infections [63, 98]. While solid scientific evidence is lacking, some clinicians believe that the immunostimulatory effects of *Echinacea* could result in the activation of CD4+ cells, thereby increasing the number of "target cells" for HIV [98]. In addition, an enriched polysaccharide extract of *E. purpurea* was shown to increase production of tumor necrosis factor (TNF) in mice, and high concentrations of TNF-alpha have been linked to HIV disease progression [99, 100]. Based on these limited data, it is unlikely that short-term ($\leq$14 days) echinacea administration for the treatment of colds and influenza presents any serious risks to patients with HIV infection. However, long-term use of *Echinacea* in patients with HIV infections should probably be avoided [98].

6.3.6 Panax ginseng

Ginseng root extract, derived from the herb *Panax ginseng*, has been used as a traditional remedy in Eastern Asia for thousands of years. Orally, *Panax ginseng* is used as an "adaptogen" for promoting resistance to environmental stress and as a tonic for improving well-being. It is also used for stimulating immune function and improving cognitive function, physical stamina, concentration, memory, and work efficiency [63]. Ginseng is administered orally in a variety of forms, including fresh-cut root, alcohol extracts, powder, capsules, and teas. Its content is standardized to percent of ginsenosides. Of note, *P. ginseng* should not be confused with Siberian ginseng or American ginseng; each belongs to the same family (Araliaceae) but forms a different genus [63, 101].

Several studies have examined the effect of *P. ginseng* on CYP activity in humans. Gurley et al. administered *P. ginseng* (5% ginsenosides, 500 mg, three times daily) for 28 days to healthy volunteers and found no effect on the metabolism of the 3A substrate midazolam [25]. In a study using a probe cocktail to assess the influence of fermented red ginseng on the activity of CYP2C9, CYP1A2, CYP3A, and CYP2C19, 15 healthy volunteers received probe cocktails before and after 14 days of fermented red ginseng administration given as one pouch (70 mL) daily. The cocktail drugs and their respective CYP enzymes were dextromethorphan 30 mg (CYP2D6), caffeine 200 mg (CYP1A2), omeprazole 20 mg (CYP2C19), midazolam 7.5 mg (CYP3A), and losartan 50 mg (CYP2C9). The GMRs did not differ significantly pre- and post-fermented red ginseng administration except for midazolam, which showed mild CYP3A inhibition (GMR, 0.816 [90% CI, 0.673–0.990]) that is not likely to be clinically significant [102].

Anderson et al. investigated the potential of *P. ginseng* to induce CYP3A4 by measuring the urinary metabolic ratio of cortisol and 6-hydroxycortisol in 20 healthy volunteers given 24 days of ginseng extract (4% ginsenosides, 100 mg twice daily) [52]. Results from this study found that *P. ginseng* did not induce CYP3A4, although the ability of urinary cortisol metabolic ratios to predict CYP3A4 activity is questionable [103]. A third in vivo study found a modest increase (29%; P value not reported) in nifedipine C_{max} in healthy volunteers after an 18-day course of ginseng (200 mg/day) [94]. In vitro investigations have found varying extents of CYP inhibition, depending on the methodology and concentrations of *P. ginseng* used [40, 89, 90, 104, 105]. A study conducted in rats showed significant increases in the hepatic CYP content of rats fed with *Panax* root, suggesting the possibility of enzyme induction [106].

Due to general inconsistencies in results from the above studies, we determined the influence of *P. ginseng* (500 mg twice daily for 28 days) on CYP3A and P-gp activity in 12 healthy volunteers using midazolam and fexofenadine probes, respectively [107]. Midazolam oral clearance increased in 11 of the 12 study subjects by an average of 51% after *P. ginseng* administration ($P = 0.01$). These data suggest that *P. ginseng* has the potential to increase CYP3A activity and lower the plasma concentrations of anti-infective medications metabolized by this pathway (Table 6.1). Conversely, *P. ginseng* had no effect on fexofenadine pharmacokinetics, suggesting that *P. ginseng* is unlikely to alter the pharmacokinetics of coadministered medications via modulation of P-gp.

Lastly, we assessed the impact of *P. ginseng* on the pharmacokinetic profile of the HIV protease inhibitor and CYP3A4 substrate lopinavir, when given in combination with the CYP3A4 inhibitor, ritonavir as the combination product, Kaletra® to 12 healthy volunteers [108]. The same *P. ginseng* formulation and dosage regimen were used as the *P. ginseng*/midazolam interaction study discussed above [107]. The GMR (90% CI), post-ginseng/pre-ginseng, for lopinavir $AUC_{0-\infty}$ was 0.95 (0.85–1.05). The GMRs for C_{max} and T ½ were 0.94 (0.84–1.04) and 1.19 (0.92–1.46), respectively. None of these changes were statistically significant ($P > 0.05$ for all comparisons). The lack of an observed effect of *P. ginseng* on lopinavir disposition is likely the result of CYP3A4 inhibition by ritonavir, which prevented the induction effects of *P. ginseng* on lopinavir metabolism via CYP3A4. These data suggest that the presence of a CYP3A4 inhibitor, such as RTV, can cancel out the induction effects of *P. ginseng*. It is unclear whether other CYP3A4 inhibitors used as pharmacokinetic boosters, such as cobicistat, produce this same effect; however, this is likely the case.

Results from in vitro drug interaction studies with *P. ginseng* have been largely inconclusive due to the use of different ginseng products and variations in study design and methodology. Similarly, drug interaction studies in humans have been conflicting and have largely yielded negative results or results that suggest weak induction of CYP3A. As a result, *P. ginseng* is unlikely to interact with anti-infectives metabolized by routes other than CYP3A.

6.3.7 African Potato (Hypoxis hemerocallidea)

Hypoxis hemerocallidea (African potato) has been used by traditional Zulu healers for hundreds of years for the treatment of bladder and urinary disorders including cystitis; it has also been used for the treatment of benign prostatic hypertrophy, prostate cancer, and lung diseases [63, 109]. The South African community is currently using *Hypoxis* as an immunostimulating agent in patients with HIV infection [110].

Mills et al. first provided in vitro evidence suggesting that *Hypoxis* is capable of modulating CYP3A4 and P-gp activity and binding to PXR [110]. *Hypoxis* inhibited CYP3A4 activity by 86% and P-gp activity to a lesser degree (i.e., *Hypoxis* showed 42–51% of the inhibitory strength of verapamil, a potent P-gp inhibitor). In addition, *Hypoxis* produced an approximate twofold dose-dependent activation of PXR. Because the PXR nuclear receptor controls the activation of CYP3A4 and P-gp, these findings suggest that *Hypoxis* administration may result in initial inhibition of CYP3A4 and P-gp, followed by induction with prolonged administration [110]. Thus, *Hypoxis* may alter the metabolism and transport of antiretroviral agents that are metabolized by CYP3A4 (i.e., the HIV protease inhibitors and NNRTIs) and/or transported by P-gp. Of note, a separate series of in vitro investigations showed that hypoxoside-induced P-gp in Caco-2 cells and stigmasterol (another ingredient in the African potato) strongly inhibited CYP3A4, CYP3A5, and CYP2C19 [111].

Based upon the in vitro data above, Mogatle et al. examined the influence of the African potato on single-dose efavirenz pharmacokinetics [61]. Ten healthy male volunteers received a single 600 mg dose of efavirenz before and after 14 days of a freshly prepared African potato decoction (15 mg/kg/day of hypoxoside). In contrast with previous in vitro findings, which suggest that *Hypoxis* modulates CYP3A4 and P-gp activity, African potato administration did not alter efavirenz pharmacokinetics in this investigation. The GMRs of C_{max} and AUC_{0-48} were 97.3 and 102.8. Potential reasons for the differences between in vivo and in vitro results discussed above are (1) relatively high concentrations of *Hypoxis* used in the in vitro investigations, which may not be applicable in human studies, (2) the fact that hypoxoside is quickly metabolized to rooperol following oral administration and is not absorbed systemically, and (3) the fact that efavirenz is largely metabolized by CYP2B6, which the African potato has not yet been shown to modulate [61].

A final in vitro study showed that the African potato ingredient *Hypoxis hemerocallidea* significantly decreased the P-gp-mediated efflux of nevirapine across Caco-2 cell monolayers ($P < 0.05$) [112]. The authors concluded that the African potato could increase the oral bioavailability of nevirapine, potentially resulting in higher plasma concentrations and increased toxicity. However, when one considers that the absolute bioavailability of nevirapine exceeds 90%, the potential increase in nevirapine absorption in the presence of the African potato would be expected to be minimal.

6.3.8 *Sutherlandia (*Sutherlandia frutescens*)*

Sutherlandia frutescens is an African herb that has been used for numerous maladies including cancer, tuberculosis, chronic fatigue syndrome, diabetes, influenza, osteoarthritis, rheumatoid arthritis, gastritis, clinical depression, anxiety, and HIV infection [109]. The bioactive constituents of *Sutherlandia* include L-canavanine, GABA, and D-pinitol [101].

Similar to their experiments with *Hypoxis*, Mills and coworkers examined the influence of *Sutherlandia* on CYP3A4 and P-gp activity and PXR activation [110]. *Sutherlandia* produced near complete (96%) inhibition of CYP3A4, while its effects on P-gp activity were less potent (*Sutherlandia* showed 19–31% of the inhibitory strength of verapamil on P-gp activity). Similar to what was observed with *Hypoxis*, *Sutherlandia* produced an approximate twofold dose-dependent activation of PXR. To this end, *Sutherlandia* administration may result in initial inhibition of CYP3A4 and – to a lesser degree – P-gp followed by induction with prolonged administration.

Muller et al. investigated the impact of *Sutherlandia frutescens* (a single dose given twice daily for 13 days) on the bioavailability of a single dose of atazanavir 400 mg in 12 healthy male subjects [113]. The GMRs (90% CI) for atazanavir plus *Sutherlandia* compared to atazanavir alone were 0.78 (0.61–1.00) and 0.80 (0.63–1.01) for atazanavir C_{max} and AUC_{0-24}, respectively. Because the confidence intervals for C_{max} and AUC_{0-24} fell below the predefined bioequivalence boundary (0.80–1.25), *Sutherlandia* plus atazanavir was not bioequivalent to atazanavir administration alone. Of note, atazanavir AUC was not extrapolated to infinity in this study as required by bioequivalence testing guidance, and atazanavir exposure was not assessed under steady-state conditions [114]. As a result, it is challenging to interpret the results of this investigation. Nonetheless, it is likely prudent for clinicians to maintain a high level of suspicion of an interaction between *Sutherlandia* and atazanavir and to instruct HIV-infected patients to avoid taking these compounds together.

6.3.9 *Black Cohosh (*Actaea racemosa*)*

Black cohosh is used to treat premenstrual syndrome (PMS), dysmenorrhea, symptoms of menopause, anxiety, dyspepsia, fever, sore throat, and cough [63].

An in vitro investigation found that six triterpene glycosides fractionated from black cohosh exhibited potent CYP3A4 inhibition as assessed by nifedipine oxidation [115]. However, in 12 healthy volunteers, 1090 mg of black cohosh (standardized to 0.2% triterpene glycosides) given twice daily for 28 days did not alter CYP3A activity using a midazolam probe [43]. In this same study, black cohosh had no significant effect on CYP1A2 and CYP2E1 activity using caffeine and chlorzoxazone probes, respectively. Similarly, two studies in healthy volunteers failed to find clinically meaningful changes in debrisoquine 8 h. urinary recovery ratios as a measure of CYP2D6 activity after 28 days of black cohosh administration [43, 116]. Lastly, the same researchers assessed the influence of black cohosh (20 mg twice daily for 14 days) on P-gp activity using digoxin as a probe; again, black cohosh did not alter the activity of this protein [33]. Based on these data, black cohosh is unlikely to interact with anti-infective medications via modulation of CYP or P-gp activity.

Due to concern that black cohosh may be linked to cases of liver failure and autoimmune hepatitis, it should not be taken by individuals receiving other hepatotoxic drugs as this may increase the risk of liver damage [63]. Anti-infective agents known to cause liver toxicity include itraconazole, voriconazole, ketoconazole, isoniazid, rifampin, efavirenz, nevirapine, delavirdine, nitrofurantoin, terbinafine, trovofloxacin, and tipranavir-ritonavir.

6.3.10   Goldenseal (Hydrastis canadensis)

Goldenseal is used to treat upper respiratory tract infections including the common cold, nasal congestion, allergic rhinitis, and a host of other maladies [63]. Goldenseal is often combined with echinacea in products touted for the treatment and prevention of the common cold. The active components of goldenseal are presumed to be the alkaloids berberine and hydrastine [117].

Data are conflicting with regard to goldenseal's ability to modulate CYP3A [43, 44]. Several in vitro investigations have identified goldenseal extracts, as well as individual isoquinoline alkaloids, as potent CYP3A4 inhibitors [75, 89, 118]. However, when goldenseal (570 mg capsules; administered as two capsules twice daily for 14 days) was given in combination with the CYP3A4 substrate indinavir, it did not alter any of indinavir's pharmacokinetic parameters [44]. Of note, the goldenseal product used in this investigation was analyzed for standard alkaloid content (2% hydrastine and 2.5% berberine) prior to the study and found to meet the US Pharmacopeia (USP) standards. Conversely, Gurley and coworkers observed strong CYP3A inhibition with goldenseal (900 mg three times daily for 28 days) using serum ratios of 1-hydroxymidazolam/midazolam determined 1 h after midazolam dosing [43]. The reason(s) for the apparent discrepancy in these two studies with regard to goldenseal's ability to inhibit CYP3A are not immediately clear. One possibility raised by authors from both studies is that goldenseal may alter the oral bioavailability of drugs that are subject to high first-pass metabolism by CYP3A in the gut wall. Since indinavir is not appreciably metabolized by intestinal CYP3A4, this may explain why goldenseal did not alter indinavir absorption and disposition. Hence, goldenseal's potential to interact with coadministered CYP3A substrates may depend on the comparative degree of intestinal versus hepatic metabolism involved in the biotransformation of the coadministered compound [43].

Separate in vitro and in vivo studies noted that goldenseal significantly inhibited CYP2D6 activity [43, 116, 118]. As a result, goldenseal should be avoided by individuals taking medications metabolized by CYP2D6. Fortunately, no commonly used antivirals, antifungals, or antibacterial agents use CYP2D6 as primary metabolic route.

In addition to drug-metabolizing enzymes, goldenseal was evaluated for its influence of P-gp-mediated drug transport [45]. Preliminary data in rats showed that the goldenseal constituent berberine produced a dose-dependent increase in the bioavailability of digoxin and cyclosporine A via inhibition of intestinal P-gp [119]. However, an in vitro investigation found data to suggest that berberine upregulates P-gp expression [120]. Contrary to data from these in vitro experiments, goldenseal (3210 mg daily for 14 days) did not significantly alter the systemic exposure of the P-gp substrate digoxin in 20 healthy volunteers [45]. Based on the lack of P-gp modulation in vivo by goldenseal, it is unlikely that this herbal preparation will alter the pharmacokinetics of coadministered P-gp substrates.

6.3.11 *Kava Kava* (**Piper methysticum**)

Kava is used to treat anxiety, stress, insomnia, and restlessness. It is also used in a variety of other conditions including attention deficit hyperactivity disorder (ADHD), depression, headache, chronic fatigue syndrome (CFS), respiratory tract infections, tuberculosis, and urinary tract infection (UTI) [63]. The active constituents of kava extracts include a number of kava lactones.

Of the kava lactones assessed, methysticin, dihydromethysticin, and desmethoxyyangonin appear to have the greatest inhibitory effect on CYP enzymes, with all three inhibiting CYP3A4 [121, 122]. Indeed, preliminary evidence from in vitro investigations suggest that kava is a significant inhibitor of CYP3A4, CYP2D6, CYP1A2, and P-gp [121, 123, 124]. However, subsequent studies in humans did not find kava to be an inhibitor of any of these proteins [43, 45, 116]. Kava was found to inhibit CYP2E1 activity by approximately 40% using chlorzoxazone as a probe [43]; however, other than several anesthetics, relatively few medications (and no anti-infectives to our knowledge) are metabolized by this isoform [21]. Several in vitro studies have observed inhibition of CYP2C9 and CYP2C19 by kava extracts; however, no data in humans are available [121, 123]. Nonetheless, aside from nelfinavir (HIV protease inhibitor), voriconazole (azole antifungal), and proguanil (prophylactic antimalarial agent), the CYP2C subfamily is not routinely involved in the metabolism of anti-infective agents [21].

There is concern that kava can cause hepatotoxicity and liver failure in patients taking recommended doses for relatively short time periods [63]. Indeed, the use of kava for as little as 3 months or less has resulted in the need for liver transplantation and death [125–129]. As a result, kava preparations should not be taken in combination with previously mentioned anti-infective agents known to cause liver toxicity.

Lastly, kava preparations have been associated with drowsiness, dizziness, and disturbances of oculomotor equilibrium and accommodation [63]. As a result, kava should be avoided by individuals taking anti-infective medications with CNS-related side effects such as efavirenz and minocycline.

6.3.12   Valerian (Valeriana officinalis)

Valerian is primarily used to treat insomnia, anxiety, and restlessness [63]. Other uses for valerian include depression, attention deficit hyperactivity disorder (ADHD), and chronic fatigue syndrome (CFS) [63].

Preliminary data from in vitro investigations suggest that valerian may inhibit CYP3A4 and P-gp [89, 130, 131]. However, two separate studies in healthy volunteers reported no statistically significant effect of valerian at 375 mg/day for 28 days and 1000 mg/day for 14 days, on CYP3A activity using 1-hydroxymidazolam/midazolam ratios and alprazolam AUC, respectively, as CYP3A probes [43, 46]. In addition, valerian (375 mg/day for 28 days) had no effect on CYP1A2, CYP2D6, and CYP2E1 activity in healthy volunteers [43]. No studies in humans have assessed the influence of valerian on P-gp activity.

Since valerian can cause drowsiness and insomnia, it should probably be avoided or at least used with caution in patients taking efavirenz, which can also cause sleep disturbances and drowsiness in some individuals [132].

6.3.13   Devil's Claw (Harpagophytum procumbens)

Devil's claw is used for nonspecific lower back pain, osteoarthritis, gout, myalgia, tendonitis, and rheumatoid arthritis [63]. Devil's claw contains the iridoid glycoside constituents harpagoside, harpagide, and procumbide, which appear to have anti-inflammatory effects [133].

Preliminary data from a single in vitro investigation suggest that devil's claw may inhibit CYP3A4, CYP2C9, and CYP2C19; it was not shown to inhibit CYP2D6 [123]. However, the influence of devil's claw on these or other CYP enzymes has not been evaluated in humans. Due to the frequent disparity in data from in vitro versus in vivo studies assessing the ability of an herbal formulation to modulate CYP activity, it is not possible to predict, with any degree of certainty, whether devil's claw will increase the systemic concentrations of anti-infectives metabolized by CYP3A4, CYP2C9, and CYP2C19; clinical studies are necessary to explore this possibility.

6.3.14   Grape Seed (Vitis vinifera)

Grape seed is primarily used for preventing cardiovascular disease, hemorrhoids, varicose veins, hypertension, and peripheral vascular disease [63]. Grape seed has also been used to treat diabetic complications such as retinopathy and neuropathy [63]. Flavonoids found in grape products exhibit a variety of effects that may prevent cardiac disease; these include antioxidant, antiplatelet, and vasodilating properties as well as anti-lipoperoxidant activity [134–136].

Grape seed extract was shown to inhibit the activities of CYP2C9, CYP2D6, and CYP3A4 in human liver microsomes [137]. Conversely, another study conducted in human hepatocytes found that grape seed extract increased CYP3A4 mRNA expression by nearly 300% versus control, thereby suggesting that grape seed extract is capable of inducing CYP3A4 activity [138]. A study in rats failed to find an appreciable effect of grape seed administration on intestinal and hepatic microsomal activity nor midazolam pharmacokinetics [137]. Studies in humans are necessary before any conclusions can be reached regarding the potential for grape seed to interact with anti-infective medications via modulation of CYP2C9, CYP2D6, and CYP3A4. A study in healthy subjects showed that grape juice appeared to induce CYP1A2 activity as evidenced by a 43% reduction in the AUC of the CYP1A2 substrate phenacetin [47]. However, as noted previously, CYP1A2 is not routinely involved in the metabolism of any anti-infective medications.

6.3.15 Green Tea (Camellia sinensis)

Green tea is used to improve mental alertness and enhance cognitive performance. It is also used to treat vomiting, diarrhea, and headache. In addition, green tea has been reported to promote weight loss and possess antioxidant, anticancer, and anti-inflammatory properties [63, 101]. Many of the purported therapeutic effects of green tea are thought to be due to the presence of catechins, polyphenols, and phytoestrogens [63]. Green tea also contains 2–4% caffeine [63].

In vitro studies in human liver microsomes and rat hepatic and intestinal microsomes and a pharmacokinetic study in rats have yielded conflicting results with regard to the influence of green tea on CYP3A activity [137, 139]. In healthy volunteers, green tea extract (844 mg catechins/day for 14 days) had no effect on CYP3A4 or CYP2D6 using alprazolam and dextromethorphan as CYP3A4 and CYP2D6 probes, respectively [48]. One study in human liver microsomes found that green tea extract inhibited CYP2C9 activity; however, the influence of green tea on CYP2C9 has not been evaluated in humans [137]. Collectively, these data do not suggest that green tea is likely to alter the metabolism of medications metabolized through CYP. Nonetheless, green tea may still interact with certain antibiotics and antifungals through alternate mechanisms.

As mentioned, green tea contains caffeine (10–80 mg per cup) whose clearance via CYP1A2 is reduced by fluoroquinolone antibiotics [21]. As a result, side effects due to excessive caffeine exposure such as anxiety, insomnia, and headache might be expected when green tea is ingested with quinolone antibiotics such as ciprofloxacin and norfloxacin [21]. In addition, green tea has been noted to cause liver toxicity. At least 14 cases of hepatotoxicity, mainly linked to green tea extracts in pill form, have been reported [140, 141]. Due to potentially additive hepatotoxic effects, green tea should be avoided by patients receiving those anti-infective medications mentioned earlier that produce liver toxicity.

6.3.16 Ginger (Zingiber officinale)

Ginger is used for motion sickness, nausea and vomiting, morning sickness during pregnancy, migraine headache, and a host of other ailments [63]. Active components of ginger include gingerdione, shogaol, gingerol, and sesquiterpene and monoterpene volatile oils [142, 143]. These constituents produce a number of pharmacologic properties including analgesic, antitussive antipyretic, sedative, anti-inflammatory, antibiotic, and weak antifungal activities [142, 144].

Relatively few studies have examined ginger for its drug interaction potential, and most of these have focused on warfarin, the S-isomer of which is metabolized through CYP2C9 [35, 42, 63]. Ginger did not alter warfarin pharmacokinetics or pharmacodynamics in healthy volunteers [35, 63]. As a result, ginger is unlikely to interact with medications metabolized by CYP2C9. Until more data are available, it is not possible to predict the interaction potential between ginger and medications metabolized through other CYP pathways.

6.3.17 Hawthorn (Crataegus monogyna, Crataegus laevigata)

Hawthorn is primarily used for the treatment of congestive heart failure, angina pectoris, hypertension, and dysrhythmias [63]. The constituents of hawthorn preparations that are responsible for its pharmacologic activities include flavonoids and oligomeric proanthocyanidins (OPCs) such as epicatechin and procyanidins [63].

Neither preclinical nor clinical studies have assessed the influence of CYP-mediated drug interaction with hawthorn. However, one study in healthy volunteers showed that 3 weeks of hawthorn and digoxin coadministration did not alter digoxin pharmacokinetics, thereby indicating that hawthorn is unlikely to modulate the systemic exposure of medications that are P-gp substrates [49]. Until more data are available, it is not possible to predict the interaction potential between hawthorn and drugs metabolized by CYP.

6.3.18 Saw Palmetto (Serenoa repens)

Saw palmetto is mainly used to treat symptoms of benign prostatic hyperplasia (BPH) [63]. Additionally, saw palmetto is used as a sedative, anti-inflammatory, mild diuretic, and antiseptic agent [63]. Saw palmetto products are frequently standardized based on their fatty acid content. Most saw palmetto extracts used in clinical studies for the treatment of BPH are berry extracts prepared with lipophilic solvents containing 80–90% free fatty acids [63].

Two studies conducted in vitro reported that saw palmetto inhibited CYP2C9, CYP2D6, and CYP3A4 activity [89]. However, a study in healthy volunteers found that 14 days of saw palmetto (197 mg) administration did not alter CYP3A4 or CYP2D6 activity in 12 healthy volunteers [50]. Confirming these results, Gurley et al. observed that saw palmetto supplementation (160 mg twice daily for 28 days) had no significant effect on CYP3A and CYP2D6 activity in 12 healthy volunteers [32]. Based on these results, saw palmetto is unlikely to interact with medications metabolized by CYP3A4 and CYP2D6. Studies in humans are necessary to determine whether saw palmetto modulates other CYP isoforms and/or drug transport proteins such as P-gp.

6.3.19 *Soy* (Glycine max)

Soy is used for the treatment of menopausal symptoms, hyperlipidemia, the prevention of osteoporosis and cardiovascular disease, and numerous other maladies [63]. The active components of soybeans include the phytoestrogens (isoflavones and lignans), phytosterols, and stigmasterol [63]. Soy constituents potentially associated with alterations in drug metabolism include genistein and daidzein [52].

In one study, genistein and daidzein were shown to inhibit UDP-glucuronosyl-transferase in rat liver extract, while genistein was shown to inhibit P-gp activity in another [145, 146]. In human liver microsomes, unhydrolized soy extract produced weak inhibition of CYP1A2 and CYP2D6 [52]. Of note, the majority of soy isoflavones in plasma occur in their unhydrolized form [52, 147]. In a series of in vitro experiments using human liver microsomes, hydrolyzed soy extract inhibited CYP3A4, CYP2C9, CYP1A2, and CYP2D6, with CYP3A4 and CYP2C 9 inhibition being the strongest [52]. In contrast to these in vitro findings, the same researchers showed that administration of soy extract (50 mg) to 20 healthy females did not alter CYP3A4 activity using 6-β-hydroxycortisol/cortisol ratios as an indicator of CYP3A4 activity [52]. The authors highlighted the lack of agreement between their in vitro and in vivo findings with regard to soy's ability to induce CYP3A, and they call into question the degree of usefulness of in vitro screening studies to detect interactions between herbs and prescription medications. Further supporting this assertion, Wang et al. showed that soy extract had no effect of CYP2C9 activity as evidenced by a lack of an interaction with the CYP2C9 substrate losartan in healthy female volunteers [51]. To this end, soy extracts are unlikely to interact with medications via CYP3A4 or CYP2C9 modulation; whether soy extract inhibits or induces other CYP pathways or transport proteins will need to be determined through studies conducted in humans.

In addition to metabolic interactions, there may be an interaction between isoflavones in soy and antibiotics. Isoflavones are converted by intestinal bacteria to their active forms, and this process may be impeded by antibiotics, which interfere with the bacteria's ability to transform isoflavones into their active moiety [148, 149]. While the occurrence of this interaction is probable, it is unlikely to be clinically relevant [63].

6.3.20 *Evening Primrose* (Oenothera biennis)

Evening primrose is used to treat premenstrual syndrome (PMS), endometriosis, chronic mastalgia, and symptoms of menopause [63]. It is also used orally for atopic eczema, psoriasis, rheumatoid arthritis, and osteoporosis [63]. Evening primrose oil contains 2–16% gamma-linolenic acid (GLA), 65–80% linoleic acid, and vitamin E [150, 151]. GLA is thought to be responsible for the anti-inflammatory effects of evening primrose oil [63].

A purified component of evening primrose oil, *cis*-linoleic acid, was assessed for its ability to inhibit the catalytic activity of cDNA-expressed CYP isoforms in a series in vitro experiments [73]. *cis*-linoleic acid was found to be a potent inhibitor (IC50 value $\leq$10 μM) of CYP2C9 and a moderate inhibitor of CYP1A2, CYP2C19, CYP2D6, and CYP3A4 (IC50 values 10–50 10 μM). Unfortunately, no drug interaction studies with evening primrose have been conducted in humans. As a result, it is not possible to predict the potential of evening primrose to interact with CYP-metabolized medications or medications transported by ATP-binding cassette proteins such as P-gp.

6.4 Issues and Concerns Related to the Use of Herbal Supplements

6.4.1 *Product Content*

Assessing herbal preparations for their potential to interact with prescription medications is wrought with a number of difficulties. First is a general lack of quality control. There is significant variability in manufacturing techniques and storage of herbal products between manufacturers, which can lead to wide variability in content within and between products. In one study of ephedra-containing dietary supplements, half of the 20 products tested contained alkaloids that differed by more than 20% of the amount listed on the label [152]. Substantial differences between content and product label claims have also been noted for dehydroepiandrosterone, ginseng, feverfew, and kava [153–157]. Even more concerning is the contamination of herbal products with heavy metals, pharmaceuticals, and prohibited animal and plant ingredients [158, 159]. Indeed, adulteration of herbal preparations with antibiotics, nonsteroidal anti-inflammatory drugs, heavy metals, and hormones is not uncommon. To this end, it is difficult, and in many cases impossible, to predict potential drug interactions with herbal products that contain suspect ingredient content and/or adulterant compounds.

Due to the variability in ingredient content between (and in some cases within) brands of herbal products, drug interaction studies should be reviewed carefully. Ideally drug interaction studies should include an independent content analysis of all herbal products used in an investigation to confirm the presence of the putative interacting ingredient(s). In cases where such a content analysis is not performed, the study should use an herbal product that is manufactured by a reputable company, preferably one whose products have been previously analyzed and used in herb-drug interaction studies. In addition, manufacturers of the herbal product(s) under study should offer evidence that the US Pharmacopeia-endorsed quality control standards were followed during the manufacturing process of the herbal preparation.

6.4.2 Study Design

Studies assessing herb-drug interactions are typically conducted in human liver microsomes, cDNA expressed CYP isoforms, rat liver microsomes, rats, and humans. The most robust of these scientific approaches are studies conducted in humans. The literature is replete with examples of conflicting data obtained from in vitro versus in vivo (human) studies. A prime example is seen with St. John's wort. A series of in vitro microsomal experiments reported that crude extracts of St. John's wort inhibited CYP1A2, CYP2C9, CYP2C19, CYP2D6, and CYP3A4 [147, 160]. Conversely, studies in humans have clearly shown that St. John's wort is a potent inducer of these enzymes [25–30] (Table 6.1). Reasons for disparity between these in vitro and in vivo findings is likely multifactorial, including the specific herbal extract under study, methodology used in preparing the extract, concentration of the constituent(s) being tested, presence of concurrent herbal constituents or adulterant pharmaceuticals that may contribute to a positive interaction, and limitations of certain in vitro systems that cannot readily assess drug transport, enzymatic induction, or phase 2 metabolism. Therefore, clinical studies in humans need to be conducted to identify those herbal preparations that have the potential to significantly interact with prescription medications.

In clinical studies, the duration that an herbal product is administered is an important consideration. Enzymatic induction is dependent on the half-life of the substrate and the rate of enzymatic turnover; therefore it is a gradual process that requires multidose administration [161]. As a result, studies that do not administer an herbal preparation for at least 2 weeks should be interpreted with caution, as results may not be indicative of those that occur with prolonged administration. In addition to treatment duration, sample sizes for herb-drug interaction studies must be sufficiently large to detect relatively small differences in the exposure of coadministered medications, as the magnitude of most herb-drug interaction studies tends to be mild.

In addition to formal studies, a number of herb-drug interactions have been described in case reports [63]. However, case reports of drug interactions involving herbal preparations are often plagued by the following problems: anecdotal data usually in a single patient, confounding medications, missing information, lack of clarity regarding the temporal association between when the herbal product was started in relation to the putative interacting drug, and lack of formal content analysis of the herbal product. As a result, data from case reports should be interpreted as either (1) hypothesis-generating, alerting clinical researchers to potential drug interaction studies that might be profitable to conduct in the future, or (2) as confirmatory evidence of a previously conducted herb-drug interaction study.

6.4.3 Patient Management Issues

Despite frequent use of CAM, many patients fail to disclose this information to their health-care provider. In one study, 70% of CAM users did not inform their primary care provider of their CAM use [162]. Patients may neglect to inform their clinicians about their CAM use since they are unlikely to attribute health problems to an herbal supplement that they assume to be "safe" and "natural" [2]. Patients may also fear disapproval from their health-care provider if they disclose their use of herbal supplements. For these reasons, clinicians should perform a complete medication history at each clinic visit to determine whether a patient has initiated treatment with a new herbal preparation; often patients will not share such information unless specifically prompted [163]. It is important that clinicians remain nonjudgmental and supportive when interacting with patients who use CAM. Indeed, in addition to potential health-related benefits from CAM, the use of CAM therapy may provide patients with HIV infection a sense of empowerment as they take an active role in managing their own health [164].

Patients who insist on using CAM should be encouraged to use brands that are USP verified, have been used in clinical trials, or are at least manufactured by a reputable company. Once information regarding CAM use is elicited from patients, it should be recorded in detail in their medical record. Specific information regarding CAM use should include start and stop dates, dosages, and name and manufacturer of the product. This information may be useful in the future when assessing a potential drug interaction between CAM and an anti-infective medication.

CAM therapy should be considered in patients who experience unexplained toxicity or lack of efficacy from a particular anti-infective agent. For example, if a patient with HIV infection had a viral load <50 copies/mL and was tolerating their antiretroviral medications well, then suddenly experienced a large increase in viral load or a new toxicity, the possibility that the patient initiated herbal therapy should be considered.

Determining whether an herbal product is likely to interact with a particular medication is oftentimes not straightforward and requires a familiarity with several quality resources. A number of Web sites are extremely valuable in helping clinicians identify potential herb-drug interactions (Table 6.3). While information may not be available with regard to a specific herb-drug interaction, interactions can often be predicted by knowing which CYP pathways an herb modulates and which CYP pathways are used by concurrently administered medications. Information contained in Tables 6.1 and 6.2 of this chapter may also be useful in predicting such herb-drug interactions.

Table 6.3 Selected internet resources for drug interactions involving herbal preparations

Source (Web address)	Description	Accessibility
Natural medicines comprehensive database (http://www.naturaldatabase.com)	Includes evidence-based monographs for nearly 1100 individual natural ingredients and a searchable herb-drug interaction calculator. Primary references with links to PubMed are included for all interactions	Paid subscription required
Natural standard: The authority on integrative medicine (www.naturalstandard.com)	Includes monographs with "interactions" section and PubMed links to primary references	Paid subscription required
American botanical council (www.herbalgram.org)	Includes monographs with "interactions" section and primary references. Provides access to the complete German commission E monographs online as well as *HerbalGram* online	Level of access is dependent on membership level; some content is free
Office of Dietary Supplements (http://www.ods.od.nih.gov)	Provides link to international bibliographic information on dietary supplements (IBIDS), which lists bibliographic citations and abstracts from published, international, and scientific literature on dietary supplements; access to additional databases is also provided	Free
Dietary supplement verification program (http://www.usp.org/USPVerified/)	Includes information on USP-verified dietary supplements and participating manufacturers along with an explanation of the verification process	Free
Micromedex (http://www.micromedex.com/)	Drug interaction calculator recognizes herbal products in addition to over-the-counter and prescription medications. Monographs for alternative medications include specific information on drug interactions. Includes ratings for risk and documentation, mechanism of drug interactions, pharmacokinetic data, and dosing recommendations. Primary references are included	Paid subscription required
Medscape (http://www.medscape.com/druginfo/druginterchecker)	Drug interaction calculator recognizes herbal products in addition to over-the-counter and prescription medications. Includes severity rating, pharmacokinetic data, mechanism of drug interactions, and dosing recommendations. Includes primary references.	Free registration required
Lexi-comp (www.lexi-comp.com)	Allows for interaction reviews of specific medications as well as patient-specific regimens; natural products are included. Assigned risk ratings and patient management information are included	Paid subscription required
The University of Liverpool (www.hiv-druginteractions.org/)	Includes drug interaction charts for antiretroviral medications in combination with other agents, including 13 herbal supplements/vitamins	Free

(continued)

Table 6.3 (continued)

Source (Web address)	Description	Accessibility
Facts & Comparisons (http://online. factsandcomparisons. com/)	Interactive tool that allows for interaction searches between herbs and prescription and over-the-counter medications. Includes severity, pharmacokinetic data, and mechanism of drug interactions	Paid subscription required
Stockley's herbal medicines interactions; available online through: (http://www. medicinescomplete)	Available as an online subscription, book and CD-ROM package, personal user CD-ROM, and book. Includes clinical and experimental interaction data on over 150 common herbs, dietary supplements, and nutraceuticals. Fully referenced and very detailed monographs; updates posted annually	All formats require purchase

In addition to identifying potential herb-drug interactions from a qualitative stand point, it is also important to appreciate the quantitative nature of these putative interactions. For example, St. John's wort is a potent inducer of several CYP enzymes and has the potentially to markedly reduce plasma concentrations of coadministered CYP substrates. As a result, St. John's wort should be avoided by patients receiving interacting medications. On the contrary, the majority of drug interactions with herbs other than St. John's wort tend to be of a mild nature, where coadministered drug concentrations are not increased/decreased by more than $\cong 35\%$. In these cases, only medications with narrow therapeutic indices are likely to be altered to a clinically significant degree. Fortunately, most anti-infective agents do not fit this description; other medications that do include cyclosporine, tacrolimus, irinotecan, sildenafil, and sirolimus.

6.5 Conclusions

CAM use is common in patients with infectious diseases, particularly those with HIV infection. Predicting herb-drug interactions in this population is often difficult, as in vitro studies frequently fail to accurately predict the ability/inability of herbal preparations to interact with medications in humans. Therefore, future studies should be conducted in humans, employ a solid study design, and use herbal products that are USP or otherwise independently verified. Financial support for such studies should be a priority among private and public funding agencies.

References

1. Eisenberg DM, Davis RB, Ettner SL et al (1998) Trends in alternative medicine use in the United States, 1990–1997. JAMA 280:1569–1575
2. Maria FA, Jahangir V (2010) Use of herbal products and potential interactions in patients with cardiovascular diseases. J Am Coll Cardiol 55:515–525

3. Jordan SA, Cunningham DG, Marles RJ (2010) Assessment of herbal medicinal products: challenges, and opportunities to increase the knowledge base for safety assessment. Toxicol Appl Pharmacol 243:198–216

4. Dietary Supplement Health and Education Act 1994, Pub. L. 103–417, Sec,1(a), 108 Stat. 4325 (Oct. 25, 1994)

5. US Pharmacopeial Convention For Manufacturers. Available at: http://www.usp.org/ verification-services/ manufacturers. Accessed on: 20 Oct 2016

6. http://www.usp.org/USPVerified/dietarySupplements/. Accessed 10 June 2010

7. Bent S, Ko R (2004) Commonly used herbal medicines in the United States. Am J Med 116:478–485

8. Michelle LD, Roger BD, Anthony JL, Gloria YY (2014) Complementary and alternative medicine use by US adults with gastrointestinal conditions: results from the 2012 National Health Interview Survey. Am J Gastroenterol 109:1705–1711

9. Maureen G, Maxim T (2013) Complementary and alternative medicine for asthma self-management. Nurs Clin N Am 48:53–149

10. Dreikorn K (2005) Complementary and alternative medicine in urology. BJU Int 96:1177–1184

11. Wu T, Yang X, Zeng X, Poole P (2008) Traditional Chinese medicine in the treatment of acute respiratory tract infections. Respir Med 102:1093–1098

12. Stickel F, Schuppan D (2007) Herbal medicine in the treatment of liver diseases. Dig Liver Dis 39:293–304

13. Chichon PG (2000) Herbs and the common cold. Adv Nurse Pract 8:31–32

14. Littlewood RA, Vanable PA (2008) Complementary and alternative medicine use among HIV-positive people: research synthesis and implications for HIV care. AIDS Care 20:1002–1018

15. Fairfield KM, Eisenberg DM, Davis RB, Libman H, Phillips RS (1998) Patterns of use, expenditures, and perceived efficacy of complementary and alternative therapies in HIV-infected patients. Arch Intern Med 158:2257–2264

16. Hsiao AF, Wong MD, Kanouse DE et al (2003) Complementary and alternative medicine use and substitution for conventional therapy by HIV-infected patients. J Acquir Immune Defic Syndr 33:157–165

17. Lesley AB, Catherine AF, Mathew DM et al (2016) Complementary medicine use by people living with HIV in Australia- a national survey. Int J STD AIDS 27(1):33–38

18. Armando H, Minerva J, Sara H, Jesús O, Eduardo S (2009) Use of alternative / complementary therapy in HIV seropositive patients. Rev Med Inst Mex Seguro Soc 47(6):651–658

19. Shayesta D, Chan KJ, Montaner JSG, Hogg RS (2006) Complementary and alternative medicine use in British Columbia -a survey of HIV positive people on antiretroviral therapy. Complement Ther Clin Pract 12:242–248

20. Tanaka E, Hisawa S (1999) Clinically significant pharmacokinetic drug interactions with psychoactive drugs: antidepressants and antipsychotics and the cytochrome P450 system. J Clin Pharmacy Ther 24:7–16

21. Indiana University School of Medicine. Cytochrome P450 Drug Interaction Table. Available at: http://medicine.iupui.edu/clinpharm/ddis/table.asp. Accessed 10 June 2010

22. Marchetti S, Mazzanti R, Beijnen JH, Schellens JH (2007) Concise review: clinical relevance of drug drug and herb drug interactions mediated by the ABC transporter ABCB1 (MDR1, P-glycoprotein). Oncologist 12:927–941

23. Garrovo C, Rosati A, Bartoli F, Decorti G (2006) St John's wort modulation and developmental expression of multidrug transporters in the rat. Phytother Res 20:468–473

24. Harris RZ, Jang GR, Tsunoda S (2003) Dietary effects on drug metabolism and transport. Clin Pharmacokinet 42:1071–1088

25. Gurley BJ, Gardner SF, Hubbard MA et al (2002) Cytochrome P450 phenotypic ratios for predicting herb-drug interactions in humans. Clin Pharmacol Ther 72:276–287

26. Gurley BJ, Swain A, Williams DK, Barone G, Battu SK (2008) Gauging the clinical significance of P-glycoprotein-mediated herb-drug interactions: comparative effects of St. John's wort,

Echinacea, clarithromycin, and rifampin on digoxin pharmacokinetics. Mol Nutr Food Res 52:772–779

27. Wang LS, Zhou G, Zhu B (2004) St John's wort induces both cytochrome P450 3A4-catalyzed sulfoxidation and 2C19-dependent hydroxylation of omeprazole. Clin Pharmacol Ther 75:191–197

28. Jiang X, Williams KM, Liauw WS et al (2004) Effect of St. John's wort and ginseng on the pharmacokinetics and pharmacodynamics of warfarin in healthy subjects. Br J Clin Pharmacol 57:592–599

29. Xu H, Liauw KM, Murray M, Day RO, McLachlan AJ (2008) Effects of St. John;s wort and CYP2C9 genotype on the pharmacokinetics and pharmacodynamics of gliclazide. Br J Pharmacol 153:1579–1586

30. Hafner V, Jäger M, Matthée A-K et al (2010) Effect of simultaneous induction and inhibition of CYP3A by St. John's wort and ritonavir an CYP3A activity. Clin Pharmacol Ther 87:191–196

31. Markowitz JS, Devane CL, Chavin KD, Taylor RM, Ruan Y, Donovan JL (2003) Effects of garlic (Allium Sativum L.) supplementation on cytochrome P450 2D6 and 3A4 activity in healthy volunteers. Clin Pharmacol Ther 74:170–177

32. Gurley BJ, Gardner SF, Hubbard MA (2004) Vivo assessment of botanical supplementation on human cytochrome P450 phenotypes: Citrus Aurantium, Echinacea Purpurea, milk thistle, and saw palmetto. Clin Pharmacol Ther 76:428–440

33. Gurley BJ, Barone GW, Williams DK (2006) Effect of milk thistle (Silybum marianum) and black cohosh (Cimicifuga racemosa) supplementation on digoxin pharmacokinetics in humans. Drug Metab Dispos 34:69–74

34. Robertson SM, Davey RT, Voell J, Formentini E, Alfaro RM, Penzak SR (2008) Effect of *Ginkgo biloba* extract on lopinavir, midazolam, and fexofenadine pharmacokinetics in healthy subjects. Curr Med Res Opin 24:591–599

35. Jiang X, Williams KM, Liauw WS et al (2005) Effect of gingko and ginger on the pharmacokinetics and pharmacodynamics of warfarin in healthy subjects. Br J Clin Pharmacol 59:425–432

36. Mahutsky MA, Anderson GD, Miller JW, Elmer GW (2006) Ginkgo biloba: evaluation of CYP2C9 drug interactions in vitro and in vivo. Am J Ther 13:24–31

37. Yin OQP, Tomlinson B, Waye MMY, Chow AHL, Chow MSS (2004) Pharmacogenetics and herb-drug interactions: experience with *Ginkgo biloba* and omeprazole. Pharmacogenetics 14:841–850

38. Yasui-Furukori N, Furukori H, Kaneda A, Kaneko S, Tateishi T (2004) The effects of Ginkgo biloba extracts on the pharmacokinetics and pharmacodynamics of donepezil. J Clin Pharmacol 44:538–542

39. Penzak SR, Robertson SM, Hunt JD et al (2010) *Echinacea Purpurea* significantly induces cytochrome P450 3A (CYP3A) but does not alter lopinavir-ritonavir exposure in healthy subjects. Pharmacotherapy 30(8):797–805

40. Gorski JC, Huang SM, Pinto A et al (2004) The effect of echinacea (Echinacea purpurea root) on cytochrome P450 activity in vivo. Clin Pharmacol Ther 75:89–100

41. Freeman C, Spelman K (2008) A critical evaluation of drug interactions with Echinacea spp. Mol Nutr Food Res 52:789–798

42. Jiang X, Blair EY, McLachlan AJ (2006) Investigation of the effects of herbal medicines on warfarin response in healthy subjects: a population pharmacokinetic-pharmacodynamic modeling approach. J Clin Pharmacol 46:1370–1378

43. Gurley BJ, Gardner SF, Hubbard MA et al (2005) In vivo effects of goldenseal, kava kava, black cohosh, and valerian on human cytochrome P450 1A2, 2D6, 2E1, and 3A4/5 phenotypes. Clin Pharmacol Ther 77:415–426

44. Sandhu RS, Prescilla RP, Simonelli TM, Edwards DJ (2003) Influence of goldenseal root on the pharmacokinetics of indinavir. J Clin Pharmacol 43:1283–1288

45. Gurley BJ, Swain A, Barone GW et al (2007) Effect of goldenseal (Hydrastis canadensis) and kava kava (Piper methysticum) supplementation on digoxin pharmacokinetics in humans. Drug Metab Dispos 35:240–245

46. Donovan JL, DeVane CL, Chavin KD et al (2004) Multiple night-time doses of valerian (Valeriana officinalis) had minimal effects on CYP3A4 activity and no effect on CYP2D6 activity in healthy volunteers. Drug Metab Dispos 32:1333–1336

47. Dong SX, Ping ZZ, Xiao WZ et al (1999) Possible enhancement of the first-pass metabolism of phenacetin by ingestion of grape juice in Chinese subjects. Br J Clin Pharmacol 48:638–640

48. Donovan JL, Chavin KD, Devane CL et al (2004) Green tea (Camellia Sinensis) extract does not alter cytochrome p450 3A4 or 2D6 activity in healthy volunteers. Drug Metab Dispos 32:906–908

49. Tankanow R, Tamer HR, Streetman DS et al (2003) Interaction study between digoxin and a preparation of hawthorn (Crataegus Oxyacantha). J Clin Pharmacol 43:637–642

50. Markowitz JS, Donovan JL, Devane CL et al (2003) Multiple doses of saw palmetto (Serenoa repens) did not alter cytochrome P450 2D6 and 3A4 activity in normal volunteers. Clin Pharmacol Ther 74:536–542

51. Wang G, Xiao CQ, Li Z et al (2009) Effect of soy extract administration on losartan pharmacokinetics in healthy female volunteers. Ann Pharmacother 43:1045–1049

52. Anderson GD, Rosito G, Mohustsy MA, Elmer GW (2003) Drug interaction potential of soy extract and Panax ginseng. J Clin Pharmacol 43:643–648

53. Piscitelli SC, Burstein AH, Chaitt D, Alfaro RM, Falloon J (2000) Indinavir concentrations and St. John's wort. Lancet 355:547–548

54. Rengelshausen J, Banfield M, Riedel KD et al (2005) Opposite effects of short-term and long-term St. John's wort intake on voriconazole pharmacokinetics. Clin Pharmacol Ther 78:25–33

55. Piscitelli SC, Burstein AH, Welden N, Gallicano KD, Falloon J (2002) The effect of garlic supplements on the pharmacokinetics of saquinavir. Clin Infect Dis 34:234–238

56. Gallicano K, Foster B, Choudhri S (2003) Effect of short-term administration of garlic supplements on single-dose ritonavir pharmacokinetics in healthy volunteers. Br J Clin Pharmacol 55:199–202

57. Piscitelli SC, Formentini E, Burstein AH, Alfaro R, Jagannatha S, Falloon J (2002) Effect of milk thistle on the pharmacokinetics of indinavir in healthy volunteers. Pharmacotherapy 22:551–556

58. DiCenzo R, Shelton M, Jordan K et al (2003) Coadministration of milk thistle and indinavir in healthy subjects. Pharmacotherapy 23:866–870

59. Mills E, Wilson K, Clarke M (2005) Milk thistle and indinavir: a randomized controlled pharmacokinetics study and meta-analysis. Eur J Clin Pharmacol 61:1–7

60. Rajnarayana K, Reddy MS, Vidyasagar J, Krishna DR (2004) Study on the influence of silymarin pretreatment on metabolism and disposition of metronidazole. Arzneimittelforschung 54:109–113

61. Mogatle S, Skinner M, Mills E, Kanfer I (2008) Effect of African potato (Hypoxis hemerocallidea) on the pharmacokinetics of efavirenz. S Afr Med J 98:945–949

62. Cristiano C (2010) Herbal interactions on absorption of drugs: mechanisms of action and clinical risk assessment. Pharmacol Res 62:207–227

63. Jellin JM (2010) Natural medicines comprehensive database. Therapeutic Research Faculty, Stockton, CA. Available at www.naturaldatabase.com. Accessed 11 June 2010

64. Dürr D, Stieger B, Kullak-Ublick GA et al (2000) St. John's wort induces intestinal P-glycoprotein/MDR1 and intestinal and hepatic CYP3A4. Clin Pharmacol Ther 68:598–604

65. Staudinger JL, Ding X, Lichti K (2006) Pregnane X receptor and natural products: beyond drug-drug interactions. Expert Opin Drug Metab Toxicol 2:847–857

66. de Maat MM, Hoetelmans RM, Math t RA et al (2001) Drug interaction between St John's wort and nevirapine. AIDS 15:420–421

67. Hyland R, Jones BC, Smith DA (2003) Identification of the cytochrome P450 enzymes involved in the N-oxidation of voriconazole. Drug Metab Dispos 31:540–547

68. Moore LB, Goodwin B, Jones SA et al (2009) St. John's wort induces hepatic drug metabolism through activation of the pregnane X receptor. Proc Natl Acad Sci U S A 97:7500–7502

69. Zhou SF, Xue CC, XQ Y, Wang G (2007) Metabolic activation of herbal and dietary constituents and its clinical and toxicological implications: an update. Curr Drug Metab 8:526–553

70. Brady JF, Isizaki H, Fukuto JM et al (1991) Inhibition of cytochrome P-450 2E1 by diallyl sulfide and its metabolites. Chem Res Toxicol 4:642–647

71. Foster BC, Foster MS, Vandenhoek S (2001) An in vitro evaluation of human cytochrome P450 3A4 and P-glycoprotein inhibition by garlic. J Pharm Pharm Sci 4:176–184

72. Arnault I, Haffner T, Siess MH, Vollmar A, Kahane R, Auger J (2005) Analytical method for appreciation of garlic therapeutic potential and for validation of a new formulation. J Pharm Biomed Anal 37:963–970

73. Zou L, Harkey GL, Henderson GL (2002) Effects of herbal components on cDNA-expressed cytochrome P450 enzyme catalytic activity. Life Sci 71:1579–1589

74. Laroche M, Choudhri S, Gallicano K, Foster B (1998) Severe gastrointestinal toxicity with concomitant ingestion of ritonavir and garlic. Can J Infect Dis 9(Suppl A):471P

75. Budzinski JW, Trudeau VL, Drouin CE, Panahi M, Arnason JT, Foster BC (2007) Modulation of human cytochrome P450 3A4 (CYP3A4) and P-glycoprotein (P-gp) in Caco-2 cell monolayers by selected commercial-source milk thistle and goldenseal products. Can J Physiol Pharmacol 85:966–978

76. Sridar C, Goosen TC, Kent UM, Williams JA, Hollenberg PF (2004) Silybin inactivates cytochromes P450 3A4 and 2C9 and inhibits major hepatic glucuronosyltransferases. Drug Metab Dispos 32:587–594

77. Molto J, Valle M, Miranda C, Cedeno S, Negredo E, Cloteta B (2012) Effect of milk thistle on the pharmacokinetics of Darunavir-ritonavir in HIV-infected patients. Antimicrob Agents Chemother 56(6):2837–2841

78. Van Erp NP, Baker SD, Zhao M et al (2005) Effect of milk thistle (Silybum Marianum) on the pharmacokinetics of irinotecan. Clin Cancer Res 11:7800–7806

79. Sidelmann UG, Cornett C, Tjørnelund J, Hansen SHA (1996) Comparative study of precision cut liver slices, hepatocytes, and liver microsomes from the wistar rats using metronidazole as a model substrate. Xenobiotica 26:709–722

80. Beckmann-Knopp S, Rietbrock S, Weyhenmeyer R et al (2000) Inhibitory effects of silibinin on cytochrome P-450 enzymes in human liver microsomes. Pharmacol Toxicol 86:250–256

81. Kim DH, Jin YH, Park JB, Kobashi K (1994) Silymarin and its components are inhibitors of beta-glucuronidase. Biol Pharm Bull 17:443–445

82. Venkataramanan R, Ramachandran V, Komoroski BJ et al (2000) Milk thistle, a herbal supplement, decreases the activity of CYP3A4 and uridine diphosphoglucuronosyl transferase in human hepatocyte cultures. Drug Metab Dispos 28:1270–1273

83. Deng JW, Shon JH, Shin HJ et al (2008) Effect of silymarin supplement on the pharmacokinetics of rosuvastatin. Pharm Res 25:1807–1814

84. Blumenthal M (1998) The Complete German Commission E Monographs, Therapeutic Guide to Herbal Medicines. American Botanical Council, Austin

85. Kubota Y, Kobayashi K, Tanaka N et al (2003) Interaction of Ginkgo biloba extract (GBE) with hypotensive agent, nicardipine, in rats. In Vivo 17:409–412

86. Shinozuka K, Umegaki K, Kubota Y et al (2002) Feeding of Ginkgo biloba extract (GBE) enhances gene expression of hepatic cytochrome P-450 and attenuates the hypotensive effect of nicardipine in rats. Life Sci 70:2783–2792

87. Umegaki K, Saito K, Kubota Y, Sanada H, Yamada K, Shinozuka K (2002) Ginkgo biloba extract markedly induces pentoxyresorufin O-dealkylase activity in rats. Jpn J Pharmacol 90:345–351

88. Sugiyama T, Kubota Y, Shinozuka K, Yamada S, Yamada K, Umegaki K (2004) Induction and recovery of hepatic drug metabolizing enzymes in rats treated with Ginkgo biloba extract. Food Chem Toxicol 42:953–957

89. Budzinski JW, Foster BC, Vandenhoek S, Arnason JT (2000) An in vitro evaluation of human cytochrome P450 3A4 inhibition by selected commercial herbal extracts and tinctures. Phytomedicine 7:273–282

90. He N, Edeki TI (2003) Effects of ginseng and Ginkgo biloba components on CYP3A4 medi-
    ated testosterone 6beta-hydroxylation in human liver microsomes [abstract]. Clin Pharmacol
    Ther 73:50. Abstract PII-81
91. Yoshioka M, Ohnishi N, Koishi T et al (2004) Studies on interactions between functional
    foods or dietary supplements and medicines. IV. Effects of ginkgo biloba leaf extract on the
    pharmacokinetics and pharmacodynamics of nifedipine in healthy volunteers. Biol Pharm
    Bull 27(12):2006–2009
92. Morse GD, Rosenkranz S, Para MF et al (2005) Amprenavir and Efavirenz pharmacokinetics
    before and after the addition of nelfinavir, Indinavir, ritonavir, or Saquinavir in seronegative
    individuals. Antimicrob Agents Chemother 49:3373–3381
93. Gallicano K, Khaliq Y, Carignan G, Tseng A, Walmsley S, Cameron DW (2001) A pharmaco-
    kinetic study of intermittent rifabutin dosing with a combination of ritonavir and saquinavir
    in patients infected with human immunodeficiency virus. Clin Pharmacol Ther 70:149–158
94. Wiegman DJ, Brinkman K, Franssen EJ (2009) Interaction of Ginkgo Biloba with efavirenz.
    AIDS 23:1184–1185
95. Kassahun K, McIntosh I, Cui D et al (2007) Metabolism and disposition in humans of
    Raltegravir (MK-0518), an anti-AIDS drug targeting the HIV-1 integrase enzyme. Drug
    Metab Dispos 35:1657–1663
96. Blonk M, Colbers A, Poirters A, Schouwenberg B, Burgera D (2012) Effect of Ginkgo biloba
    on the pharmacokinetics of Raltegravir in healthy volunteers. Antimicrob Agents Chemother
    56(10):5070–5075
97. Modarai M, Gertsch J, Suter A, Heinrich M, Kortenkamp A (2007) Cytochrome P450 inhibi-
    tory action of Echinacea preparations differs widely and co-varies with alkylamide content.
    J Pharm Pharmacol 59:567–573
98. The New Mexico AIDS Education and Training Center at the University of New Mexico
    Health Sciences Center. Echinacea. Fact Sheet Number 726. Available at: www.aidsinfonet.
    org. Accessed 11 June 2010
99. Al S, Laba JG, Moore JA, Lee TDG (2008) *Echinacea*-induced macrophage activation.
    Immunopharmacol Immunotoxicol 30:553–574
100. Morlat P, Pereira E, Clayette P et al (2008) Early evolution of plasma soluble TNF-alpha p75
    receptor as a marker of progression in treated HIV-infected patients. AIDS Res Hum Retrovir
    24:1383–1389
101. van den Bout-van den Beukel CJ, Koopmans PP, van der Ven AJ, De Smet PA, Burger DM
    (2006) Possible drug-metabolism interactions of medicinal herbs with antiretroviral agents.
    Drug Metab Rev 38:477–514
102. Kim M-G, Kim Y, Jeon J-Y, Kim D-S (2016) Effect of fermented red ginseng on cyto-
    chrome P450 and P-glycoprotein activity in healthy subjects, as evaluated using the cocktail
    approach. Br J Clin Pharmacol 82:1580–1590
103. Streetman DS, Bertino JS, Nafziger AN (2000) Phenotyping of drug-metabolizing enzymes
    in adults: a review of in-vivo cytochrome P450 phenotyping probes. Pharmacogenetics
    10(3):187–216
104. Smith M, Lin KM, Zheng YP (2001) An open trial of nifedipine-herb interactions: nifedip-
    ine with St. John's wort, ginseng, or ginkgo biloba [abstract]. Clin Pharmacol Ther 69:86.
    Abstract PIII-89
105. Henderson GL, Harkey MR, Gershwin ME, Hackman RM, Stern JS, Stresser DM (1999)
    Effects of ginseng components on c-DNA-expressed cytochrome P450 enzyme catalytic
    activity. Life Sci 65:209–214
106. Furutsu M, Koyama Y, Kusakabe M, Takahashi S (1997) Preventative effect of the extract
    of Du-zhong (Tochu) leaf and ginseng root on acute toxicity of chlorpyrifos. Jpn J Toxicol
    Environ Health 43:92–100
107. Malati CY, Robertson SM, Hunt JD et al (2012) Influence of Panax ginseng on cytochrome
    P450 (CYP)3A and P-glycoprotein (P-gp) activity in healthy participants. J Clin Pharmacol
    52:932–939

108. Calderón MM, Chairez CL, Gordon LA, Alfaro RM, Kovacs JA, Penzak SR (2014) Influence of Panax ginseng on the steady state pharmacokinetic profile of lopinavir-ritonavir in healthy volunteers. Pharmacotherapy 34:1151

109. Mills E, Cooper C, Seely D (2005) Kanfer. African herbal medicines in the treatment of HIV: Hypoxis and Sutherlandia. An overview of evidence and pharmacology. Nutr J 4:19

110. Mills E, Foster BC, van Heeswijk R et al (2005) Impact of African herbal medicines on anti-retroviral metabolism. AIDS 19:95–97

111. Nair VD, Foster BC, Thor Arnason J, Mills EJ, Kanfer I (2007) In vitro evaluation of human cytochrome P450 and P-glycoprotein-mediated metabolism of some phytochemicals in extracts and formulations of African potato. Phytomedicine 14:498–507

112. Brown L, Heyneke O, Brown D, van Wyk JP, Hamman JH (2008) Impact of traditional medicinal plant extracts on antiretroviral drug absorption. J Ethnopharmacol 119:588–592

113. Müller AC, Skinner MF, Kanfer I (2013) Effect of the African traditional medicine, Sutherlandia frutescens, on the bioavailability of the antiretroviral protease inhibitor, Atazanavir. Evid Based Complement Alternat Med 2013:324618

114. US FDA (2016) Guidance for Industry Bioavailability and Bioequivalence Studies for Orally Administered Drug Products — General Considerations.Available at: http://www.fda.gov/ohrms/dockets/ac/03/briefing/3995B1_07_GFI-BioAvail-BioEquiv.pdf. Accessed on: 04 Nov 2016

115. Tsukamoto S, Aburatani M, Ohta T (2005) Isolation of CYP3A4 inhibitors from the Black Cohosh (Cimicifuga racemosa). Evid Based Complement Alternat Med 2:223–226

116. Gurley BJ, Swain A, Hubbard MA et al (2008) Clinical assessment of CYP2D6-mediated herb-drug interactions in humans: effects of milk thistle, black cohosh, goldenseal, kava kava, St. John's wort, and Echinacea. Mol Nutr Food Res 52:755–763

117. Abourashed EA, Khan IA (2001) High-performance liquid chromatography determination of hydrastine and berberine in dietary supplements containing goldenseal. J Pharm Sci J Pharm Sci 90:817–822

118. Etheridge AS, Black SR, Patel PR, So J, Mathews JM (2007) An in vitro evaluation of cyto-chrome P450 inhibition and P-glycoprotein interaction with goldenseal, Ginkgo biloba, grape seed, milk thistle, and ginseng extracts and their constituents. Planta Med 73:731–741

119. Qiu W, Jiang XH, Liu CX, Ju Y, Jin JX (2009) Effect of berberine on the pharmacokinetics of substrates of CYP3A and P-gp. Phytother Res 23:1553–1558

120. Lin HL, Liu TY, Lui WY, Chi CW (1999) Up-regulation of multidrug resistance transporter expression by berberine in human and murine hepatoma cells. Cancer 85(9):1937–1942

121. Mathews JM, Etheridge AS, Black SR (2002) Inhibition of human cytochrome P450 activi-ties by kava extract and kavalactones. Drug Metab Dispos 30:1153–1157

122. Zou L, Henderson GL, Harkey MR, Sakai Y, Li A (2004) Effects of kava (Kava-kava, 'Awa, Yaqona, Piper methysticum) on c-DNA-expressed cytochrome P450 enzymes and human cryopreserved hepatocytes. Phytomedicine 11:285–294

123. Unger M, Frank A (2004) Simultaneous determination of the inhibitory potency of herbal extracts on the activity of six major cytochrome P450 enzymes using liquid chromatogra-phy/mass spectrometry and automated online extraction. Rapid Commun Mass Spectrom 18:2273–2281

124. Weiss J, Sauer A, Frank A, Unger M (2005) Extracts and kavalactones of Piper methysticum G. Forst (kava-kava) inhibit P-glycoprotein in vitro. Drug Metab Dispos 33:1580–1583

125. Escher M, Desmeules J, Giostra E, Mentha G (2001) Drug points: hepatitis associated with kava, a herbal remedy for anxiety. BMJ 322:139

126. Russmann S, Lauterberg BH, Hebling A (2001) Kava Hepatotoxicity [letter]. Ann Intern Med 135:68

127. Liver Toxicity with kava (2001) Pharmacist's letter/Prescriber's letter. Ann Intern Med 18(1):180115

128. Consultation letter MLX 286: Proposals to prohibit the herbal ingredient Kava-Kava (Piper methysticum) in unlicensed medicines. Medicines Control Agency, United Kingdom, July 19, 2002

129. Li XZ, Ramzan I (2010) Role of ethanol in kava hepatotoxicity. Phytother Res 24:475–480
130. Lefebvre T, Foster BC, Drouin CE et al (2004) In vitro activity of commercial valerian root extracts against human cytochrome P450 3A4. J Pharm Pharmaceut Sci 7:265–273
131. Hellum BH, Nilsen OG (2008) In vitro inhibition of CYP3A4 metabolism and P-glycoprotein-mediated transport by trade herbal products. Basic Clin Pharmacol Toxicol 102:466–475
132. Sustiva Prescribing Information (2010) Bristol Myers Squibb Pharma. Sustiva Prescribing Information, Princeton
133. Chantre P, Cappelaere A, Leblan D et al (2000) Efficacy and tolerance or Harpagophytum procumbens versus diacerhein in treatment of osteoarthritis. Phytomedicine 7:177–184
134. Chisholm A, Mann J, Skeaff M et al (1998) A diet rich in walnuts favourably influences plasma fatty acid profile in moderately hyperlipidaemic subjects. Eur J Clin Nutr 52:12–16
135. Freedman JE, Parker C, Li L et al (2001) Select flavonoids and whole juice from purple grapes inhibit platelet function and enhance nitric oxide release. Circulation 103:2792–2798
136. Nuttall SL, Kendall MJ, Bombardelli E, Morazzoni P (1998) An evaluation of the antioxidant activity of a standardized grape seed extract, Leucoselect. J Clin Pharm Ther 23:385–389
137. Nishikawa M, Ariyoshi N, Kotani A et al (2004) Effects of continuous ingestion of green tea or grape seed extracts on the pharmacokinetics of midazolam. Drug Metab Pharmacokinet 19:280–289
138. Raucy JL (2003) Regulation of CYP3A4 expression in human hepatocytes by pharmaceuticals and natural products. Drug Metab Dispos 31:533–539
139. Sohn OS, Surace A, Fiala ES et al (1994) Effects of green and black tea on hepatic xenobiotic metabolizing systems in the male F344 rat. Xenobiotica 24:119–127
140. Bonkovsky HL (2006) Hepatotoxicity associated with supplements containing Chinese green tea (Camellia sinensis). Ann Intern Med 144:68–71
141. Jimenez-Saenz M, Martinez-Sanchez MDC (2006) Acute hepatitis associated with the use of green tea infusions. J Hepatol 44:616–619
142. Suekawa M, Ishige A, Yuasa K et al (1984) Pharmacological studies on ginger. I. Pharmacological actions of pungent constitutents, (6)-gingerol and (6)-shogaol. J Pharmacobiodyn 7:836–848
143. Pongrojpaw D, Somprasit C, Chanthasenanont A (2007) A randomized comparison of ginger and dimenhydrinate in the treatment of nausea and vomiting in pregnancy. J Med Assoc Thail 90:1703–1709
144. Langner E, Greifenberg S, Gruenwald J (1998) Ginger: history and use. Adv Ther 15:25–44
145. Castro AF, Altenberg GA (1997) Inhibition of drug transport by genistein in multidrug-resistant cells expressing P-glycoprotein. Biochem Pharmacol 53:89–93
146. Jäger W, Zembsch B, Wolschann P et al (1998) Metabolism of the anticancer drug flavopiridol, a new inhibitor of cyclin dependent kinases, in rat liver. Life Sci 62:1861–1873
147. South African Development Committee. Ministerial consultative meeting on nutrition and HIV/AIDS, Johannesburg, South Africa, January 20, 2002
148. Morito K, Hirose T, Kinjo J et al (2001) Interaction of phytoestrogens with estrogen receptors alpha and beta. Biol Pharm Bull 24:351–356
149. Halm BM, Franke AA, Ashburn LA, Hebshi SM, Wilkens LR (2008) Oral antibiotics decrease urinary isoflavonoid excretion in children after soy consumption. Nutr Cancer 60:14–22
150. Kleijnen J (1994) Evening primrose oil. BMJ 309:824–825
151. Belch J, Hill A (2000) Evening primrose oil and borage oil in rheumatologic conditions. Am J Clin Nutr 71:352S–356S
152. Gurley BJ, Gardner SF, Hubbard MA (2000) Content versus label claims in ephedra-containing dietary supplements. Am J Health-Syst Pharm 57:963–969
153. Parasrampuria J, Schwartz K (1998) Quality control of dehydroepiandrosterone dietary supplement products. JAMA 280:1565. [Letter]
154. Liberti LE, Der Marderosian A (1978) Evaluation of commercial ginseng products. J Pharm Sci 67:1487–1489
155. Cui J, Garle M, Eneroth P, Björkhem I (1994) What do commercial ginseng preparations contain? Lancet 344:134. [Letter]

156. Heptinstall S, Awang DV, Dawson BA, Kindack D, Knight DW, May J (1992) Parthenolide content and bioactivity of feverfew (*Tanacetum parthenium* (L.) Schulkz-Bip). Estimation of commercial and authenticated feverfew products. J Pharm Pharmacol 44:391–395

157. Webb G (1997) "fX" chemically adulterated product does not contain kava. HerbalGram 39:9

158. Saper RB, Phillips RS, Sehgal A et al (2008) Lead, mercury, and arsenic in US- and Indian-manufactured Ayurvedic medicines sold via the internet. JAMA 300:915–923

159. Corns CM (2003) Herbal remedies and clinical biochemistry. Ann Clin Biochem 40(Pt 5):489–507

160. Obach RS (2000) Inhibition of human cytochrome P450 enzymes by constituents of St. John's wort, an herbal preparation used in the treatment of depression. J Pharmacol Exp Ther 294:88–95

161. Kashuba ADM, Bertino JS Jr (2005) Mechanisms of drug interactions I: absorption, metabolism and excretion. In: Piscitelli SC, Rodvold KA (eds) Drug interactions in infectious diseases, 2nd edn. Human Press, Totowa, pp 13–39

162. Weber K, Schneider M, Sacks H et al (2002) Trends in complementary/alternative medicine (CAM) use in a large cohort of HIV-infected women in the US from 1994–2001. XIV International AIDS Conference, Barcelona Spain. 7–12 July 2002: Abstract WePeB6004

163. Behm Dillon DM, Penzak SR, Bailey Klepser T (2004) The use of herbals by patients with HIV. Adv Pharm 2:41–60

164. Lorenc A, Robinson N (2013) A review of the use of complementary and alternative medicine and HIV: issues for patient care. AIDS Patient Care STDs 27(9):503–510

Chapter 7
In Vitro Modeling of Drug-Drug Interactions

Grant T. Generaux

7.1 Introduction

The last two decades have witnessed an impressive growth in the use and standardization of in vitro tools to investigate and characterize the mechanisms responsible for the absorption, distribution, metabolism, and excretion (ADME) of pharmaceutical agents. In parallel with the increasing use of in vitro tools to understand mechanism, there have been an ever-increasing number of researchers helping to grow our understanding of how these individual ADME mechanisms can be integrated with human (patho)physiology using a combination of in vitro-in vivo extrapolation (IVIVE) approaches and mechanistic, physiologically based pharmacokinetic (PBPK) models. This increase in the use of IVIVE and PBPK models has resulted in a significant uptick in quantitative predictions of human pharmacokinetics. The growth of these two fundamental areas – mechanistic PBPK modeling and in vitro tools for ADME – when combined with an increase in the access to and use of convenient PBPK and scientific software packages has greatly increased our ability to accurately characterize and predict drug-drug interactions (DDI) based on in vitro data.

G.T. Generaux (✉)
New Hope, PA, USA
e-mail: ggeneraux@gmail.com

7.2   General Principles Affecting the Magnitude of Drug-Drug Interactions

Until recently, it was common to evaluate the potential for a DDI solely by a qualitative comparison to a known clinical DDI. In the past, if Drug A (a CYP3A4 inhibitor) produced a threefold increase in exposure of a substrate of interest, the anticipated effect of a comparator Drug B would be evaluated by measuring its CYP3A4 IC_{50}. An educated guess regarding the effect of Drug B coadministration on our substrate of interest's exposure would be made by determining whether its IC_{50} value was higher or lower than that of Drug A. With such an approach, the in vitro data was generally put into very little context, and there was little, if any, quantitative integration of the in vitro data with other information pertaining to the drug or patient population (such as pharmacokinetic or pathophysiology) in order for researchers to give concrete recommendations to clinical study teams.

Taking a mechanistic and integrative approach when embarking on the prediction of a potential DDI has become more common, as it is quite useful to integrate the information and data necessary to conduct mechanistic predictions systematically based on in vitro data. What types of prior information and data are used depends somewhat on whether the focus of the in vitro modeling is a victim (e.g., substrate of a metabolic enzyme or a transporter substrate) or a perpetrator (e.g., inhibitor or inducer). However, the perpetrator and victim are present in each DDI prediction, so the systematic approach outlined below is useful for either case. In general, the evaluation of DDI potential boils down to two key questions that represent two sides of the same coin:

1. What is the capacity of the perpetrator to affect various ADME processes of the victim drug?
2. Which ADME mechanisms of the victim drug are most important from the perspective of safety and efficacy?

There is nothing new or novel about these questions. What has recently changed, and what is important for the application and success of in vitro DDI modeling, is the ability to quantify these questions in terms of individual mechanisms and then integrate the resulting information in the form of a model that results in accurate predictions.

7.2.1   Interaction Potential of the Perpetrator Compound

7.2.1.1   Bioavailability

Bioavailability, often represented as F or % F, is the fraction of administered dose that ends up in systemic circulation. From the perspective of classical pharmacokinetics, bioavailability is calculated according to the equation below:

$$F = \frac{\text{AUC}_{\text{PO}}}{\text{AUC}_{\text{IV}}} \times \frac{\text{Dose}_{\text{IV}}}{\text{Dose}_{\text{PO}}}$$

This is useful information for understanding how much of the administered dose is delivered into systemic circulation; however, bioavailability is a measure (or "parameter") which can be further broken down into several processes, each of which may affect DDI in a different way. If we think about bioavailability from the perspective of distinct processes, then it can also be defined as follows:

$$F = F_{\text{a}} \times F_{\text{g}} \times F_{\text{h}}$$

where F_{a}, or the fraction absorbed, is the fraction of dose that crosses the apical membrane of the gastrointestinal tract and F_{g} and F_{h} are the fractions that escape metabolism during the first pass through the gut and liver, respectively. For a drug acting as a perpetrator, all of these processes affect the resultant hepatic and systemic exposure of the perpetrator and thus can significantly influence the potential for the perpetrator to perturb ADME processes. How each of these processes affects the DDI potential of the perpetrator can be ascertained by measuring and/or estimating some of these mechanisms, either by in vitro studies or in silico calculations. DDI models that demonstrate how to break each of these processes down into their mechanistic pieces are discussed below.

7.2.1.2 Distribution of Perpetrator

Another factor that influences the DDI potential of a perpetrator drug is the manner in which it is distributed throughout the body. The distribution of a compound within the body is a complicated topic that involves many processes. Distribution may differ significantly based on whether the compound has a high passive permeability or is a drug transporter substrate, which may allow it to accumulate in tissues in an unanticipated manner, compared to what would be expected from physicochemical properties alone. Based upon concepts from classical pharmacokinetics, a compound's volume of distribution (V_{D}) is described as

$$V_{\text{D}} = \frac{\text{Dose}_{\text{IV}}}{C_0}$$

In this equation, Dose_{IV} is the administered IV dose, and C_0 is the initial systemic concentration following administration. Conceptually, like bioavailability, multiple processes comprise volume of distribution, namely, the relative binding of a compound to tissue and to plasma proteins:

$$V_{\text{D}} \propto \frac{fb_{\text{T}}}{fb_{\text{P}}}$$

Intuitively, this relationship makes sense – if a compound has high nonspecific binding in the tissues, then the resultant plasma concentrations will be lower relative to a given dose. Thus, based on the above equation, the estimate of V_D will be higher. This relationship also shows how a classical V_D measurement can be misleading. For example, a compound with overall low nonspecific tissue binding but high specific binding in one particular tissue can appear to have a large V_D, when there is really a disproportionate accumulation in that tissue. In order to integrate in vitro measures of potency for different ADME mechanisms that occur in different tissues, knowing only whether a perpetrator has a small or large V_D is insufficient; the extent to which the perpetrator is distributed to the tissues of interest must also be estimated.

In the absence of active transport processes, how a compound distributes to the different tissues of the body is a function of its physicochemical properties and the lipid composition of different tissues. Researchers have developed ways to predict a compound's distribution using in silico approaches. One of the more widely used approaches has been published by Rogers et al. and allows for the prediction of tissue partition coefficients (K_p) based on a compound's $logP$, pKa values, and plasma protein binding [1, 2]. The use of K_p allows for using plasma concentration to predict what a given tissue concentration would be by multiplying the plasma concentration by K_p. For the purposes of DDI, and in the absence of measured data on tissue exposure, the K_p value can be used to predict tissue concentrations of the perpetrator, which may be particularly relevant for perpetrators that affect metabolizing enzymes, efflux transporters, or nuclear receptors – processes where the concentration within the cell is most relevant to DDI potential.

In the case of uptake transporters such as the OATPs or OATs, the most relevant perpetrator concentration is the unbound plasma concentration for non-hepatic tissues; for hepatic uptake transporters, it is appropriate to use an estimate of portal vein concentration.

7.2.1.3 Hepatic and Intestinal Perpetrator Concentrations

Oral administration of a compound can lead to significantly elevated concentrations in both the gut as well as the liver, particularly while the compound is being absorbed. The magnitude and duration of elevated concentrations depends on the dose of the compound, as well as the fraction absorbed (F_a) and first-order absorption rate constant (ka). The oral absorption of a compound is a complex process, which may involve multiple steps including tablet disintegration, drug dissolution, drug precipitation, or saturation of drug transporters residing on the lumen of the gastrointestinal tract. Ideally, this process is modeled using a model that has a sophisticated PBPK absorption model (e.g. ACAT or ADAM absorption models). However, in the absence of such sophisticated PBPK models, there are widely used steady-state approximations for estimating gut and hepatic concentrations, which are suitable for use with static DDI prediction models [3, 4]. For liver, the equation is shown below:

where C_{max} is the maximum plasma concentration, Q_h is the hepatic blood flow, and Fa and ka are as defined above.

$$I_{hepatic,inlet} = C_{max} + \frac{Dose \times F_a \times ka}{Q_h}$$

For prediction of gut concentration, the following equation is used:

$$I_{gut} = \frac{Dose \times F_a \times ka}{Q_g}$$

where Q_g is the enterocytic blood flow, which is reported by Galetin et al. to be between 2 and 10% cardiac output [4]. In the equations above, Dose $\times F_a \times$ ka determines the input rate of the compound into the liver or gut, and Q_h or Q_g determines the compound's clearance from the liver and gut, respectively.

7.2.1.4 Potency of the Perpetrator Compound

In terms of in vitro modeling of DDI, a compound's potency refers to its strength of binding or association with various enzymes, transporters, or receptors. A variety of in vitro experimental systems for generating data that are suitable for predicting DDI exist. Chapter 7 David Rodrigues Drug-Drug Interactions (2nd edition) has a detailed overview of the different experimental systems that can be used to evaluate DDIs involving metabolism [5]. For transporter-based DDIs, Zamek-Gliszczynski et al. provide a good overview of the experimental systems that can be used to generate transport-related parameters and recommendations on study design in order to generate high-quality data [6].

Generally, suitable potency values for making quantitative predictions related to the mechanism of interest are generated using an in vitro system that isolate the mechanism in some way. However, the degree to which the mechanism needs to be isolated will depend upon the question being addressed.

7.2.2 Interaction Potential of the Victim Compound

7.2.2.1 Sensitivity of Substrate

One of the most important determinants of the magnitude of DDI is the degree to which the victim substrate depends on a particular ADME process for its systemic clearance (f_{CL}). The relationship between f_{CL} and the increased victim exposure (AUC$_i$/AUC) for victim substrates is shown below and is essentially the same equation that will be discussed later when different static DDI models are addressed:

$$\frac{\text{AUC}_i}{\text{AUC}} = \frac{1}{\left(\dfrac{f_{\text{CL}}}{1+\dfrac{[I]}{K_i}}\right)+\left(1-f_{\text{CL}}\right)}$$

To make the calculations easier for exploring how AUC_i/AUC depends on f_{CL}, set $[I] = K_i$, thus capturing the situation when 50% of the enzyme or transporter activity is inhibited. This simplifies the equation to

$$\frac{\text{AUC}_i}{\text{AUC}} = \frac{1}{\left(\dfrac{f_{\text{CL}}}{2}\right)+\left(1-f_{\text{CL}}\right)}$$

With this simplification in hand, a few values can be checked to illustrate the equation's behavior. For instance, setting f_{CL} to 0.1, 0.5, and 1, the resultant exposure increases are 1.05-fold, 1.34-fold, and 2-fold, respectively. This fits the expectation that exposure should increase by twofold when all clearance processes are inhibited by 50%. Similarly using a more significant inhibition of 90% (i.e., $[I]/K_i = 9$)

$$\frac{\text{AUC}_i}{\text{AUC}} = \frac{1}{\left(\dfrac{f_{\text{CL}}}{10}\right)+\left(1-f_{\text{CL}}\right)}$$

The resultant exposure increases are 1.1-fold, 1.82-fold, and 10-fold for f_{CL} values of 0.1, 0.5, and 1, respectively. Just from this small sampling of f_{CL} values, it is evident that the victim's exposure increase is very sensitive to the value of f_{CL}. This fact is even more evident if AUC_i/AUC is plotted as a function of f_{CL} for several different levels of inhibition. Figure 7.1 illustrates that the AUC_i/AUC values increase exponentially as f_{CL} approaches 1, particularly when inhibition levels are above ~90%.

Fraction cleared is typically characterized in terms of fraction of systemic clearance due to metabolism. However, the principle also applies to any ADME process that is a rate-limiting contributor to systemic clearance and occurs in parallel with other systemic clearance pathways. For instance, if a compound undergoes hepatic clearance and the rate-limiting step is uptake into hepatocytes via a combination of OATP1B1 and passive diffusion across the sinusoidal hepatocyte membrane, then the f_{CL} due to OATP1B1 will be calculated as

$$f_{\text{CL}} = \frac{\text{CL}_{\text{OATP1B1}}}{\text{CL}_{\text{OATP1B1}} + \text{CL}_{\text{passive diffusion}}}$$

Fig. 7.1 Impact of fraction cleared (f_{CL}) on substrate AUC ratio at different levels of enzyme or transporter inhibition

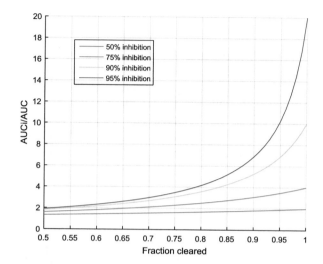

The above equations are useful for evaluating the DDI potential when one of two or more parallel pathways is inhibited, as is indicated by the separation of the denominator into f_{CL} and $1-f_{CL}$ terms. However, in cases where the enzymes and/or transporters that play a role in the DDI occur in series, it may be more appropriate to address such a question using a dynamic DDI model (e.g., PBPK or mechanistic PK model), particularly when the enzyme or transporter of interest is not the rate-limiting step. One example of this scenario is the HMG-CoA reductase atorvastatin, which is transported into the hepatocyte via OATPs and subsequently metabolized by hepatic CYP3A4. Although atorvastatin is metabolized exclusively by CYP3A4 ($fm_{CYP3A4} \sim 1$), plasma levels of oral atorvastatin increased by only 47% following the coadministration of clarithromycin, and systemic exposure of atorvastatin following an IV microdose did not change after coadministration with itraconazole [7, 8]. These results suggest that the rate-limiting uptake clearance of atorvastatin into the hepatocyte is masking the effect of CYP3A4 inhibition from being observed in the systemic circulation. In this example, the prediction of atorvastatin exposure change based on CYP3A4 alone, without taking into consideration hepatocyte uptake as a rate-limiting step, would greatly overestimate the increase in plasma exposure observed clinically.

7.2.2.2 Oral Bioavailability of the Substrate

The discussion thus far of victim DDI potential has focused on perturbations to systemic clearance; however, for orally administered compounds, DDIs involving the bioavailability of a victim compound can have a profound impact on the magnitude of a DDI. Earlier F was described as the successive multiplication of drug escaping through the different barriers (e.g., apical enterocyte membrane, enterocyte, and hepatocyte) that it must overcome prior to reaching systemic circulation:

$$F = F_a \times F_g \times F_h$$

Of particular interest is the inhibition of gut CYP3A4 following administration of an oral CYP3A4 inhibitor. Because CYP3A4 is highly expressed in the gut, intestinal CYP3A4 metabolism plays a significant role in the first-pass extraction of many orally administered drugs. Similar to the above analysis regarding the impact that f_{CL} has on victim exposure, how different values of F_g affect DDI potential can also be evaluated. The following equation describes how inhibition of intestinal metabolism increases systemic exposure:

$$\frac{\mathrm{AUC}_i}{\mathrm{AUC}} = \frac{1}{\left(F_g + \dfrac{1 - F_g}{1 + \dfrac{[I]_{gut}}{K_i}} \right)}$$

For the above equation, F_g can be estimated from in vitro data by using the following equation:

$$F_g = \frac{Q_{gut}}{Q_{gut} + \mathrm{fu}_{gut} \times \mathrm{CLu}_{int,gut}}$$

where Q_{gut} signifies a clearance term representing the effect of both enterocytic blood flow (Q_{ent}) as well as clearance due to passive permeability (CL_{perm}), fu_{gut} represents the unbound fraction of drug in the enterocytes, and $\mathrm{CLu}_{int,gut}$ represents the unbound intrinsic gut clearance [4]. As Q_{gut} is a parameter which is derived from both Q_{ent} and CL_{perm}, it needs to be calculated from the following equation:

$$Q_{gut} = \frac{Q_{ent} \times CL_{perm}}{Q_{ent} + CL_{perm}}$$

Those with some background or experience working with the well-stirred model of hepatic clearance will recognize the two above equations as analogous to the well-stirred equations for bioavailability (F) and hepatic clearance (CL_h). Additional details, including derivation, of the Q_{gut} model can be found in Galetin et al. [4]. Figure 7.2 illustrates the relationship between $\mathrm{AUC}_i/\mathrm{AUC}$ and F_g for various levels of intestinal extraction and degrees of gut CYP3A4 inhibition.

As expected, for compounds which are not subject to significant first-pass extraction in the gut (i.e., $F_g \sim 1$), even potent inhibition of CYP3A4 does not lead to an appreciable increase in systemic exposure. It is important to note that the increase in systemic exposure due to inhibition of first-pass extraction is multiplied by any

Fig. 7.2 Impact of fraction escaping gut metabolism (Fg) on substrate AUC ratio at different levels of enzyme or transporter inhibition

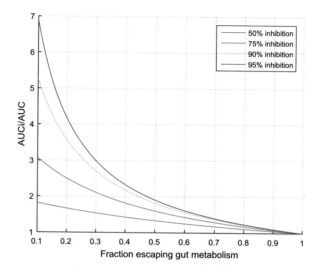

increase in systemic exposure due to inhibition of systemic clearance. This is intuitive as inhibition of first-pass extraction increases bioavailability, and thus the amount entering systemic circulation, whereas the inhibition of systemic clearance affects the rate at which compound is removed from systemic circulation. In contrast, compounds with an F_g value near 1 will be more susceptible to large decreases in bioavailability in the presence of CYP3A4 enzyme inducers.

7.2.2.3   Kinetics and Saturation

A key characteristic for ADME processes that contribute to the victim's systemic clearance is the assumption of linearity. With the exception of passive permeability across a membrane, ADME processes generally follow Michaelis-Menten kinetics and are therefore subject to saturation. What saturation means for DDI potential depends on the degree of saturation, whether the compound is a perpetrator or a victim, and which ADME process is being affected. Saturation of a perpetrator's clearance mechanism may increase its exposure and its interaction potential, if there are no parallel, unsaturated clearance pathways that can act as a relief valve for the additional exposure.

For victim compounds, how exposure is affected depends on the nature of the interaction with the perpetrator as well as the ADME process affected. The inhibition of saturated enzymes or transporters involved in systemic clearance can make the victim less sensitive to exposure increases compared to in their unsaturated state. This phenomenon may be most likely observed in the case where saturated enzymes or transporters are involved in limiting bioavailability. In this case, the inhibition of intestinal P-gp or CYP3A4 may be of little consequence unless the victim dose is lowered below a saturating concentration.

7.3   Types of DDI Prediction Models

Models used to predict drug-drug interactions from in vitro data fall into two basic categories: static and dynamic.

7.3.1   *Mechanistic Static Models*

To predict the magnitude of a DDI, static DDI models use a constant perpetrator concentration, potency measures for the perpetrator, and information on the sensitivity of the victim to the mechanism the perpetrator is affecting. Static models are particularly useful during drug discovery and early drug development, where they are used to evaluate the risk of DDI based on in vitro data and an early prediction of compound exposure. Such a DDI prediction, guided by the sensitivity, therapeutic index, and frequency of co-medications, can go a long way toward de-risking a compound and informing future clinical studies.

7.3.1.1   Direct Inhibition Models

The most commonly used mechanistic static model is shown below. It is used to predict an increase in AUC for cytochrome P450 (CYP) substrates (excluding CYP3A4 substrates) following direct inhibition of their biotransformation [9]:

$$\frac{\text{AUC}_i}{\text{AUC}} = \frac{1}{\left(\dfrac{\text{fm}_{\text{CYP}}}{1 + \dfrac{[I]}{K_i}} + \left(1 - \text{fm}_{\text{CYP}}\right) \right)}$$

In this equation, fm_{CYP} is the fraction of systemic clearance due to biotransformation by the CYP of interest, $[I]$ is the estimated inhibitor concentration available to interact with the CYP of interest, and K_i is the inhibitor constant of the perpetrator for a given CYP. The degree to which a victim drug relies on a single pathway for its systemic clearance is the primary factor that determines its sensitivity to a DDI. At the biochemical level, the degree to which a given substrate relies on an individual CYP for its biotransformation depends on both the relative expression of all CYPs within a given individual, as well as the relative affinity of the substrate for the various CYPs. Sophisticated PBPK modeling platforms, such as Simcyp and GastroPlus, have the ability to use a population distribution of CYP expression levels and affinities, thus allowing for the prediction of variability in exposure change due to DDI. In the case of mechanistic static DDI models, approaches can be used to incorporate population variability into the predictions from static models. However, most commonly, average values are used for the relative expression of the CYPs, and the relative affinity of the victim substrates for the individual CYPs. Additionally, it is necessary to ensure that any contributions of non-CYP systemic clearance routes (e.g., direct conjugation via UGTs, renal clearance) are captured in the fm_{CYP} parameter.

As discussed above, with CYP3A4 substrates, inhibition of gut CYP3A4 is another potential source of DDI. An orally administered CYP3A4 inhibitor can significantly increase the F_g, and thus exposure, of an orally administered CYP3A4 substrate by inhibiting first-pass metabolism within the enterocytes. Therefore, for predicting interactions involving substrates of CYP3A4, the equation above is modified to incorporate an additional term that accounts for the inhibition of intestinal metabolism:

$$\frac{AUC_i}{AUC} = \frac{1}{\left(\dfrac{fm_{CYP}}{1+\dfrac{[I]}{K_i}}\right)+\left(1-fm_{CYP}\right)} \times \frac{1}{Fg_{CYP}+\dfrac{\left(1-Fg_{CYP}\right)}{1+\dfrac{[I]_{gut}}{K_i}}}$$

7.3.1.2 Metabolism-Dependent Inhibition Models

For perpetrators that result in metabolism-dependent inhibition, the equations above for direct inhibition require a minor modification in order to account for the different enzyme inhibition kinetics occurring with irreversible binding:

$$\frac{AUC_i}{AUC} = \frac{1}{\left(\dfrac{fm_{CYP}}{1+\dfrac{[I]\times k_{inact}}{k_{deg}\times\left([I]+K_I\right)}}\right)+\left(1-fm_{CYP}\right)}$$

This equation is identical to the one used above for direction inhibition, with the exception that the $1+\dfrac{[I]\times k_{inact}}{k_{deg}\times\left([I]+K_I\right)}$ term replaces the simpler $1+\dfrac{[I]}{K_i}$ term that is used in the static equation for direct inhibition. In this term, K_I is the inhibitor concentration required for half-maximal inactivation, k_{inact} is the maximal rate of inactivation, and k_{deg} is the rate constant for enzyme degradation. The values of $[I]$, K_I, and k_{inact} are all parameters which are estimated or measured in vitro for the inhibitor of interest. K_{deg}, however, is what is often referred to as a "system parameter," because it represents a physiological process or anatomical state that does not change from compound to compound. System parameters only rarely show up in static DDI models, but they are a key feature of dynamic PBPK DDI models, which are discussed in Sect. 7.3.2.

Similar to the case of direct inhibition, there is an extended version of the metabolism-dependent inhibition model that includes the impact on victim exposure of inactivating gut CYP3A4:

$$
\frac{AUC_i}{AUC} = \cfrac{1}{\left(\cfrac{fm_{CYP}}{1+\cfrac{[I] \times k_{inact}}{k_{deg} \times \left([I]+K_I\right)}} + \left(1-fm_{CYP}\right)\right)} \times \cfrac{1}{\left(\cfrac{\left(1-Fg_{CYP}\right)}{1+\cfrac{[I]_{gut} \times k_{inact}}{k_{deg(gut)} \times \left([I]_{gut}+K_I\right)}} + Fg_{CYP}\right)}
$$

A subtle but notable difference between the two k_{deg} values in this equation is that the k_{deg} used in the hepatic term represents an estimate of the true degradation rate for the CYP of interest, whereas $k_{deg(gut)}$ represents the rate at which enterocytes are shed from the gut wall. The latter is a process that occurs more rapidly than the turnover of the CYP itself [9].

7.3.1.3 Enzyme Induction Models

The DDI models described so far have all focused on enzyme inhibition, where the concern is an increased frequency of toxicity or other off-target effects driven by increased exposure. For anti-infective drugs, a DDI which results in reduced exposure is also a significant concern because a patient receiving subtherapeutic exposure over a prolonged time period can experience therapeutic failure, or even worse, resistance can develop and eventually compromise drug efficacy in future patients. As with the models used for inhibition, there exist static DDI models that help predict and thus reduce the risk of therapeutic failure with DDIs involving enzyme inducers:

$$
\frac{AUC_i}{AUC} = \cfrac{1}{\left(1+\cfrac{d \times E_{max} \times [I]}{[I]+EC_{50}}\right) \times fm_{CYP} + \left(1-fm_{CYP}\right)} \times
$$

$$
\cfrac{1}{\left(1+\cfrac{d \times E_{max} \times [I]_{gut}}{[I]_{gut}+EC_{50}}\right) \times \left(1-F_g\right) + F_g}
$$

where fm_{CYP} is the fraction of systemic clearance due to biotransformation by the CYP of interest, $[I]$ is the estimated inducer concentration available to interact with the CYP of interest, EC_{50} is the concentration of inducer that gives half-maximal induction, E_{max} is the maximum fold induction observed in vitro (typically in cultured human hepatocytes), and d is the calibration, or scaling, factor [10].

One parameter of note that was not present in the inhibition static models is the scaling factor, d. Enzyme inhibition is a phenomenon acting directly on the enzyme and for which the percent inhibition calculated translates directly to the in vivo scenario. In contrast, enzyme induction is an indirect process involving the binding of an inducer to various transcription factors, which in turn increase the transcription of the target enzyme. Consequently, the expression of transcription factors, and thus transcription rate, can differ significantly between different in vitro systems, as well as between the in vitro and in vivo scenario. A calibration, or scaling, factor is thus introduced into the equation above to account for this difference in expression between in vitro and in vivo. One approach for determining such a calibration factor is illustrated by Fahmi et al., who utilized the above equation to estimate the value of d *via* least-squares regression, using the known clinical reduction in AUC for a set of known clinical inducers, in combination with the AUC reductions predicted by the static induction model [9]. For additional details on the calculation of the calibration factor, as well as the use of static induction models for predicting DDI, please refer to Fahmi et al. [10].

7.3.2 Physiologically Based Pharmacokinetic (PBPK) DDI Models

In the past decade, the use of PBPK models to predict DDI has evolved from an approach used mainly within academia or evaluated on an exploratory basis within a handful of pharmaceutical companies to an approach that is now used to predict DDI on a regular basis throughout the pharmaceutical industry. As such, it is beginning to mature with respect to regulatory acceptance [11]. PBPK models are conceptually straightforward – multiple compartments consisting of physiological volumes represent the major tissues of the body, with the compartments being connected by physiological blood flows. In contrast to traditional compartmental pharmacokinetic models, PBPK models are created by linking drug-specific parameters (i.e., in vitro and/or in silico data) with system parameters (i.e., physiology and anatomy). The strength of PBPK models lies in this separation of drug and system parameters, allowing different physiological states, such as disease or genetics, to be represented in the model and any interaction between physiology and the drug to be evaluated in an integrated fashion. An additional strength of PBPK models compared to static DDI models is they are dynamic. Thus, in addition to predicting exposure change following continuous, long-term coadministration between a perpetrator and victim drug, a PBPK model would be able to investigate what impact a change in perpetrator dosing schedule would have on a given DDI.

From the perspective of DDI prediction, one major advantage of PBPK modeling is related to the separation of system and drug parameters described above. During drug development, DDI studies are often performed in healthy volunteers, and while many anti-infective agents will be administered to otherwise healthy individuals, they need to be given to individuals with a variety of health conditions. The ability to incorporate pathophysiology into a PBPK model allows for the simulation of DDI outcomes with particular diseases or in specific sets of patients for which we might otherwise lack the resources to evaluate in clinical trials.

Another advantage of PBPK models relates to polypharmacy and involves the ability to predict and evaluate complex DDI. The static DDI models discussed above in Sect. 7.3.1 all have in common the evaluation of a single mechanism at a time (e.g., direct inhibition, induction) and a focus on DDIs involving parallel, but not sequential, clearance pathways in the absence of enzyme or transporter saturation. The dynamic nature of PBPK models provides a natural platform with which to investigate complex DDIs involving multiple mechanisms and organs. Taken in combination with the effect of disease on DDI, the ability to predict DDI outcomes involving multiple perpetrators and/or victims has the potential to significantly reduce or even eliminate the need to run multiple clinical trials for evaluating key co-medications within the target population.

7.4 Summary

The use of in vitro modeling to predict and understand DDIs has and will increasingly continue to positively affect the prediction and management of DDIs during the development of new anti-infective agents. Static in vitro DDI models are useful during drug discovery and early drug development, where they can be used to evaluate the risk of DDI based on in vitro data and an early prediction of compound exposure. Such a DDI prediction, guided by the sensitivity, therapeutic index, and frequency of co-medications, can go a long way toward de-risking a compound and informing future clinical studies. The usefulness of using PBPK models for DDI prediction stands out for questions involving complex DDIs or DDIs involving special populations or diseases. PBPK models are also critical for exploring alternative dosing regimens and other issues related to clinical study design. Although the science of IVIVE for transporters, transporter/enzyme interplay, and special populations/diseases continues to develop at a rapid pace, in vitro modeling of DDI is a scientifically mature subject with increasing regulatory acceptance and is evolving into a key asset to help in the development of life-altering medicines for patients in need.

References

1. Rodgers T, Leahy D, Rowland M (2005) Physiologically based pharmacokinetic modeling 1: predicting the tissue distribution of moderate-to-strong bases. J Pharm Sci 94:1259–1276
2. Rodgers T, Rowland M (2006) Physiologically based pharmacokinetic modelling 2: predicting the tissue distribution of acids, very weak bases, neutrals and zwitterions. J Pharm Sci 95:1238–1257
3. Ito K, Iwatsubo T, Kanamitsu S et al (1998) Prediction of pharmacokinetic alterations caused by drug-drug interactions: metabolic interaction in the liver. Pharmacol Rev 50(3):387–411
4. Galetin A, Hinton LK, Burt H et al (2007) Maximal inhibition of intestinal first-pass metabolism as a pragmatic indicator of intestinal contribution to the drug-drug interactions for CYP3A4 cleared drugs. Curr Drug Metab 8(7):685–693
5. Oglive BW, Usuki E, Yerino P et al (2008) In vitro approaches for studying the inhibition of drug-metabolizing enzymes and identifying the drug-metabolizing enzymes responsible for the metabolism of drugs (reaction phenotyping) with emphasis a cytochrome P450. Drug-Drug Interactions, 2nd edition 231–358
6. Zamek-Gliszczynski MJ, Lee CA, Poirier A et al (2013) ITC recommendations for transporter kinetic parameter estimation and translational modeling of transport-mediated PK and DDIs in humans. Clin Pharmacol Ther 94(1):64–79
7. Jacobson TA (2004) Comparative pharmacokinetic interaction profiles of pravastatin, simvastatin, and atorvastatin when coadministered with cytochrome P450 inhibitors. Am J Cardiol 94(9):1140–1146
8. Maeda K, Ikeda Y, Fujita T et al (2011) Identification of the rate-determining process in the hepatic clearance of atorvastatin in a clinical cassette microdosing study. Clin Pharmacol Ther 90(4):575–581
9. Shardlow CE, Generaux GT, MacLauchlin CC et al (2011) Utilizing drug-drug interaction prediction tools during drug development: enhanced decision making based on clinical risk. Drug Metab Displays 39:2076–2084
10. Fahmi OA, Maurer TS, Kish M et al (2008) A combined model for predicting CYP3A4 clinical net drug-drug interaction based on CYP3A4 inhibition, inactivation, and induction determined in vitro. Drug Metab Displays 36:1698–1708
11. Rowland M, Lesko LJ, Rostami-Hodjegan A (2015) Physiologically based pharmacokinetics is impacting drug development and regulatory decision making. CPT Pharmacometrics Syst Pharmacol 4:313–315

Chapter 8
Probe Cocktail Studies

Anne N. Nafziger and Joseph S. Bertino Jr

8.1 Purpose and Use of Cocktail Studies

The conduct of drug interaction studies has been revolutionized by the ability to evaluate more than one potential drug-drug interaction (DDI) within a single study. DDI studies were formerly conducted as a group of studies to evaluate the potential of interactions through common or suspected metabolic pathways that were expected to be of clinical significance [1, 2]. Earlier DDI studies primarily used specific, approved drugs with narrow therapeutic indices (e.g., digoxin, phenytoin, theophylline, warfarin) that were likely to be co-administered and for which there could be important clinical consequences. However, these types of studies had significant limitations and were applicable only to the specific drugs studied. The studies were a surrogate for studies of a metabolic pathway. For example, theophylline is metabolized by CYP1A2, and DDI studies with theophylline were then extrapolated to predict other DDIs that might occur via CYP1A2 metabolism.

Cocktail studies provide a means to screen for DDIs through multiple metabolic pathways within a single study. A cocktail study is comprised of concurrent administration of probe substrates and assessment of biomarkers to simultaneously assess DME activities before (baseline) and during drug treatment. Evaluation of DME can be for the effect of a drug on constitutive DME (i.e., is the drug under study an

A.N. Nafziger (✉)
Bertino Consulting, Schenectady, NY, USA

School of Pharmacy & Pharmaceutical Sciences, Department of Pharmacy Practice,
University at Buffalo, State University of New York, Buffalo, NY, USA
e-mail: anne.nafziger@bertinoconsulting.com

J.S. Bertino Jr
College of Physicians & Surgeons, Columbia University, New York, NY, USA

Bertino Consulting, Schenectady, NY, USA

© Springer International Publishing AG 2018
M.P. Pai et al. (eds.), *Drug Interactions in Infectious Diseases: Mechanisms and Models of Drug Interactions*, Infectious Disease,
https://doi.org/10.1007/978-3-319-72422-5_8

259

Table 8.1 Chapter abbreviations

AUC	Area under the concentration-time curve
CI	Confidence intervals
Cmax	Maximum concentration
CYP	Cytochrome P450
DDI	Drug-drug interaction
DME	Drug-metabolizing enzymes
EM	Extensive metabolizer
EMA	European Medicines Agency
FDA	US Food and Drug Administration
IM	Intermediate metabolizer
MHLW	Ministry of Health, Labour and Welfare
NAT2	N-acetyltransferase 2
PBPK	Physiologically based pharmacokinetic modeling
PMDA	Pharmaceuticals and Medical Devices Agency
PhRMA	Pharmaceutical Research and Manufacturers of America
PM	Poor metabolizer
UGTs	UDP-glucuronosyltransferases
UM	Ultra-rapid metabolizer

inhibitor, inducer, or activator?) or to evaluate the effect of an inhibitor, inducer, or activator on the pharmacokinetics of the DME pathway for the drug in question. By observing whether changes in activity occur with co-administration of the treatment drug, the mechanistic basis of, and the qualitative potential for, drug interactions can be evaluated.

The most frequent use for a cocktail study is to determine the constitutive activity of defined DMEs and then reevaluate the DME activities after inhibition, activation, and/or induction by an investigational drug, although use has also been proposed to determine drug-therapeutic protein interactions [3]. Most often, these studies are conducted to evaluate the potential for both inhibition and induction. Cocktail studies are particularly important when there are shared metabolic pathways and the clinically relevant pharmacokinetic DDIs through these pathways are uncertain [4]. At least theoretically, cocktail studies can be used to assess DDIs involving transporter pathways, although transporter drug interactions have the potential to be more complex than simple DDIs [5]. Some validated probes and biomarkers have been identified for common transporters [5], but, to date, no cocktails have been successfully validated for transporter probes [5–7]. This chapter will therefore focus on cocktail studies for assessment of drug-metabolizing enzymes (DMEs), but the same principles apply to evaluation of transporter-related DDIs [6] or interactions that result from both DMEs and transporters. For quick reference, Table 8.1 provides a list of abbreviations used in this chapter.

A probe is a substance, typically a drug, which is a selective substrate for a specific DME or a substrate metabolized to a specific metabolite by a specific DME pathway. A biomarker is the metric used to evaluate the DME activity (or phenotype) of the given probe through the specific enzyme pathway. A variety of pharmacokinetic parameters may be used as biomarkers. These include total area under the concentration-time curve ($AUC_{0-\infty}$), systemic or partial clearance of the probe or a metabolite, or metabolic ratios of a metabolite to the parent compound [2]. Biomarkers may be measured in a variety of biologic samples, but are most often measured in serum, plasma, or urine.

Cocktail studies are usually conducted in healthy volunteers. It is important to remember that other factors in addition to the co-administered drugs can influence DME, and such factors are more likely to be present in patients than in healthy volunteers. Examples of these factors include active disease states such as cancer [8] and renal, hepatic, and cardiac failure [9–11]; increased cytokine levels [12]; human immunodeficiency virus infection [13]; environmental exposures such as tobacco smoke [14], alcohol consumption [15], fruit juice consumption [16–18], and other dietary exposures [19]; fasting [20]; age (particularly for children less than 1 year of age) [21]; and pregnancy [22]. In addition, medical interventions such as hemodialysis can alter DME activity [23]. Therefore, studies conducted in healthy volunteers reflect phenotypes and DDI potential within similar, healthy populations and may not reflect either basal DME activity or the enzyme activity changes that occur in patient populations with acute or chronic health conditions. DDI studies conducted in healthy subjects can potentially describe the worst-case scenario since inflammatory disease often results in a reduction in DME activity and thus reduces the potential for inhibitory DDIs [24]. Thus, the extent of an identified DDI may be reduced in a patient or may change over time as a disease is treated or progresses. When studies are conducted in patients, the investigator should not compromise on biomarker sampling; this has been a limitation for application of cocktail studies in the clinical setting. Because of the potentially greater variability or alteration in DME activity among patients, enhanced sampling may be necessary to accurately assess DME activity. Unfortunately, in many published probe and cocktail studies, investigators have attempted to provide limited sampling guidelines but have used incorrect statistical analyses to devise this limited sampling [25]. This type of misanalysis invalidates single sample recommendations.

In drug development, cocktail studies have numerous advantages. First, the effect of interindividual variability in DME over time is minimized by conducting one study in the same subjects rather than five or more studies in different subjects. Second, intraindividual variability is decreased by using subjects as their own controls (and thus controlling for genetic factors) [26, 27]. Third, research costs are reduced by assessing multiple enzyme systems in one study rather than during multiple studies of one enzyme system [4, 28]. Finally, combining the above factors leads to increased efficiency and a compressed timeline for drug development. Because the number of DDI studies per new molecular entity is increasing [29], efficiency becomes ever more important. In spite of this, cocktail studies are used infrequently [30]. Cocktail studies should be preferred over the use of physiologically based pharmacokinetic modeling (PBPK) since they provide real data rather than modeled data derived only from average literature values.

There are also potential limitations, but these can be addressed by proper study design. DDIs are possible among the probes. If interactions occur, they could result in findings of greater or lesser DME activity changes than those actually related to the actions of the treatment drug. For this reason, it is essential that the combination of probes has been validated as a cocktail. This validation is separate from the work required to validate individual probes and biomarkers. Individually validated probes and biomarkers cannot be assumed to make a validated cocktail when combined. Other challenges include a lack of safe probes and limited availability of some probes that are part of validated cocktails. Special requirements may be needed for sample collection and handling, and these special requirements may not be described or readily accessible in the literature, but rather personally known to investigators or laboratories. Lastly, sensitive and specific assays may be lacking for validated biomarkers. Advances in assay methodologies allow for multiple biomarkers to be assayed simultaneously using small specimen quantities [31–33], and this has increased the feasibility of conducting cocktail studies. Small quantities of blood can be collected and thereby reduce subject risk while lowering study costs. In order to obtain accurate results, it is essential that individual probes, biomarkers, and each cocktail combination be adequately validated prior to use [4] and that the exact validated cocktail methodology be followed during study conduct [34].

A cocktail study may not completely eliminate requirements for additional DDI studies, but the approach of using cocktail studies prior to more specific definitive studies has been endorsed by the US Food and Drug Administration (FDA) [35, 36] and European Medicines Agency (EMA) [37]. Investigators can anticipate that if a significant change from basal DME activity is identified during a cocktail study, the regulatory agency with oversight may require additional specific DDI studies for the investigational drug and other frequently co-administered drugs that are expected to result in clinically significant DDIs [38]. The rationale for requiring specific DDI studies is open to debate since DDIs cannot be quantitatively predicted or used to provide specific dosage adjustments for individual patients [39]. Initial cocktail studies should use the most selective probe substrates that are part of a validated cocktail. If subsequent studies are conducted, other clinically relevant validated probes can be used.

8.2 In Vitro Studies and the Conduct of Cocktail Studies

In vitro studies are the first step in prediction of DDIs. The FDA recommends that appropriate in vitro screening be done to evaluate whether therapeutic concentrations of an investigational drug are metabolized by *CYP1A2*, *CYP2B6*, *CYP2C8*, *CYP2C9*, *CYP2C19*, *CYP2D6*, and *CYP3A* isozymes [35]. While other CYP enzymes (e.g., *CYP2A6*, *CYP2J2*, *CYP4F2*, *CYP2E1*) or non-CYP phase I enzymes [40] or phase II enzymes (e.g., UDP-glucuronosyltransferases [UGTs]) are less frequently involved in clinically important DDIs, they should be considered for study when appropriate [35]. For example, if an investigational new drug is likely to be co-administered with a drug primarily metabolized by *CYP2A6*, in vitro screening for a DDI should be conducted.

The Pharmaceutical Research and Manufacturers of America (PhRMA) published recommendations for conduct of in vitro DDI studies [41] that are based on FDA guidance and a joint conference of regulatory and scientific agencies [37]. The recommended study types commonly use pooled human liver microsomes or cDNA-expressed human CYPs and may underestimate or overestimate effects that will occur in vivo [40]. Multiple factors influence the accuracy of predicted DDIs from in vitro studies. These include probe selection, determination of intrinsic clearance, choice of substrate and inhibitor concentration range, effect of organic solvents on enzyme activity, buffering of the system, and whether transcellular transporters are important in vivo [40, 42–44]. Underestimation can occur when hepatic drug concentrations are substantially higher than plasma concentrations after oral drug administration [45]. Because the free fraction of drug is generally responsible for in vivo DDIs, DDIs may be overestimated when in vivo plasma or hepatic protein binding is high. In addition, the contribution of an enzyme to the overall metabolic clearance in vivo may be either underestimated or overestimated if the metabolic pathway is partially saturated at in vivo concentrations such as those found during first-pass metabolism [46].

In vitro screening can be used to investigate whether inhibition, activation, or induction affects elimination through the DME systems. If screening assays find that an enzyme pathway does not metabolize an investigational drug, then clinical studies to evaluate CYP450 inhibitors or inducers are unlikely to be required by regulatory agencies. The FDA goes so far as to say that if no inhibition or induction is found in vitro, then no in vivo interaction studies are needed [47]. However, in vitro studies may not predict DDI in a number of circumstances. These situations include when induction or activation occurs or predominates in vivo, measurable plasma concentrations are incorrectly extrapolated to hypothetical concentrations at the site of metabolic activity, the relative contribution of CYP pathways is not accurately known, mechanism-based inactivation takes place, an improper enzyme inhibition model is chosen, interactions occur with transporters rather than DMEs, or CYP inhibitors also affect P-glycoprotein or other transporters [48, 49].

There are numerous examples of in vitro screenings that were not predictive of in vivo DDIs [7, 48]. Cocktail studies can also be used when in vitro assays may not be available or accurate (e.g., for herbal products or therapeutic proteins) [30]. Given the possible problems of bias and error with in vitro testing, and that not all factors affecting in vitro-in vivo correlations may be known, confirmation of presence or lack of in vivo DDIs may be desirable even when in vitro screening is negative. Also, although some methods are available, in vitro screening may be unable to adequately assess the potential for DDIs that occur through enzyme induction [41].

A full discussion of the proper design and application of in vitro studies is beyond the scope of this chapter but is discussed at length elsewhere [37, 41, 50, 51]. The PhRMA recommendations give specific study design guidelines to assist investigators in the conduct of in vitro studies [35, 36, 52–54].

8.3 Cocktail Study Methodology

8.3.1 Probes and Biomarkers

Probes, biomarkers, and specific cocktail combinations must each be validated. Choosing validated probes and biomarkers is essential for the acquisition of accurate and useful data. Specific recommendations for validation criteria have been published [2, 6, 55, 56]. Probes should be substrates that are specific for the elimination pathway of an individual CYP enzyme in in vitro studies [47]. If more than one metabolic pathway is involved in the metabolism of the probe, the second pathway should constitute <10% of the total clearance [55]. Although not required for validation, probes should be safe and commercially available worldwide [2]. Consistent use of validated probes allows for comparison between studies and across different populations.

Biomarkers are the metrics used to assess the metabolism of the probe drug. Biomarkers must be reproducible (i.e., have a low coefficient of variation for repeated tests). The biomarker should reflect known genetic polymorphisms and should not be dependent upon other factors unrelated to enzyme activity (e.g., urinary pH, urinary flow, renal function) [57–59]. During in vivo studies, biomarkers need to measure change from baseline to induction as well as from baseline to inhibition. They should also be able to assess enzyme activation. Sampling of the biomarker over time must be appropriate to quantitate both induction and inhibition, and this means that sampling strategies will usually differ by study phase.

Biomarkers that are direct metrics are preferred, e.g., total area under the plasma or serum concentration-time curve (AUC), total body clearance, and total AUC metabolic ratios [2, 34]. When AUC is used, the complete AUC (i.e., $AUC_{0-\infty}$) and not partial AUC (e.g., AUC_{0-last}) should be determined. Appropriate sampling duration is required in order to characterize at least 80–85% of the AUC with plasma versus time concentrations (<15–20% extrapolation) [60]. If metabolites are used as part of the biomarker, correlation of metabolite formation with the activity and content of the enzyme in subcellular fractions should have been shown [56, 61, 62]. Indirect metrics such as urinary or plasma metabolic ratios (metabolite/drug) or recovery ratios (drug + metabolite) have frequently been used, but are not recommended, and generally have not been validated [57, 63]. Simpler ratios and single-point measurements are usually not satisfactory parameters and can lead to errors in interpretation [57, 64–66]. This is particularly true when divergent primary metabolic pathways are mediated by different enzymes and lead to the formation of the same secondary metabolite [37]. Limited sampling strategies are published for many biomarkers but may introduce excessive variability and lack adequate accuracy [64, 66] when applied within the setting of cocktail studies. As listed above, many issues and limitations in published biomarkers exist. Thus, the investigator should be cognizant that just because a probe or biomarker has been used alone or as part of a cocktail does not make it validated or appropriate for use.

Table 8.2 Validated in vivo cytochrome P450 (CYP) probe substrates

CYP enzyme	Validated probe substrates
CYP1A2	Caffeine [68–70], plasma paraxanthine/caffeine AUC [68, 69], theophylline [70], tizanidine [71]
CYP2B6	Bupropion [72, 73]
CYP2C8	Rosiglitazone [74, 75], repaglinide [76]
CYP2C9	(S)-warfarin [77], tolbutamide [78]
CYP2C19	(S)-mephenytoin [79, 80], omeprazole [81–83], lansoprazole [84], [^{13}C] pantoprazole [85, 86]
CYP2D6	Debrisoquine [56], dextromethorphan [56, 87], desipramine [88], nebivolol [89]
CYP2E1	Chlorzoxazone [90, 91]
CYP3A	Midazolam (IV ± oral) [61, 62, 92], alfentanil (IV ± oral) [93–95], felodipine [96], triazolam [97, 98]

As of this writing, the following probes (noted with the enzymes that they measure) and biomarkers have been validated. Many have been used in validated cocktails and the findings published. As such, the following probes and biomarkers are appropriate for use in DDI studies. This is not an exhaustive list of validated probes or biomarkers. Recommendations of regulatory agencies may differ [35, 36, 52, 53] and are frequently made without referencing [35, 53, 54, 67]. Table 8.2 lists validated single probes by enzyme pathway. Table 8.3 lists validated cocktails that include at least four probes for CYP pathways of major importance.

CYP1A2 Caffeine is a validated probe with caffeine systemic clearance or the plasma paraxanthine-to-caffeine ratio $AUC_{0-\infty}$ as the biomarker [68, 69]. Although urinary metabolite ratios are frequently used as the biomarker [87], these are not optimal for the reasons examined elsewhere in the chapter. Chlorzoxazone inhibits in vivo caffeine metabolism [101] and therefore these two probes should not be used together. Theophylline is also a validated CYP1A2 probe [70]. Tizanidine is proposed as a sensitive probe and has been evaluated in comparison to caffeine [71, 102]. However, currently tizanidine is not part of a validated cocktail.

CYP2B6 The probe bupropion has been validated, and the (S,S)-hydroxybupropion/(S)-bupropion $AUC_{0-\infty\ ratio}$ is the validated biomarker [72, 73]. *CYP2B6* is considered an enzyme importance by the FDA [103], but currently no CYP2B6 probe is part of a validated cocktail.

CYP2C8 Currently, no validated CYP2C8 probes are part of a validated cocktail. Rosiglitazone is a selective substrate and valid probe for *CYP2C8* [74, 75], but the risk of hypoglycemia is a potential safety issue [104]. Repaglinide is a selective and sensitive substrate of *CYP2C8* but is also metabolized by *CYP3A* [76] and is a substrate for *OATP1B1* [105]. Amodiaquine (*N*-deethylation) and repaglinide are recommended for use by the EMA, with the caveat that these substrates are not validated probes, but they may be used as alternatives [67]. Pioglitazone has been recommended as a probe but is also metabolized by *CYP3A* [104, 106].

Table 8.3 Validated multidrug cocktails with at least four validated probe substrates for CYP enzymes of major importance and their preferred biomarkers

		Cooperstown cocktail 5+1	Sanofi-Aventis	Inje cocktail
Reference		Chainuvati et al [99]	Turpault et al [100]	Ryu et al [31]
CYP1A2	Probe	Oral caffeine 2 mg/kg	Oral caffeine 100 mg	Oral caffeine 93 mg
	Biomarker	Plasma paraxanthine:caffeine AUC$_{0-12\,hr}$	Plasma paraxanthine:caffeine AUC$_{0-24\,hr}$	Plasma paraxanthine:caffeine AUC$_{0-12hr}$
CYP2C9	Probe	Oral warfarin 10 mg	Oral warfarin 10 mg	Oral losartan 50 mg
	Biomarker	Plasma (S)-warfarin AUC$_{0-\infty}$	Plasma (S)-warfarin AUC$_{0-\infty}$	Plasma AUC$_{losartan}$:AUC$_{E-3174}$ or urinary 8-h AUC$_{losartan}$:AUC$_{E-3174}$
CYP2C19	Probe	Oral omeprazole 40 mg	Oral omeprazole 20 mg	Oral omeprazole 20 mg
	Biomarker	5OH-omeprazole:omeprazole AUC$_{0-10\,hr}$	Plasma omeprazole AUC$_{0-\infty}$	Plasma omeprazole:5OH-omeprazole AUC$_{0-12hr}$
CYP2D6	Probe	Oral dextromethorphan 30 mg	Oral metoprolol 100 mg	Oral dextromethorphan 30 mg
	Biomarker	Plasma dextromethorphan:dextrophan AUC$_{0-\infty}$ ratio or apparent dextromethorphan CL	Metoprolol is not a valid probe; therefore, no biomarker is recommended	Plasma dextromethorphan:dextrophan AUC$_{0-\infty}$ ratio
CYP3A	Probe	IV midazolam 0.025 mg/kg	Oral midazolam 0.03 mg/kg	Oral midazolam 2 mg
	Biomarker	plasma midazolam AUC$_{0-\infty}$	Plasma midazolam AUC$_{0-\infty}$	Plasma midazolam AUC$_{0-\infty}$
		plasma 1'OH midazolam:midazolam AUC$_{0-\infty}$	Plasma 1'OH midazolam:midazolam AUC$_{0-\infty}$	Plasma 1'OH midazolam:midazolam AUC$_{0-\infty}$

CYP2C9 (S)-warfarin is a validated CYP2C9 probe with plasma (S)-warfarin AUC$_{0-\infty}$ as the biomarker [77]. Low-dose (125 mg) tolbutamide is also a validated CYP2C9 probe with oral tolbutamide plasma clearance [81] as the biomarker [78]. Unfortunately, tolbutamide use in a cocktail has only been validated with caffeine and dextromethorphan [107], and this limits its usefulness in studies that also wish to evaluate *CYP2C19* and *CYP3A* isozymes. Losartan is used as a CYP2C9 probe in some cocktails but does not adequately distinguish between common CYP2C9 genotypes when evaluated as plasma AUC$_{losartan}$/AUC$_{E-3174}$ [108] or urinary 8-h AUC$_{losartan}$/AUC$_{E-3174}$ [109–111]. While (S)-flurbiprofen has been validated as a cocktail component, it does not correlate with other validated CYP2C9 probes (i.e., (S)-warfarin, tolbutamide) and exhibits greater variability in inhibition [77]. While flurbiprofen AUC0-∞ differs by CYP2C9 genotype and has been evaluated in vivo at baseline and after inhibition with fluconazole, it has not been fully validated [112]. For these reasons, Kumar et al. have suggested (S)-warfarin AUC$_{0-\infty}$ as the preferred CYP2C9 biomarker [77]. The urinary ratio of 4'-hydroxy-diclofenac/diclofenac has been proposed as a CYP2C9 biomarker, but correlation with CYP2C9 phenotypes is inconsistent [111, 113, 114]. Further research demonstrated that plasma 4'-hydroxy-diclofenac/diclofenac is unchanged by CYP2C9 genotype [115], and therefore this biomarker should not be used.

CYP2C19 Omeprazole is the most commonly used, validated CYP2C19 probe [81–83] with the 5-hydroxyomeprazole/omeprazole $AUC_{0\text{-}10 \text{ hr}}$ as the preferred biomarker [31, 69]. Many studies use a single 2- or 3-hr. metabolic ratio as the biomarker, but this is suboptimal as the omeprazole Cmax can vary markedly [81, 116]. Esomeprazole, the S-isomer of omeprazole, should also be valid as a CYP2C19 probe. Lansoprazole is also a validated probe [84] but is not included in a validated cocktail. (S)-mephenytoin has been proposed as a CYP2C19 probe [79, 80], but there are issues with the stability and duration of urine collection [117] as well as safety concerns [87]. In addition, mephenytoin is generally not commercially available. For these reasons, mephenytoin is not recommended. $[^{13}C]$pantoprazole shows promise as a CYP2C19 probe although it has not been validated as a cocktail component [85, 86].

CYP2D6 Oral dextromethorphan is the preferred CYP2D6 probe [56], with the use of plasma dextromethorphan/dextrorphan AUC $ratio_{0\text{-}\infty}$ or plasma dextromethorphan oral clearance as biomarkers [57]. While the 12-hr. urinary dextromethorphan/dextrorphan ratio has been validated, it should be viewed as inferior to plasma measurements because of issues related to urine specimen collection and handling, including pH considerations [56]. Debrisoquine is a validated CYP2D6 probe, and if used, the 12-hr. urinary debrisoquine/4-hydroxydebrisoquine metabolic ratio is the validated biomarker [56, 87]. However, debrisoquine is of limited usefulness because it is not available in North America or Asia. Desipramine is a validated probe [88] but has safety issues and is not part of a validated cocktail. Metoprolol is used in some validated cocktails. However, metoprolol cannot be considered a validated CYP2D6 probe because it does not correlate with the metabolic ratios of other validated probes (i.e., debrisoquine, sparteine, dextromethorphan) in non-Caucasian populations [56]. There are few situations where one would be interested in DDI data that are only applicable to Caucasian populations. Nebivolol is a validated CYP2D6 probe, but is not part of a validated cocktail [89]. A review of CYP2D6 probes has been published by Frank et al. [56].

CYP3A isozymes Midazolam is the validated, gold-standard CYP3A isozyme probe [47, 61, 62], although some researchers believe that more than one probe is needed when assessing CYP3A activity [118]. Validated CYP3A isozymes biomarkers include midazolam $AUC_{0\text{-}\infty}$ and plasma 1-hydroxymidazolam/midazolam $AUC_{0\text{-}\infty}$ ratio [61]. Single-point ratios of 1-hydroxymidazolam/midazolam have been used, but these are demonstrated to be invalid biomarkers [64, 119]. Urinary 1-hydroxymidazolam/midazolam ratios are not useful because they do not accurately reflect baseline CYP3A activity [63]. Although simvastatin is listed as a recommended CYP3A isozyme probe in the most recent FDA guidance [35], simvastatin does not correlate with CYP3A activity during inhibition or induction and therefore should not be used as a CYP3A probe [120]. Both oral and intravenous alfentanil are validated CYP3A isozyme probes with plasma alfentanil $AUC_{0\text{-}\infty}$

as the biomarker [93, 94]. Additionally, pupillometry can be used as a surrogate for alfentanil effect, obviating the need for blood sampling [121]. However, alfentanil is not part of a validated cocktail. Because quinine has not been validated, is a P-glycoprotein substrate, and inhibits *CYP2D6*, it should not be used in cocktail studies [122, 123]. Triazolam is a validated CYP3A probe but is not part of a validated cocktail [97, 98]. Felodipine is proposed as a CYP3A probe [96], but correlation with other CYP3A probes has not been done [124]. At one time, dapsone was used as a CYP3A probe, but subsequent research showed it to be suboptimal. Dapsone is metabolized by CYP3A isozymes, CYP2C9, and CYP2E1 [125, 126], lacks correlation with other CYP3A isozyme probes, and fails to accurately assess CYP3A inhibition or induction [2, 49, 127–129]. The erythromycin breath test is not specific for CYP3A isozymes (and is also a P-glycoprotein substrate) and should not be used [130]. Some authors have suggested other drugs (e.g., buspirone [124, 131] and sildenafil [47, 124]) as CYP3A probes. While these may be appropriate for in vitro assessment of CYP3A activity [132, 133], data that support use as in vivo probes are currently lacking. Additionally, the endogenous compounds ß-hydroxy-cortisol and ß-hydroxycholesterol should not be used as they have been shown to lack correlation with midazolam biomarkers [134, 135], or they have been incorrectly validated by comparison to midazolam [135–137].

Finding a selective CYP3A5 substrate has been challenging. Because CYP3A5 and CYP3A4 are structurally similar, the specificities of substrates and inhibition are also very similar [138], as are the determinants of constitutive expression [139]. No validated, specific CYP3A5 probe is currently available. Also, CYP3A5 is considered an enzyme of emerging importance rather than a major enzyme [140].

CYP2E1 Chlorzoxazone is the preferred probe for CYP2E1 and has been validated at the 250 mg dose [90, 91]. The corresponding biomarker is the plasma 6-hydroxy-chlorzoxazone/chlorzoxazone $AUC_{0-\infty}$ ratio [141, 142] or apparent chlorzoxazone clearance [90]. Chlorzoxazone inhibits CYP3A isozymes, and an interaction has been demonstrated when chlorzoxazone is dosed with oral midazolam [2, 143]. Therefore, these two probes should not be co-administered during a cocktail study, and CYP3A isozymes cannot be assessed while evaluating CYP2E1 with chlorzoxazone. As CYP2E1 is considered an enzyme of limited importance, exclusion of this enzyme from cocktail studies is unlikely to be a problem.

Miscellaneous One validated CYP probe can be used to measure the activity of a phase II enzyme, N-acetyltransferase (NAT2). The presence of NAT2 genotype variants has been evaluated with caffeine [144, 145] or dapsone [146], although the two probes were not highly correlated in an acutely ill population [147]. Usually NAT2 activity is evaluated during a cocktail study when urinary caffeine metabolite ratios are collected because caffeine or dapsone has been administered to evaluate a CYP enzyme rather than primarily to determine acetylator status. Little is known about how changes in NAT2 activity relate to changes in the measured biomarkers. For this reason, assessment of the biomarkers is primarily used to evaluate NAT2 phenotype, not the potential for DDIs via NAT2.

8.3.2   *Validated Cocktails*

Once validated probes and biomarkers have been identified, it is important to assure that the cocktail combination of probes has also been validated. First, one should evaluate whether the probes used in the cocktail study are validated. Second, one should be sure that a validation of the combination of probes in the cocktail has been published. There must be clear evidence that there are no clinical or metabolic interactions among the probe drugs when used concurrently [34].

There are a number of reasons that published cocktails may not be appropriate for use. The use of validated probes and biomarkers without appropriate evidence of a lack of interaction between the probes is insufficient. The use of validated cocktails, but choosing biomarkers that are invalid or have yet to be validated, can also lead to erroneous results. Therefore, one must evaluate a cocktail for evidence that each probe or biomarker is valid and that the combination of probes has been validated.

Unfortunately, many unvalidated cocktails are in use, and cocktail studies are frequently published that use unvalidated or invalid probes, biomarkers, and/or cocktails. For example, a cocktail may include one or more component probes that are invalid, or have been shown not to be valid (e.g., the 6β-hydroxycortisol/cortisol molar ratio [134, 148, 149], the 4-h (single-point) 1-hydroxymidazolam/midazolam ratio for measuring CYP3A isozymes activity [64]), or do not distinguish between all genotypes (e.g., losartan as a probe to evaluate CYP2C9 activity [31]). Another possibility is that the component probe has been validated, but the chosen biomarker has not [2]. For example, while midazolam is a validated probe for CYP3A isozyme activity, there are numerous midazolam biomarkers that are used but not validated. Midazolam clearance and $AUC_{0-\infty}$ are validated biomarkers [55, 61, 62]; neither single-point midazolam concentrations [119, 150] nor the 1-hydroxymidazolam/midazolam single-point ratio accurately measures CYP3A isozyme activity [64]. In other words, midazolam is a validated probe, midazolam clearance and $AUC_{0-\infty}$ are validated biomarkers, but single-point concentrations or single-point metabolic ratios are not validated biomarkers and should not be used. Substitution of validated biomarkers can be done if the individual probe and its use as part of a cocktail are validated.

Another problem is that the cocktail components may have been individually validated, but the concurrent administration of the probes has not [151, 152]. Finally, some in vitro probes and cocktails are suggested for in vivo use without supporting in vivo data [153]. Thus, it is essential that the investigator be sure that the probe drugs, biomarkers, and cocktail combination have each been validated in order to assure accurate study results.

8.4   Application of Cocktail Study Methodology

During a study, all DME polymorphisms that may be relevant (based on preclinical data) should be evaluated. When there is evidence that 30% or more of an investigational drug is cleared through CYP-mediated metabolism, the cocktail study should be designed to include CYP enzymes of major metabolic importance (i.e., *CYP1A2*, *CYP2C9*, *CYP2C19*, *CYP2D6*, and *CYP3A4*) [35, 36]. Other CYP enzymes that are considered to be of importance (e.g., *CYP2B6*, *CYP3A5*) should be included if in vitro assays suggest they play a role in metabolism of the investigational drug of interest [41]. *CYP1A1*, *CYP1B1*, *CYP2A6*, *CYP2E1*, *CYP2J2*, *CYP4A11*, and *CYP4F2* are considered to be of low importance and typically do not need to be investigated in cocktail studies.

Some validated cocktails lack the ability to evaluate an important and relevant DME pathway. For example, the validated Cooperstown cocktail did not include a CYP2C9 probe [154] although *CYP2C9* is responsible for metabolism of approximately 20% of marketed drugs [155]. This problem was overcome by addition of a CYP2C9 probe (warfarin) and validation of a new combination, the Cooperstown 5 + 1 cocktail [99]. The Cooperstown 5 + 1 cocktail evaluates all of the major DME pathways although CYP3A isozyme assessment is limited to hepatic activity. Oral midazolam has been used to assess intestinal plus hepatic CYP3A activity, but has not been validated as part of the Cooperstown 5 + 1 cocktail. However, there is little reason to believe that rapidly absorbed oral midazolam could not be used in this cocktail. Published studies have used oral midazolam as part of this cocktail [156]. The six-drug Pittsburgh cocktail is designed to assess *CYP1A2*, *CYP2C9*, *CYP2C19*, *CYP2D6*, and *CYP2E1* [26, 128]. However, the CYP2C19 probe (mephenytoin) is not valid, *CYP2E1* is a DME of low importance (and therefore is seldom of interest), and this cocktail does not contain a valid probe drug for CYP3A. Since CYP3A is responsible for approximately 50% of drug metabolism via CYP enzymes [49], the utility of the Pittsburgh cocktail is severely limited. Although the Pittsburgh cocktail is a six-drug cocktail, it only contains validated probes for three CYPs of major importance. While addition of midazolam to this cocktail would probe CYP3A isozymes [143], doing so without a complete validation of the cocktail would be scientifically unsound.

Published, validated cocktails that contain at least four validated probes for major CYPs are shown in Table 8.3. These include the Cooperstown 5 + 1 drug cocktail [99], the Sanofi-Aventis cocktail [100], and the Inje cocktail [31]. The reader should be aware that there are published investigations that use these probe cocktails but fail to evaluate validated biomarkers. It is also important to note that most investigators cannot acquire the entire set of probe drugs needed for the Pittsburgh cocktail. Currently there are no cocktails that combine validated CYP probes with a validated P-glycoprotein probe, and in fact, no validated P-glycoprotein probes exist [6].

8.5 Subject Selection

A decision should be made during study design as to whether inclusion criteria will specify extensive metabolizers (EMs) identified by a priori genotyping or prior phenotyping or will allow other metabolizer genotypes. One option is to genotype subjects and use genotype during the screening period to determine eligibility. Another is to use genotype to stratify during the analysis phase. When genotyping is done prior to the study, the additional costs of post hoc pharmacogenetic analyses may be avoided. Knowledge of subject phenotype may be required to appropriately interpret findings [157–159].

Using pharmacogenomic inclusion/exclusion criteria can increase statistical power by reducing the variability introduced by inclusion of a range of polymorphisms. By reducing intersubject variability, the sample size is reduced. Using subjects with EM genotypes to evaluate DDI potential is most common because such individuals have a greater quantity of enzyme and therefore are at greater risk for a DDI [160]. Individuals with PM genotypes have little or no enzyme production and therefore are unlikely to experience metabolic DDIs, and studies in PMs may result in underestimation of DDIs if results are not stratified by phenotype [160, 161]. Exclusion of poor metabolizers (PMs) may also increase trial safety by removing the risks associated with excessive drug exposure and limiting or avoiding the need for intensive monitoring during study drug administration. For these reasons, exclusion of subjects who are PMs makes a study more efficient. Likewise, inclusion of ultra-rapid metabolizer phenotypes (UMs) may result in overestimation of DDIs and can result in markedly different findings than if only EMs are studied [162]. Restricting a study to subjects with an EM phenotype will result in findings that are applicable to the greatest number of individuals, but will also decrease generalizability of the study.

Alternatively, cocktail studies can be specifically designed to evaluate DDI by enzyme genetic polymorphism. Some researchers include EMs and IMs (if phenotyping or genotyping has been done prior to study start). If a drug is metabolized by a polymorphic enzyme, then enrollment of adequate numbers of subjects who are PMs and EMs can allow comparison of pharmacokinetic parameters and thereby indicate the extent of the DDI that is expected with strong enzyme inhibition. In such a situation, additional interaction studies with such inhibitors would be unnecessary [47]. Enrollment of EM genotypes is encouraged when studying polymorphic DMEs. Although the focus of their statement is on pharmacogenomic studies, the Industry Pharmacogenomics Working Group (http://i-pwg.org) has endorsed the use of homogeneous populations when possible [159].

When genotype is not used to determine study eligibility, it is essential that the methods and quality of evaluating both genotype and phenotype be included in the protocol because lack of accurate phenotyping or incorrect genotype can result in spurious findings [159]. When multiple genotypes and phenotypes are included, the results should be presented by phenotype subgroup. Subgroup data presentation provides the maximal information for understanding DDI potential.

A cocktail protocol must also control environmental factors that may result in inhibition or induction of DMEs. Food-drug interactions [16, 17], cigarette smoking [14], or alcohol consumption [15] should be avoided when possible, or at a minimum, assessed and recorded [19].

8.6   Drug Dosage and DME Evaluations

The investigational drug dose and duration should be sufficient to estimate maximum induction or inhibition at clinically relevant dosages. Therefore, the investigational drug should be dosed at the highest dose likely to be employed in clinical use. The drug(s) used to inhibit or induce the enzyme pathways should also be dosed at the highest clinical dose and the drug(s) shortest dosing interval [47]. One must also consider the duration of activation or induction when determining the appropriate interval between DME evaluations [163]. Dosing the investigational, inhibitory, and induction drugs in this manner will maximize the chance of identifying an interaction.

Exposure measures (e.g., total AUC, maximum concentration [Cmax], time to Cmax) and pharmacokinetic parameters (e.g., clearance, volume of distribution) should be measured in every study. Additional measures such as pharmacodynamic parameters should be considered when appropriate. When the objective of the study is to quantify the effects on different enzymes, the complete AUC or pharmacokinetic parameter for the biomarker (not metabolic ratios) is the preferred metric [34, 57]. Simpler ratios such as metabolite-to-parent drug ratios in urine may have more confounding factors, and the magnitude of an effect may be difficult to translate into inhibition potency, induction potency, and treatment recommendations. If a study assesses single parent-to-metabolite ratios (rather than a complete AUC), further in vivo evaluation may be required to provide quantitative data on changes in exposure.

8.7   Sampling, Assays, and Sample Analyses

If in vitro data indicate CYP inhibition, induction, or activation, there should be appropriate adjustment of the specimen sampling strategy. The frequency of sampling must allow accurate determination of the relevant measures and parameters for the parent drug and the active metabolites. Baseline sampling should be performed on the same schedule as during the cocktail validation study. Further modifications to the sampling scheme can be based on baseline DME activity and genotype, the expectation of inhibition or induction, and the substrate specificity for the enzyme system.

There may be important issues related to handling specimens prior to assay. Appropriate and consistent storage of blood and urine samples during collection is essential. When metabolic ratios are dependent upon renal clearance and a drug is lipid soluble, then diurnal variation in urinary pH has the potential to affect intraindividual variability in urinary ratios (for the parent drug) and plasma ratios (for the metabolite) [58]. As such, control of the duration of specimen collections should be standardized [37]. One must also know that the timing of the specimen collection is adequate to identify changes related to either inhibition or induction.

Stability of the probe in urine or plasma is required (i.e., the biomarker should not change over time, either prior to assay or during specimen storage). Urinary pH can influence detectable metabolic ratios and lead to a marked increase in variability [58]. In some circumstances, it is necessary to stabilize the urine during collection [57, 59]. For example, when collecting urine for dextromethorphan/dextrorphan, 3 grams of ascorbic acid per 2-liter collection bottle is added to acidify the urine and standardize pH. Urine samples for dextromethorphan and its metabolites should be DE conjugated with β-glucuronidase before measurement in order to include unconjugated dextromethorphan and the 3-hydroxy methorphinan metabolite. Failure to deconjugate the urine may lead to incorrect measurement of metabolites [56].

Analytical interference should not be caused by the probe, investigational drug, or metabolites. The assay must be sensitive enough to allow determination of drugs and metabolites in the collected samples. In general LC-MS-MS is recommended as an analytical instrument due to its precision, specificity, and ability to quantitate very low concentrations of substances in body fluids. UPLC can be employed in simultaneous assay of multiple probe drugs and metabolites [164]. Deuterated drug is encouraged as the internal standard. Documentation of a lack of analytical interference between the cocktail drugs, their metabolites, and any internal standards is also important [165].

8.8 Statistical Considerations

Consideration of the desired study power, inter- and intraindividual variability in enzyme activity, and definition of a clinically important mean group difference in the measured biomarkers are all important aspects of study design. Each will influence the sample size calculations. Having an adequate number of subjects is essential, and lack of attention to sample size may result in an underpowered study [160]. Information on intraindividual variability for many CYP biomarkers can be found in the review by Zhou et al. [4] as well as the original research publications.

Sample size should be calculated for both the CYP enzyme of greatest interest based on in vitro findings and the biomarker with the greatest intraindividual variability. Calculating sample size from each of these and then using the larger sample size will provide adequate power for all of the CYPs under study. Routine use of the FDA-recommended minimal sample size of 12 [166] can lead to a study with inadequate power.

Correct statistical evaluation begins with log transformation of the data. The rationale for this is that most pharmacokinetic metrics are not normally distributed but are right skewed [34, 47]. Log transformation tends to normalize or "correct" distribution of the data. When data are normally distributed, measures of variance (e.g., confidence intervals, standard deviations, interquartile ranges) are symmetrical. In many studies, the sample size is too small to adequately evaluate for data distribution, and hence, log transformation is recommended regardless of the apparent distribution of the raw data [166].

Regulatory agencies agree that DDI studies should be analyzed using bioequivalence criteria rather than statistical testing (i.e., significance testing) [37]. Results for the biomarker metric (e.g., total AUC or Cmax) should be reported as the 90% confidence intervals (CI) around the geometric mean ratio of the biomarker measurements before and after treatment. The ratio is constructed from either the enzyme activity during investigational drug administration (treatment) to basal enzyme activity (baseline) or the enzyme activity during investigational drug plus inhibitor/inducer (treatment) to enzyme activity during investigational drug treatment alone (baseline). Confidence intervals provide an estimate of the distribution of the observed systemic exposure of treatment versus the control state and convey a probability of the magnitude of the interaction [47].

As a general rule, to meet bioequivalence criteria, the 90% CI should be within the conventional limits of 0.8–1.25 for AUC and 0.7–1.43 for Cmax [47]. However, it is recommended that these limits be flexible and dependent upon the pharmacodynamics of the investigational drug or other clinical or safety considerations [37]. If the investigator plans to report 90% CI but specify limits other than the conventional limits noted above, these should be stated prior to study conduct. If a study is intended for submission to a regulatory agency, that agency should agree to any change in the confidence interval limits before the protocol design is finalized. Significance testing (e.g., parametric tests such as Student's t-test or nonparametric tests such as the Wilcoxon signed-rank test) rather than bioequivalence testing is not appropriate because small, consistent systemic exposure differences can be statistically significant ($p < 0.05$) but not clinically relevant [34, 37, 47]. Unfortunately, not all cocktail validation studies have been analyzed with the appropriate statistical methodology.

Data presentation should include both interindividual variability and intraindividual variability (by metabolizer phenotype if appropriate). Reporting mean data with standard deviations is inadequate. For interactions in which an increase in variability is of concern (e.g., narrow therapeutic index drugs), the focus of the statistical analysis should be on measures of variability [47] rather than measures of central tendency such as the mean or median. This is because the measures of variability assist in prediction of the range of the DDIs anticipated to occur in the clinical setting. Mean or median change in enzyme activity is less useful from a clinical perspective.

8.9 Application of Cocktail Studies and Conclusions

Cocktail studies can assess the potential for DDIs and therefore assist the pharmaceutical industry with go/no-go decisions. They also allow assessment of the need for additional, specific DDI studies. Because cocktail studies assess the potential extent of DDIs, qualitative recommendations for drug dosing and use may be made. Evaluation of variability in the extent of DDIs can result in useful clinical information. For example, the presence of large interindividual variability in clearance may translate into large interindividual differences in the extent of DDIs. The importance and implications of enzyme polymorphism for different genotypes and the implications for product labeling can also be evaluated.

Both the FDA [35] and the EMA [67] endorse the use of cocktail studies as part of a systematic, comprehensive, and mechanistic approach to DDI as part of drug development. They also recommend cocktail studies to evaluate for DDIs when such studies are conducted in an adequate number of subjects and use validated biomarkers and cocktails. The FDA recommends that metabolic DDIs be explored for investigational compounds, including those that are not significantly eliminated by metabolism [67]. The FDA then works with the sponsor to determine whether further DDI studies are needed after studies with in vitro probes and early in vivo studies have been completed [67]. Specific suggestions about preferred probe substrates and study designs are provided by the FDA [53] although not all of the probe substrate recommendations are supported by review of the literature or validation studies, and many of the probes are not part of a validated cocktail.

The EMA recommends that cocktail studies use safe, validated probes and provide specific criteria that should be present in the probe drugs [67]. In addition, the EMA specifies that validated cocktails should be used, and it prefers cocktails that are supported by published validation data.

There is little published guidance available in English from Japan's Ministry of Health, Labour and Welfare (MHLW). What is available is consistent with recommendations provided by the FDA and EMA [167].

In order to predict DDIs in the clinical setting and make clinical adjustments to dosing, it is necessary to have information on substrate specificity, the extent of inhibition or induction, interindividual variability of the CYP enzyme, and whether inhibition/induction is affected by the disease state in which the drug is used. This information is often difficult to acquire from clinical studies. For this reason, conducting cocktail studies in patient populations may provide valuable data. There is a need for investigation of variability of inhibition within metabolizer phenotypes for mild-moderate inhibition and narrow therapeutic index drugs as well as investigation of variability of inhibition within enzymes such as *CYP3A4* where there are no polymorphisms but up to sevenfold interindividual variability in enzyme activity [49, 61, 168–170].

We hope that the future development of cocktails will include validation of cocktails that contain safe and validated probe drugs that are readily available worldwide and validated biomarkers that can be collected efficiently and assayed easily and concurrently.

References

1. In vivo drug metabolism/drug interaction studies – study design, data analysis, and recommendations for dosing and labeling. Food and Drug Administration, Rockville, 1999
2. Fuhr U, Jetter A, Kirchheiner J (2007) Appropriate phenotyping procedures for drug metabolizing enzymes and transporters in humans and their simultaneous use in the "cocktail" approach. Clin Pharmacol Ther 81(2):270–283

3. Zhou H, Sharma A (2016) Therapeutic protein-drug interactions: plausible mechanisms and assessment strategies. Expert Opin Drug Metab Toxicol 12(11):1–9
4. Zhou H, Tong Z, McLeod JF (2004) "Cocktail" approaches and strategies in drug development: valuable tool or flawed science? J Clin Pharmacol 44(2):120–134
5. Ebner T, Ishiguro N, Taub ME (2015) The use of transporter probe drug cocktails for the assessment of transporter-based drug-drug interactions in a clinical setting-proposal of a four component transporter cocktail. J Pharm Sci 104(9):3220–3228
6. Ma JD, Tsunoda SM, Bertino JS Jr, Trivedi M, Beale KK, Nafziger AN (2010) Evaluation of in vivo p-glycoprotein phenotyping probes: a need for validation. Clin Pharmacokinet 49(4):223–237
7. Stopfer P, Giessmann T, Hohl K, Sharma A, Ishiguro N, Taub ME et al (2016) Pharmacokinetic evaluation of a drug transporter cocktail consisting of digoxin, furosemide, metformin, and rosuvastatin. Clin Pharmacol Ther 100(3):259–267
8. Baker SD, van Schaik RH, Rivory LP, Ten Tije AJ, Dinh K, Graveland WJ et al (2004) Factors affecting cytochrome P-450 3A activity in cancer patients. Clin Cancer Res 10(24):8341–8350
9. Frye RF, Zgheib NK, Matzke GR, Chaves-Gnecco D, Rabinovitz M, Shaikh OS et al (2006) Liver disease selectively modulates cytochrome P450--mediated metabolism. Clin Pharmacol Ther 80(3):235–245
10. Frye RF, Schneider VM, Frye CS, Feldman AM (2002) Plasma levels of TNF-alpha and IL-6 are inversely related to cytochrome P450-dependent drug metabolism in patients with congestive heart failure. J Card Fail 8(5):315–319
11. McConn DJ, Lin YS, Mathisen TL, Blough DK, Xu Y, Hashizume T et al (2009) Reduced duodenal cytochrome P450 3A protein expression and catalytic activity in patients with cirrhosis. Clin Pharmacol Ther 85(4):387–393
12. Lee CM, Pohl J, Morgan ET (2009) Dual mechanisms of CYP3A protein regulation by proinflammatory cytokine stimulation in primary hepatocyte cultures. Drug Metab Dispos 37(4):865–872
13. Jones AE, Brown KC, Werner RE, Gotzkowsky K, Gaedigk A, Blake M et al (2010) Variability in drug metabolizing enzyme activity in HIV-infected patients. Eur J Clin Pharmacol 66:475
14. Hukkanen J, Vaisanen T, Lassila A, Piipari R, Anttila S, Pelkonen O et al (2003) Regulation of CYP3A5 by glucocorticoids and cigarette smoke in human lung-derived cells. J Pharmacol Exp Ther 304(2):745–752
15. He P, Court MH, Greenblatt DJ, Von Moltke LL (2005) Genotype-phenotype associations of cytochrome P450 3A4 and 3A5 polymorphism with midazolam clearance in vivo. Clin Pharmacol Ther 77(5):373–387
16. Saari TI, Laine K, Neuvonen M, Neuvonen PJ, Olkkola KT (2008) Effect of voriconazole and fluconazole on the pharmacokinetics of intravenous fentanyl. Eur J Clin Pharmacol 64(1):25–30
17. Culm-Merdek KE, von Moltke LL, Gan L, Horan KA, Reynolds R, Harmatz JS et al (2006) Effect of extended exposure to grapefruit juice on cytochrome P450 3A activity in humans: comparison with ritonavir. Clin Pharmacol Ther 79(3):243–254
18. Ngo N, Yan Z, Graf TN, Carrizosa DR, Kashuba AD, Dees EC et al (2009) Identification of a cranberry juice product that inhibits enteric CYP3A-mediated first-pass metabolism in humans. Drug Metab Dispos 37(3):514–522
19. Zhou SF (2008) Drugs behave as substrates, inhibitors and inducers of human cytochrome P450 3A4. Curr Drug Metab 9(4):310–322
20. Lammers LA, Achterbergh R, de Vries EM, van Nierop FS, Klumpen HJ, Soeters MR et al (2015) Short-term fasting alters cytochrome P450-mediated drug metabolism in humans. Drug Metab Dispos 43(6):819–828
21. Blake MJ, Gaedigk A, Pearce RE, Bomgaars LR, Christensen ML, Stowe C et al (2007) Ontogeny of dextromethorphan O- and N-demethylation in the first year of life. Clin Pharmacol Ther 81(4):510–516

22. Tracy TS, Venkataramanan R, Glover DD, Caritis SN (2005) Temporal changes in drug metabolism (CYP1A2, CYP2D6 and CYP3A activity) during pregnancy. Am J Obstet Gynecol 192(2):633–639
23. Nolin TD, Appiah K, Kendrick SA, Le P, McMonagle E, Himmelfarb J (2006) Hemodialysis acutely improves hepatic CYP3A4 metabolic activity. J Am Soc Nephrol 17(9):2363–2367
24. Morgan ET, Goralski KB, Piquette-Miller M, Renton KW, Robertson GR, Chaluvadi MR et al (2008) Regulation of drug-metabolizing enzymes and transporters in infection, inflammation, and cancer. Drug Metab Dispos 36(2):205–216
25. Bosilkovska M, Samer CF, Deglon J, Rebsamen M, Staub C, Dayer P et al (2014) Geneva cocktail for cytochrome p450 and P-glycoprotein activity assessment using dried blood spots. Clin Pharmacol Ther 96(3):349–359
26. Zgheib NK, Frye RF, Tracy TS, Romkes M, Branch RA (2006) Validation of incorporating flurbiprofen into the Pittsburgh cocktail. Clin Pharmacol Ther 80(3):257–263
27. Ozdemir V, Kalowa W, Tang BK, Paterson AD, Walker SE, Endrenyi L et al (2000) Evaluation of the genetic component of variability in CYP3A4 activity: a repeated drug administration method. Pharmacogenetics 10(5):373–388
28. Schellens JH, van der Wart JH, Brugman M, Breimer DD (1989) Influence of enzyme induction and inhibition on the oxidation of nifedipine, sparteine, mephenytoin and antipyrine in humans as assessed by a "cocktail" study design. J Pharmacol Exp Ther 249(2):638–645
29. Lesko LJ, Lagishetty CV (2016) Are we getting the best return on investment from clinical drug-drug interaction studies? J Clin Pharmacol 56(5):555–558
30. Zhang L (2010) A regulatory perspective on the utilization of cocktail approach in assessing drug interactions. FIP Pharmaceutical Sciences World Congress – American Association of Pharmaceutical Scientists Annual meeting, New Orleans, 16 November 2010
31. Ryu JY, Song IS, Sunwoo YE, Shon JH, Liu KH, Cha IJ et al (2007) Development of the "Inje cocktail" for high-throughput evaluation of five human cytochrome P450 isoforms in vivo. Clin Pharmacol Ther 82(5):531–540
32. Rezk NL, Brown KC, Kashuba AD (2006) A simple and sensitive bioanalytical assay for simultaneous determination of omeprazole and its three major metabolites in human blood plasma using RP-HPLC after a simple liquid-liquid extraction procedure. J Chromatogr B Analyt Technol Biomed Life Sci 844(2):314–321
33. Liu Y, Jiao J, Zhang C, Lou J (2009) A simplified method to determine five cytochrome p450 probe drugs by HPLC in a single run. Biol Pharm Bull 32(4):717–720
34. CHMP Efficacy Working Party Therapeutic Subgroup on Pharmacokinetics. London: European Medicines Agency; 2009 22 January. Contract No.: EMEA/618604/2008
35. Drug interaction studies —study design, data analysis, implications for dosing, and labeling recommendations In: Pharmacology C (ed) US Food and Drug Administration, Silver Spring, MD, 2012, pp 1–79
36. Bjornsson TD, Callaghan JT, Einolf HJ, Fischer V, Gan L, Grimm S et al (2003) The conduct of in vitro and in vivo drug-drug interaction studies: a pharmaceutical research and manufacturers of America (PhRMA) perspective. Drug Metab Dispos 31(7):815–832
37. Tucker GT, Houston JB, Huang SM (2001) Optimizing drug development: strategies to assess drug metabolism/transporter interaction potential – toward a consensus. Pharm Res 18(8):1071–1080
38. Huang SM, Temple R, Throckmorton DC, Lesko LJ (2007) Drug interaction studies: study design, data analysis, and implications for dosing and labeling. Clin Pharmacol Ther 81(2):298–304
39. Bertino JS Jr, Nafziger AN (2007) Labeling of drug interactions. Is change needed? Clin Pharmacol Ther 81(Suppl):S90
40. Liu X, Jia L (2007) The conduct of drug metabolism studies considered good practice (I): analytical systems and in vivo studies. Curr Drug Metab 8(8):815–821
41. Bjornsson T, Callaghan J, Einolf H, Fischer V, Gan L, Grimm S et al (2003) The conduct of in vitro and in vivo drug-drug interaction studies: a PhRMA perspective. J Clin Pharmacol 43:443–469

42. Jia L, Liu X (2007) The conduct of drug metabolism studies considered good practice (II): in vitro experiments. Curr Drug Metab 8(8):822–829
43. Venkatakrishnan K, von Moltke LL, Obach RS, Greenblatt DJ (2003) Drug metabolism and drug interactions: application and clinical value of in vitro models. Curr Drug Metab 4(5):423–459
44. Wienkers LC, Heath TG (2005) Predicting in vivo drug interactions from in vitro drug discovery data. Nat Rev Drug Discov 4(10):825–833
45. Krosser S, Neugebauer R, Dolgos H, Fluck M, Rost KL, Kovar A (2006) Investigation of sarizotan's impact on the pharmacokinetics of probe drugs for major cytochrome P450 isoenzymes: a combined cocktail trial. Eur J Clin Pharmacol 62(4):277–284
46. Rostami-Hodjegan A, Tucker G (2004) 'In silico' simulations to assess the 'in vivo' consequences of 'in vitro' metabolic drug–drug interactions. Drug Discov Today Technol 1(4):441–448
47. Drug interaction studies-study design, data analysis, and implications for dosing and labeling. U.S. Food and Drug Administration, Rockville, 2006
48. Davit B, Reynolds K, Yuan R, Ajayi F, Conner D, Fadiran E et al (1999) FDA evaluations using in vitro metabolism to predict and interpret in vivo metabolic drug-drug interactions: impact on labeling. J Clin Pharmacol 39(9):899–910
49. Weaver RJ (2001) Assessment of drug-drug interactions: concepts and approaches. Xenobiotica 31(8-9):499–538
50. Lu C, Hatsis P, Berg C, Lee FW, Balani SK (2008) Prediction of pharmacokinetic drug-drug interactions using human hepatocyte suspension in plasma and cytochrome P450 phenotypic data. II. In vitro-in vivo correlation with ketoconazole. Drug Metab Dispos 36(7):1255–1260
51. Lu C, Miwa GT, Prakash SR, Gan LS, Balani SK (2007) A novel model for the prediction of drug-drug interactions in humans based on in vitro cytochrome p450 phenotypic data. Drug Metab Dispos 35(1):79–85
52. Grimm SW, Einolf HJ, Hall SD, He K, Lim HK, Ling KH et al (2009) The conduct of in vitro studies to address time-dependent inhibition of drug-metabolizing enzymes: a perspective of the pharmaceutical research and manufacturers of America. Drug Metab Dispos 37(7):1355–1370
53. Drug development and drug interactions: table of substrates, inhibitors and inducers. US Food and Drug Administration, 2015
54. Table 2-1: Examples of clinical index substrates for P450-mediated metabolism (for use in index clinical DDI studies) (9/26/2016): US Food and Drug Administration; 2016 [updated 26 September 2016. Available from: http://www.fda.gov/Drugs/DevelopmentApprovalProcess/DevelopmentResources/DrugInteractionsLabeling/ucm093664.htm – table2-1
55. Watkins P (1994) Noninvasive tests of CYP3A enzymes. Pharmacogenetics 4:171–184
56. Frank D, Jaehde U, Fuhr U (2007) Evaluation of probe drugs and pharmacokinetic metrics for CYP2D6 phenotyping. Eur J Clin Pharmacol 63(4):321–333
57. Borges S, Li L, Hamman MA, Jones DR, Hall SD, Gorski JC (2005) Dextromethorphan to dextrorphan urinary metabolic ratio does not reflect dextromethorphan oral clearance. Drug Metab Dispos 33(7):1052–1055
58. Ozdemir M, Crewe KH, Tucker GT, Rostami-Hodjegan A (2004) Assessment of in vivo CYP2D6 activity: differential sensitivity of commonly used probes to urine pH. J Clin Pharmacol 44(12):1398–1404
59. Labbe L, Sirois C, Pilote S, Arseneault M, Robitaille NM, Turgeon J et al (2000) Effect of gender, sex hormones, time variables and physiological urinary pH on apparent CYP2D6 activity as assessed by metabolic ratios of marker substrates. Pharmacogenetics 10(5):425–438
60. Bioavailability and bioequivalence requirements; Abbreviated applications; Final Rule. In: Food and Drug Administration H (ed), U.S. Department of Health and Human Services, 2002, pp 77668–77675
61. Thummel KE, Shen DD, Podoll TD, Kunze KL, Trager WF, Bacchi CE et al (1994) Use of midazolam as a human cytochrome P450 3A probe: II. Characterization of inter- and intraindividual hepatic CYP3A variability after liver transplantation. J Pharmacol Exp Ther 271(1):557–566

62. Thummel KE, Shen DD, Podoll TD, Kunze KL, Trager WF, Hartwell PS et al (1994) Use of midazolam as a human cytochrome P450 3A probe: I. In vitro-in vivo correlations in liver transplant patients. J Pharmacol Exp Ther 271(1):549–556
63. Streetman DS, Kashuba AD, Bertino JS Jr, Kulawy R, Rocci ML Jr, Nafziger AN (2001) Use of midazolam urinary metabolic ratios for cytochrome P450 3A (CYP3A) phenotyping. Pharmacogenetics 11(4):349–355
64. Lee LS, Bertino JS Jr, Nafziger AN (2006) Limited sampling models for oral midazolam: midazolam plasma concentrations, not the ratio of 1-hydroxymidazolam to midazolam plasma concentrations, accurately predicts AUC as a biomarker of CYP3A activity. J Clin Pharmacol 46(2):229–234
65. Ma JD, Nafziger AN, Kashuba AD, Kim MJ, Gaedigk A, Rowland E et al (2004) Limited sampling strategy of S-warfarin concentrations, but not warfarin S/R ratios, accurately predicts S-warfarin AUC during baseline and inhibition in CYP2C9 extensive metabolizers. J Clin Pharmacol 44(6):570–576
66. Rogers JF, Nafziger AN, Kashuba AD, Streetman DS, Rocci ML Jr, Choo EF et al (2002) Single plasma concentrations of 1′-hydroxymidazolam or the ratio of 1′-hydroxymidazolam:midazolam do not predict midazolam clearance in healthy subjects. J Clin Pharmacol 42(10):1079–1082
67. Guideline on the investigation of drug interactions. In: Products CfHM, editor. CPMP/EWP/560/95/Rev. 1 Corr. 2 ed. European Medicines Agency, London, 2012, pp 1–59
68. Fuhr U, Rost KL (1994) Simple and reliable CYP1A2 phenotyping by the paraxanthine/caffeine ratio in plasma and in saliva. Pharmacogenetics 4(3):109–116
69. Johnson BM, Song IH, Adkison KK, Borland J, Fang L, Lou Y et al (2006) Evaluation of the drug interaction potential of aplaviroc, a novel human immunodeficiency virus entry inhibitor, using a modified Cooperstown 5 + 1 cocktail. J Clin Pharmacol 46(5):577–587
70. Sarkar MA, Jackson BJ (1994) Theophylline N-demethylations as probes for P4501A1 and P4501A2. Drug Metab Dispos 22(6):827–834
71. Granfors MT, Backman JT, Neuvonen M, Neuvonen PJ (2004) Ciprofloxacin greatly increases concentrations and hypotensive effect of tizanidine by inhibiting its cytochrome P450 1A2-mediated presystemic metabolism. Clin Pharmacol Ther 76(6):598–606
72. Faucette SR, Hawke RL, Lecluyse EL, Shord SS, Yan B, Laethem RM et al (2000) Validation of bupropion hydroxylation as a selective marker of human cytochrome P450 2B6 catalytic activity. Drug Metab Dispos 28(10):1222–1230
73. Kharasch ED, Mitchell D, Coles R (2008) Stereoselective bupropion hydroxylation as an in vivo phenotypic probe for cytochrome P4502B6 (CYP2B6) activity. J Clin Pharmacol 48(4):464–474
74. Totah RA, Rettie AE (2005) Cytochrome P450 2C8: substrates, inhibitors, pharmacogenetics, and clinical relevance. Clin Pharmacol Ther 77(5):341–352
75. Niemi M, Backman JT, Neuvonen PJ (2004) Effects of trimethoprim and rifampin on the pharmacokinetics of the cytochrome P450 2C8 substrate rosiglitazone. Clin Pharmacol Ther 76(3):239–249
76. Bidstrup TB, Bjornsdottir I, Sidelmann UG, Thomsen MS, Hansen KT (2003) CYP2C8 and CYP3A4 are the principal enzymes involved in the human in vitro biotransformation of the insulin secretagogue repaglinide. Br J Clin Pharmacol 56(3):305–314
77. Kumar V, Wahlstrom JL, Rock DA, Warren CJ, Gorman LA, Tracy TS (2006) CYP2C9 inhibition: impact of probe selection and pharmacogenetics on in vitro inhibition profiles. Drug Metab Dispos 34(12):1966–1975
78. Jetter A, Kinzig-Schippers M, Skott A, Lazar A, Tomalik-Scharte D, Kirchheiner J et al (2004) Cytochrome P450 2C9 phenotyping using low-dose tolbutamide. Eur J Clin Pharmacol 60(3):165–171
79. Kupfer A, Preisig R (1984) Pharmacogenetics of mephenytoin: a new drug hydroxylation polymorphism in man. Eur J Clin Pharmacol 26(6):753–759
80. Xie HG, Stein CM, Kim RB, Wilkinson GR, Flockhart DA, Wood AJ (1999) Allelic, genotypic and phenotypic distributions of S-mephenytoin 4′-hydroxylase (CYP2C19) in healthy

Caucasian populations of European descent throughout the world. Pharmacogenetics 9(5):539–549

81. Chang M, Tybring G, Dahl ML, Gotharson E, Sagar M, Seensalu R et al (1995) Interphenotype differences in disposition and effect on gastrin levels of omeprazole – suitability of omeprazole as a probe for CYP2C19. Br J Clin Pharmacol 39(5):511–518

82. Lasker JM, Wester MR, Aramsombatdee E, Raucy JL (1998) Characterization of CYP2C19 and CYP2C9 from human liver: respective roles in microsomal tolbutamide, S-mephenytoin, and omeprazole hydroxylations. Arch Biochem Biophys 353(1):16–28

83. Karam WG, Goldstein JA, Lasker JM, Ghanayem BI (1996) Human CYP2C19 is a major omeprazole 5-hydroxylase, as demonstrated with recombinant cytochrome P450 enzymes. Drug Metab Dispos 24(10):1081–1087

84. Miura M, Tada H, Yasui-Furukori N, Uno T, Sugawara K, Tateishi T et al (2004) Pharmacokinetic differences between the enantiomers of lansoprazole and its metabolite, 5-hydroxylansoprazole, in relation to CYP2C19 genotypes. Eur J Clin Pharmacol 60(9):623–628

85. Thacker DL, Modak AS, Lemler SM, Flockhart DA, Desta Z (2010) Cytochrome P450 (CYP) 2C19 specific breath test using (+)-[13C]-pantoprazole as a phenotype probe. Clin Pharmacol Ther 87(Suppl 1):S51

86. Desta Z, Modak A, Nguyen PD, Lemler SM, Kurogi Y, Li L et al (2009) Rapid identification of the hepatic cytochrome P450 2C19 activity using a novel and noninvasive [13C]pantoprazole breath test. J Pharmacol Exp Ther 329(1):297–305

87. Streetman D, Bertino J Jr, Nafziger A (2000) Phenotyping of drug-metabolizing enzymes in adults: a review of in-vivo cytochrome P450 phenotyping probes. Pharmacogenetics 10:187–216

88. Spina E, Avenoso A, Campo GM, Scordo MG, Caputi AP, Perucca E (1997) Effect of ketoconazole on the pharmacokinetics of imipramine and desipramine in healthy subjects. Br J Clin Pharmacol 43(3):315–318

89. Lindamood C, Ortiz S, Shaw A, Rackley R, Gorski JC (2011) Effects of commonly administered agents and genetics on nebivolol pharmacokinetics: drug-drug interaction studies. J Clin Pharmacol 51(4):575–585

90. Frye RF, Adedoyin A, Mauro K, Matzke GR, Branch RA (1998) Use of chlorzoxazone as an in vivo probe of cytochrome P450 2E1: choice of dose and phenotypic trait measure. J Clin Pharmacol 38(1):82–89

91. Mishin VM, Rosman AS, Basu P, Kessova I, Oneta CM, Lieber CS (1998) Chlorzoxazone pharmacokinetics as a marker of hepatic cytochrome P4502E1 in humans. Am J Gastroenterol 93(11):2154–2161

92. Tsunoda S, Velez R, von Moltke L, Greenblatt D (1999) Differentiation of intestinal and hepatic cytochrome P450 3A activity with use of midazolam as an in vivo proble: effect of ketoconazole. Clin Pharmacol Ther 66:461–471

93. Chaobal HN, Kharasch ED (2005) Single-point sampling for assessment of constitutive, induced, and inhibited cytochrome P450 3A activity with alfentanil or midazolam. Clin Pharmacol Ther 78(5):529–539

94. Kharasch ED, Walker A, Hoffer C, Sheffels P (2004) Intravenous and oral alfentanil as in vivo probes for hepatic and first-pass cytochrome P450 3A activity: noninvasive assessment by use of pupillary miosis. Clin Pharmacol Ther 76(5):452–466

95. Kharasch ED, Hoffer C, Walker A, Sheffels P (2003) Disposition and miotic effects of oral alfentanil: a potential noninvasive probe for first-pass cytochrome P4503A activity. Clin Pharmacol Ther 73(3):199–208

96. Jalava KM, Olkkola KT, Neuvonen PJ (1997) Itraconazole greatly increases plasma concentrations and effects of felodipine. Clin Pharmacol Ther 61(4):410–415

97. von Moltke LL, Greenblatt DJ, Harmatz JS, Duan SX, Harrel LM, Cotreau-Bibbo MM et al (1996) Triazolam biotransformation by human liver microsomes in vitro: effects of metabolic inhibitors and clinical confirmation of a predicted interaction with ketoconazole. J Pharmacol Exp Ther 276(2):370–379

98. Perloff MD, von Moltke LL, Court MH, Kotegawa T, Shader RI, Greenblatt DJ (2000) Midazolam and triazolam biotransformation in mouse and human liver microsomes: relative contribution of CYP3A and CYP2C isoforms. J Pharmacol Exp Ther 292(2):618–628

99. Chainuvati S, Nafziger AN, Leeder JS, Gaedigk A, Kearns GL, Sellers E et al (2003) Combined phenotypic assessment of cytochrome p450 1A2, 2C9, 2C19, 2D6, and 3A, N-acetyltransferase-2, and xanthine oxidase activities with the "Cooperstown 5+1 cocktail". Clin Pharmacol Ther 74(5):437–447

100. Turpault S, Brian W, Van Horn R, Santoni A, Poitiers F, Donazzolo Y et al (2009) Pharmacokinetic assessment of a five-probe cocktail for CYPs 1A2, 2C9, 2C19, 2D6 and 3A. Br J Clin Pharmacol 68(6):928–935

101. Berthou F, Goasduff T, Lucas D, Dreano Y, Le Bot MH, Menez JF (1995) Interaction between two probes used for phenotyping cytochromes P4501A2 (caffeine) and P4502E1 (chlorzoxazone) in humans. Pharmacogenetics 5(2):72–79

102. Backman JT, Granfors MT, Neuvonen PJ (2006) Rifampicin is only a weak inducer of CYP1A2-mediated presystemic and systemic metabolism: studies with tizanidine and caffeine. Eur J Clin Pharmacol 62(6):451–461

103. Zhang L, Zhang Y, Zhao P, Huang S-M (2009) Predicting drug-drug interactions: an FDA perspective. AAPS J 11(2):300–306

104. Backman JT, Filppula AM, Niemi M, Neuvonen PJ (2016) Role of cytochrome P450 2C8 in drug metabolism and interactions. Pharmacol Rev 68(1):168–241

105. Niemi M, Backman JT, Juntti-Patinen L, Neuvonen M, Neuvonen PJ (2005) Coadministration of gemfibrozil and itraconazole has only a minor effect on the pharmacokinetics of the CYP2C9 and CYP3A4 substrate nateglinide. Br J Clin Pharmacol 60(2):208–217

106. Daily EB, Aquilante CL (2009) Cytochrome P450 2C8 pharmacogenetics: a review of clinical studies. Pharmacogenomics 10(9):1489–1510

107. Bruce MA, Hall SD, Haehner-Daniels BD, Gorski JC (2001) In vivo effect of clarithromycin on multiple cytochrome P450s. Drug Metab Dispos 29(7):1023–1028

108. Yasar U, Forslund-Bergengren C, Tybring G, Dorado P, Llerena A, Sjoqvist F et al (2002) Pharmacokinetics of losartan and its metabolite E-3174 in relation to the CYP2C9 genotype. Clin Pharmacol Ther 71(1):89–98

109. Babaoglu MO, Yasar U, Sandberg M, Eliasson E, Dahl ML, Kayaalp SO et al (2004) CYP2C9 genetic variants and losartan oxidation in a Turkish population. Eur J Clin Pharmacol 60(5):337–342

110. Allabi AC, Gala JL, Horsmans Y, Babaoglu MO, Bozkurt A, Heusterspreute M et al (2004) Functional impact of CYP2C95, CYP2C96, CYP2C98, and CYP2C911 in vivo among black Africans. Clin Pharmacol Ther 76(2):113–118

111. Yasar U, Eliasson E, Forslund-Bergengren C, Tybring G, Gadd M, Sjoqvist F et al (2001) The role of CYP2C9 genotype in the metabolism of diclofenac in vivo and in vitro. Eur J Clin Pharmacol 57(10):729–735

112. Kumar V, Brundage RC, Oetting WS, Leppik IE, Tracy TS (2008) Differential genotype dependent inhibition of CYP2C9 in humans. Drug Metab Dispos 36(7):1242–1248

113. Dorado P, Berecz R, Norberto MJ, Yasar U, Dahl ML, Llerena A (2003) CYP2C9 genotypes and diclofenac metabolism in Spanish healthy volunteers. Eur J Clin Pharmacol 59(3):221–225

114. Dorado P, Cavaco I, Caceres MC, Piedade R, Ribeiro V, Llerena A (2008) Relationship between CYP2C8 genotypes and diclofenac 5-hydroxylation in healthy Spanish volunteers. Eur J Clin Pharmacol 64(10):967–970

115. Kirchheiner J, Meineke I, Steinbach N, Meisel C, Roots I, Brockmoller J (2003) Pharmacokinetics of diclofenac and inhibition of cyclooxygenases 1 and 2: no relationship to the CYP2C9 genetic polymorphism in humans. Br J Clin Pharmacol 55(1):51–61

116. Hassan-Alin M, Andersson T, Niazi M, Rohss K (2005) A pharmacokinetic study comparing single and repeated oral doses of 20 mg and 40 mg omeprazole and its two optical isomers, S-omeprazole (esomeprazole) and R-omeprazole, in healthy subjects. Eur J Clin Pharmacol 60(11):779–784

117. Tybring G, Bottiger Y, Widen J, Bertilsson L (1997) Enantioselective hydroxylation of omeprazole catalyzed by CYP2C19 in Swedish white subjects. Clin Pharmacol Ther 62(2):129–137
118. Masica AL, Mayo G, Wilkinson GR (2004) In vivo comparisons of constitutive cytochrome P450 3A activity assessed by alprazolam, triazolam, and midazolam. Clin Pharmacol Ther 76(4):341–349
119. Kim JS, Nafziger AN, Tsunoda SM, Choo EE, Streetman DS, Kashuba AD et al (2002) Limited sampling strategy to predict AUC of the CYP3A phenotyping probe midazolam in adults: application to various assay techniques. J Clin Pharmacol 42(4):376–382
120. Chung E, Nafziger AN, Kazierad DJ, Bertino JS Jr (2006) Comparison of midazolam and simvastatin as cytochrome P450 3A probes. Clin Pharmacol Ther 79(4):350–361
121. Kharasch ED, Thummel KE, Watkins PB (2005) CYP3A probes can quantitatively predict the in vivo kinetics of other CYP3A substrates and can accurately assess CYP3A induction and inhibition. Mol Interv 5(3):151–153
122. Christensen M, Andersson K, Dalen P, Mirghani RA, Muirhead GJ, Nordmark A et al (2003) The Karolinska cocktail for phenotyping of five human cytochrome P450 enzymes. Clin Pharmacol Ther 73(6):517–528
123. Mirghani RA, Ericsson O, Tybring G, Gustafsson LL, Bertilsson L (2003) Quinine 3-hydroxylation as a biomarker reaction for the activity of CYP3A4 in man. Eur J Clin Pharmacol 59(1):23–28
124. Foti RS, Rock DA, Wienkers LC, Wahlstrom JL (2010) Selection of alternative CYP3A4 probe substrates for clinical drug interaction studies using in vitro data and in vivo simulation. Drug Metab Dispos 38(6):981–987
125. Mitra AK, Thummel KE, Kalhorn TF, Kharasch ED, Unadkat JD, Slattery JT (1995) Metabolism of dapsone to its hydroxylamine by CYP2E1 in vitro and in vivo. Clin Pharmacol Ther 58(5):556–566
126. Gill HJ, Tingle MD, Park BK (1995) N-Hydroxylation of dapsone by multiple enzymes of cytochrome P450: implications for inhibition of haemotoxicity. Br J Clin Pharmacol 40(6):531–538
127. Gass RJ, Gal J, Fogle PW, Detmar-Hanna D, Gerber JG (1998) Neither dapsone hydroxylation nor cortisol 6beta-hydroxylation detects the inhibition of CYP3A4 by HIV-1 protease inhibitors. Eur J Clin Pharmacol 54(9-10):741–747
128. Frye RF, Matzke GR, Adedoyin A, Porter JA, Branch RA (1997) Validation of the five-drug "Pittsburgh cocktail" approach for assessment of selective regulation of drug-metabolizing enzymes. Clin Pharmacol Ther 62(4):365–376
129. Sharma A, Pilote S, Belanger PM, Arsenault M, Hamelin BA (2004) A convenient five-drug cocktail for the assessment of major drug metabolizing enzymes: a pilot study. Br J Clin Pharmacol 58(3):288–297
130. Kinirons MT, O'Shea D, Kim RB, Groopman JD, Thummel KE, Wood AJ et al (1999) Failure of erythromycin breath test to correlate with midazolam clearance as a probe of cytochrome P4503A. Clin Pharmacol Ther 66(3):224–231
131. Ragueneau-Majlessi I, Boulenc X, Rauch C, Hachad H, Levy RH (2007) Quantitative correlations among CYP3A sensitive substrates and inhibitors: literature analysis. Curr Drug Metab 8(8):810–814
132. Zhu M, Zhao W, Jimenez H, Zhang D, Yeola S, Dai R et al (2005) Cytochrome P450 3A-mediated metabolism of buspirone in human liver microsomes. Drug Metab Dispos 33(4):500–507
133. HY K, Ahn HJ, Seo KA, Kim H, Oh M, Bae SK et al (2008) The contributions of cytochromes P450 3A4 and 3A5 to the metabolism of the phosphodiesterase type 5 inhibitors sildenafil, udenafil, and vardenafil. Drug Metab Dispos 36(6):986–990
134. Galteau MM, Shamsa F (2003) Urinary 6beta-hydroxycortisol: a validated test for evaluating drug induction or drug inhibition mediated through CYP3A in humans and in animals. Eur J Clin Pharmacol 59(10):713–733

135. Tomalik-Scharte D, Lutjohann D, Doroshyenko O, Frank D, Jetter A, Fuhr U (2009) Plasma 4beta-hydroxycholesterol: an endogenous CYP3A metric? Clin Pharmacol Ther 86(2):147–153
136. Kasichayanula S, Boulton DW, Luo WL, Rodrigues AD, Yang Z, Goodenough A et al (2014) Validation of 4beta-hydroxycholesterol and evaluation of other endogenous biomarkers for the assessment of CYP3A activity in healthy subjects. Br J Clin Pharmacol 78(5):1122–1134
137. Bjorkhem-Bergman L, Backstrom T, Nylen H, Ronquist-Nii Y, Bredberg E, Andersson TB et al (2013) Comparison of endogenous 4beta-hydroxycholesterol with midazolam as markers for CYP3A4 induction by rifampicin. Drug Metab Dispos 41(8):1488–1493
138. Xie HG, Wood AJ, Kim RB, Stein CM, Wilkinson GR (2004) Genetic variability in CYP3A5 and its possible consequences. Pharmacogenomics 5(3):243–272
139. Lin YS, Dowling AL, Quigley SD, Farin FM, Zhang J, Lamba J et al (2002) Co-regulation of CYP3A4 and CYP3A5 and contribution to hepatic and intestinal midazolam metabolism. Mol Pharmacol 62(1):162–172
140. Daly AK (2006) Significance of the minor cytochrome P450 3A isoforms. Clin Pharmacokinet 45(1):13–31
141. Lucas D, Ferrara R, Gonzalez E, Bodenez P, Albores A, Manno M et al (1999) Chlorzoxazone, a selective probe for phenotyping CYP2E1 in humans. Pharmacogenetics 9(3):377–388
142. Girre C, Lucas D, Hispard E, Menez C, Dally S, Menez JF (1994) Assessment of cytochrome P4502E1 induction in alcoholic patients by chlorzoxazone pharmacokinetics. Biochem Pharmacol 47(9):1503–1508
143. Palmer JL, Scott RJ, Gibson A, Dickins M, Pleasance S (2001) An interaction between the cytochrome P450 probe substrates chlorzoxazone (CYP2E1) and midazolam (CYP3A). Br J Clin Pharmacol 52(5):555–561
144. Jetter A, Kinzig M, Rodamer M, Tomalik-Scharte D, Sorgel F, Fuhr U (2009) Phenotyping of N-acetyltransferase type 2 and xanthine oxidase with caffeine: when should urine samples be collected? Eur J Clin Pharmacol 65(4):411–417
145. Kalow W, Tang BK (1993) The use of caffeine for enzyme assays: a critical appraisal. Clin Pharmacol Ther 53(5):503–514
146. Rothman N, Hayes RB, Bi W, Caporaso N, Broly F, Woosley RL et al (1993) Correlation between N-acetyltransferase activity and NAT2 genotype in Chinese males. Pharmacogenetics 3(5):250–255
147. O'Neil WM, Drobitch RK, MacArthur RD, Farrough MJ, Doll MA, Fretland AJ et al (2000) Acetylator phenotype and genotype in patients infected with HIV: discordance between methods for phenotype determination and genotype. Pharmacogenetics 10(2):171–182
148. Luo X, Li XM, ZY H, Cheng ZN (2009) Evaluation of CYP3A activity in humans using three different parameters based on endogenous cortisol metabolism. Acta Pharmacol Sin 30(9):1323–1329
149. Chen M, Nafziger AN, Bertino JS Jr (2006) Drug-metabolizing enzyme inhibition by ketoconazole does not reduce interindividual variability of CYP3A activity as measured by oral midazolam. Drug Metab Dispos 34(12):2079–2082
150. Penzak SR, Busse KH, Robertson SM, Formentini E, Alfaro RM, Davey RT Jr (2008) Limitations of using a single postdose midazolam concentration to predict CYP3A-mediated drug interactions. J Clin Pharmacol 48(6):671–680
151. Wang Z, Gorski JC, Hamman MA, Huang SM, Lesko LJ, Hall SD (2001) The effects of St John's wort (Hypericum perforatum) on human cytochrome P450 activity. Clin Pharmacol Ther 70(4):317–326
152. Tomalik-Scharte D, Jetter A, Kinzig-Schippers M, Skott A, Sorgel F, Klaassen T et al (2005) Effect of propiverine on cytochrome P450 enzymes: a cocktail interaction study in healthy volunteers. Drug Metab Dispos 33(12):1859–1866
153. Videau O, Delaforge M, Levi M, Thevenot E, Gal O, Becquemont L et al (2010) Biochemical and analytical development of the CIME cocktail for drug fate assessment in humans. Rapid Commun Mass Spectrom 24(16):2407–2419

154. Streetman DS, Bleakley JF, Kim JS, Nafziger AN, Leeder JS, Gaedigk A et al (2000) Combined phenotypic assessment of CYP1A2, CYP2C19, CYP2D6, CYP3A, N-acetyltransferase-2, and xanthine oxidase with the "Cooperstown cocktail". Clin Pharmacol Ther 68(4):375–383

155. Rendic S (2002) Summary of information on human CYP enzymes: human P450 metabolism data. Drug Metab Rev 34(1-2):83–448

156. Shelepova T, Nafziger AN, Victory J, Kashuba AD, Rowland E, Zhang Y et al (2005) Effect of a triphasic oral contraceptive on drug-metabolizing enzyme activity as measured by the validated Cooperstown 5+1 cocktail. J Clin Pharmacol 45(12):1413–1421

157. Furman KD, Grimm DR, Mueller T, Holley-Shanks RR, Bertz RJ, Williams LA et al (2004) Impact of CYP2D6 intermediate metabolizer alleles on single-dose desipramine pharmacokinetics. Pharmacogenetics 14(5):279–284

158. Kim MJ, Bertino JS Jr, Gaedigk A, Zhang Y, Sellers EM, Nafziger AN (2002) Effect of sex and menstrual cycle phase on cytochrome P450 2C19 activity with omeprazole used as a biomarker. Clin Pharmacol Ther 72(2):192–199

159. Bromley CM, Close S, Cohen N, Favis R, Fijal B, Gheyas F et al (2009) Designing pharmacogenetic projects in industry: practical design perspectives from the Industry Pharmacogenomics Working Group. Pharmacogenomics J 9(1):14–22

160. Williams JA, Johnson K, Paulauskis J, Cook J (2006) So many studies, too few subjects: establishing functional relevance of genetic polymorphisms on pharmacokinetics. J Clin Pharmacol 46(3):258–264

161. Eap CB, Lessard E, Baumann P, Brawand-Amey M, Yessine MA, O'Hara G et al (2003) Role of CYP2D6 in the stereoselective disposition of venlafaxine in humans. Pharmacogenetics 13(1):39–47

162. Kirchheiner J, Schmidt H, Tzvetkov M, Keulen JT, Lotsch J, Roots I et al (2007) Pharmacokinetics of codeine and its metabolite morphine in ultra-rapid metabolizers due to CYP2D6 duplication. Pharmacogenomics J 7(4):257–265

163. Inui N, Akamatsu T, Uchida S, Tanaka S, Namiki N, Karayama M et al (2013) Chronological effects of rifampicin discontinuation on cytochrome P450 activity in healthy Japanese volunteers, using the cocktail method. Clin Pharmacol Ther 94(6):702–708

164. De Bock L, Boussery K, Colin P, De Smet J, T'Jollyn H, Van Bocxlaer J (2012) Development and validation of a fast and sensitive UPLC-MS/MS method for the quantification of six probe metabolites for the in vitro determination of cytochrome P450 activity. Talanta 89:209–216

165. Scott RJ, Palmer J, Lewis IA, Pleasance S (1999) Determination of a 'GW cocktail' of cytochrome P450 probe substrates and their metabolites in plasma and urine using automated solid phase extraction and fast gradient liquid chromatography tandem mass spectrometry. Rapid Commun Mass Spectrom 13(23):2305–2319

166. Statistical approaches to establishing bioequivalence. U.S. Department of Health and Human Services, Food and Drug Administration, Rockville, 2001

167. Group Mr. Drug interaction guideline for drug development and labeling recommendations (draft for public comment). Japan: Ministry of Health, Labour and Welfare, 2014, pp 75–76

168. Lin YS, Lockwood GF, Graham MA, Brian WR, Loi CM, Dobrinska MR et al (2001) In-vivo phenotyping for CYP3A by a single-point determination of midazolam plasma concentration. Pharmacogenetics 11(9):781–791

169. Watkins PB, Wrighton SA, Maurel P, Schuetz EG, Mendez-Picon G, Parker GA et al (1985) Identification of an inducible form of cytochrome P-450 in human liver. Proc Natl Acad Sci U S A 82(18):6310–6314

170. Kronbach T, Fischer V, Meyer UA (1988) Cyclosporine metabolism in human liver: identification of a cytochrome P-450III gene family as the major cyclosporine-metabolizing enzyme explains interactions of cyclosporine with other drugs. Clin Pharmacol Ther 43(6):630–635

Chapter 9
Design and Data Analysis in Drug Interaction Studies

David E. Nix and Keith Gallicano

9.1 Study Rationale

Drug interaction studies should be considered for drugs that are likely to be administered concomitantly to large numbers of patients. The drugs may be indicated for the same disease process, and their use in combination is considered therapeutically rational. Alternatively, the drugs may have different indications, but the two disease processes occur frequently in the same population. Drugs involved in interactions are divided into precipitant drugs (drugs that cause a change in the pharmacokinetics and/or pharmacodynamics of another drug) and object drugs (drugs affected by the precipitant drug). A drug can act as a precipitant drug and an object drug at the same time when two drugs affect each other during concomitant administration.

To study large numbers of potential interactions routinely for all drugs is not feasible or desirable. Consequently, screening methods are required to identify drugs that are likely to interact. A chemist who is knowledgeable about drug interactions affecting gastrointestinal absorption may be able to identify potential interactions involving chelation, physical binding, or other incompatibility. Metabolism of object drugs may be studied using in vitro cytochrome P450 (CYP) enzyme preparations to identify enzymes involved in the metabolism [1, 2]. Databases are available that list drugs that inhibit or induce various CYP subtypes. Once metabolism is determined to be a major elimination pathway and the responsible enzyme subtypes are known, these databases can be used to identify potential precipitant drugs [3]. Preliminary interaction studies of substrates with metabolic inhibitors and inducers can be performed using the same in vitro enzyme preparations as those used to

D.E. Nix (✉)
The University of Arizona College of Pharmacy, Tucson, AZ, USA
e-mail: nix@pharmacy.arizona.edu

K. Gallicano
Novum Pharmaceutical Research Services, Murrieta, CA, USA

© Springer International Publishing AG 2018
M.P. Pai et al. (eds.), *Drug Interactions in Infectious Diseases: Mechanisms and Models of Drug Interactions*, Infectious Disease,
https://doi.org/10.1007/978-3-319-72422-5_9

determine metabolic pathways of substrates [2, 4]. Similar methods have been adapted to investigate drug interactions involving intestinal metabolism and drug transport [5–7].

Interactions involving protein binding displacement are not usually clinically significant. However, protein binding interactions should be examined for drugs that [1] exhibit high binding to plasma proteins (>90%), [2] have a narrow therapeutic index, [3] occupy most of the available plasma protein binding sites at clinically relevant concentrations, and [4] have a small volume of distribution (<10 L/70 kg). Drugs that are the most important candidates for drug interaction studies are those that are restrictively cleared by an elimination organ; a concern is also apparent for drugs that are nonrestrictively cleared, have a narrow therapeutic index and a small volume of distribution, and are administered intravenously. In the former case, a transient increase in unbound concentration could produce harmful adverse effects [8, 9]. Preliminary protein binding studies can be carried out in vitro, recognizing that metabolites may contribute to protein binding displacement interactions. Interactions involving renal clearance changes may be expected for drugs that are mainly eliminated by renal excretion. For these drugs, the presence of significant tubular secretion or reabsorption suggests possible interactions. Pharmacodynamic interactions should be suspected for drugs that have similar pharmacologic or toxicologic effects.

9.2   Study Design: General Issues

Current regulatory guidances provide some insight into designs for in vivo drug interaction studies [10, 11]. These guidances recommend three designs: (1) randomized crossover, (2) one-sequence crossover, or (3) parallel. A position paper by Pharmaceutical Research and Manufacturers of America (PhRMA) Drug Metabolism and Clinical Pharmacology Technical Working Groups has defined a minimal best practice for in vitro and in vivo pharmacokinetic drug-drug interaction studies targeted to drug development, with the goal of harmonizing approaches by regulatory agencies and industry sponsors [12]. The US Food and Drug Administration maintains a web page that covers many issues regarding drug-drug interactions in drug development from lists of prototype inhibitors and substrates to decision trees, labeling, and dose adjustment (https://www.fda.gov/Drugs/DevelopmentApprovalProcess/DevelopmentResources/DrugInteractionsLabeling/ucm080499.htm; Accessed 3/1/2017).

Drug interaction studies involve the measurement of pharmacokinetics or a specific pharmacodynamic effect in the presence and absence of an interacting drug. Such studies typically employ a within-subject design in which individuals receive both treatments in either fixed or random order. A fixed-order design (single sequence) denotes a longitudinal study in which the treatments are administered sequentially over two or more time periods. Longitudinal studies are often conducted in patients who are receiving long-term therapy of the object drug or those undergoing treatment with drugs possessing long elimination half-lives (>24 h). A two-period, longitudinal study involves the administration of the object drug alone followed by measurement of the pharmacokinetics or effect parameter(s) in period 1. A washout period may or may not be necessary. Then, the object and suspected precipitant drugs are concomitantly administered in period 2. Measurements of the pharmacokinetics or effect parameters are then repeated following administration of the combination treatment. In the longitudinal design, potential period effects are confounded with the treatment effects. If a change in the clearance (Cl) of the object drug is observed, the change may have been caused by the precipitant drug or by some other intercurrent event. Perhaps the food intake differed between the two periods (treatment phases), or a portion of the subjects acquired a mild viral infection between the two periods. If females are included as subjects, the number of subjects in the luteal phase of the menstrual cycle may differ between the two periods.

The study must be designed with full knowledge of the pharmacokinetics of both drugs. If the study involves single doses of the object drug, then adequate washout of the first dose must be allowed before starting the second treatment phase. For the control treatment, measuring serum concentrations or effect for at least four to five half-lives is important. If reduced clearance and increased half-life are expected, the sampling time may need to be extended following concomitant treatment compared to the control period. If the study involves multiple-dose administration of the object drug, then the serum concentrations should reach steady state during both periods. Steady state may take longer during the interaction phase if the half-life is prolonged.

The major advantage of a two-period, longitudinal design is that the potential for carryover effect from prior administration of the precipitant drug is avoided. A switchback design in which the object drug is replicated at least once after the precipitant drug is discontinued is useful to determine the effects of starting and stopping a metabolic inhibitor or inducer on the baseline characteristics of the object drug. Such a design was used to establish the rebound to baseline pharmacokinetic parameters of steady-state zidovudine at 14 d after rifampin was discontinued in period 2 [13].

9.2.1 Crossover Designs

A crossover study evaluates treatments administered in two or more planned sequences with subjects randomly allocated to the different sequences. The design is characterized by T, P, and S in which T is the number of treatments, P is the number of periods, and S is the number of sequences. All of these numbers must be ≥ 2 [14]. Designs that have a single (fixed) sequence are sometimes referred to as "crossover-like," but should be considered as a longitudinal study.

There are two main types of crossover designs: nonreplicated and replicated. Nonreplicated designs have the same number of treatments as periods, and the number of possible sequences increases as the factorial of T (i.e., when $T = 3$, $S = 6$). Replicate designs have more periods than treatments, such that at least one treatment is replicated within a subject. Optimum designs are those that are balanced with equal numbers in each sequence and balanced for carryover effects and variance for the given number of treatments. A design that has each treatment followed by a different treatment the same number of times is balanced for carryover. The presence of a carryover effect is important to assess in drug interaction studies, and enough subjects in each sequence are needed to allow testing of this effect. In a variance-balanced design, each treatment appears the same number of times in each period.

The simplest nonreplicated crossover design is the 2, 2, 2 design. Suppose treatment A involves giving the object drug alone and treatment B involves giving the object drug with the precipitant drug. Subjects would receive the two treatments in one of two sequences, AB or BA, in which treatment A or B would be given during the first period and then switched to the other treatment during the second period. Carryover effects may be introduced for subjects receiving treatment B (sequence BA) in the first period if drug exposures of the object drug are increased by the precipitant drug. An adequate washout period must be planned between the two periods to prevent differential carryover in the two sequences. This may sometimes be difficult if the duration of an "adequate" washout period is not known a priori. Carryover and sequence effects, however, are confounded in the 2, 2, 2 design, and studies in which the two treatments are replicated must be conducted for optimal evaluation of carryover effects.

When nonreplicated studies involve more than two periods, the number of sequences should be carefully planned rather than testing all possible sequences. Usually a subset of sequences is chosen that defines a variance-balanced design. In a three-period, crossover pharmacokinetic study with treatments A, B, and C, six possible sequences ABC, ACB, BAC, BCA, CAB, and CBA must be included to maintain a carryover-balanced design. If carryover is a concern when the object and precipitant drugs are given together in treatments B and C, then a large sample size may be required to ensure an adequate number of subjects per sequence to test the carryover effect. A three-period crossover study may also be used to study potential bidirectional interactions. Here, treatments including drug A, drug B, and drugs A + B are required. A four-period, crossover study would have 4 or 24 possible sequences. The goal is to select four sequences from a 4 × 4 Latin square in which each treatment is administered once during each of four periods, each subject receives all four treatments, and each treatment follows the other three treatments once (balanced for carryover). An example of a "Williams design" involves the four sequences (ABCD, BDAC, CADB, and DCBA). The total number of subjects is selected as a multiple of 4 and subjects are randomized in blocks of 4 to undergo treatments in the sequence assigned [15].

There is considerable interest in replicate crossover designs for bioequivalence studies in which the test and reference treatments are administered each on two separate occasions. This allows for assessment of intraindividual variability in systemic exposure and estimation of carryover effects. The analysis of replicate designs considers that some individuals may differ from the mean response and allows for the determination of "individual bioequivalence." Optimal designs for carryover estimation of the two treatments are AA, BB, AB, and BA for two-period designs, ABB and BAA for three-period designs, and AABB, BBAA, ABBA, and BAAB for four-period designs [14, 16–18]. Switchback designs, either ABA and BAB or ABAB and BABA, are preferred to estimate the intraindividual variability [14]. Similar designs may be employed for drug interaction studies because they increase the confidence that a drug interaction detected is a true interaction.

Replicate measurements may also be obtained in more traditional study designs. As an example, the object drug may be administered as a multiple-dose regimen, and measurements can be made during more than 1 day or dosing interval before changeover to the next treatment. This was done in a randomized crossover study to investigate the interaction between cimetidine and theophylline [19]. Theophylline was administered at a subject-specific dose (concentration controlled) for 23 days. Subjects received treatment 1 (cimetidine or placebo) on days 5–11, washout on days 12–16, and treatment 2 (cimetidine or placebo) on days 17–23. The order of cimetidine and placebo treatments was randomly assigned. The pharmacokinetics of theophylline were assessed on the first, fourth, and seventh days of each treatment period. In the analysis, the data from the fourth and seventh days were treated as replicate measurements of the effect at steady state. Because theophylline exhibits large interindividual variability in clearance, doses were adjusted in a run-in phase to provide similar mean steady-state concentrations before evaluating the interaction. This example also shows how concentration control can be incorporated into the design of a drug interaction study.

9.2.2 Parallel Designs

A parallel design may be used for evaluating drug interactions. However, such designs are less desirable, because the drug variability is usually greater between individuals than within individuals. A simple parallel design study consists of two groups of subjects/patients, one group that is receiving the object drug and one that is receiving the object drug concomitantly with the suspected precipitant drug. Most studies of this type are performed in patient populations that are receiving the drug or drugs therapeutically. There may be problems with comparability of the two patient groups in terms of pharmacokinetics of the object drug regardless of the precipitant drug. The two groups may or may not be randomly selected. If random assignment is not used, additional issues of bias must be considered. When studies of this type are necessary, the use of population modeling may be used for evaluating the presence or absence of the interaction. An example of using population modeling to evaluate a drug interaction involved imipramine and alprazolam [20]. The parallel design may be advantageous for drugs with long elimination half-lives in studies where a long washout period is impractical for a crossover or longitudinal design. When there are safety concerns, randomized studies may not be ethical.

If a population can be identified that requires one or both of the drugs of interest for therapeutic reasons, a convenience sample (sparse or rich sampling) can be used with population PK modeling to study their potential interaction.

A placebo-controlled, parallel-group study can be conducted when possible inherent group differences in a parallel design or time-dependent effects in a single-sequence longitudinal design are a concern. Subjects in each group receive treatment on more than one occasion, and treatment effects are adjusted for baseline values in the first period (placebo) of each treatment group. Alternatively, the mean treatment differences are estimated within each group, and then these differences are compared between treatment groups. A placebo-controlled, parallel-group design was used to show no clinically significant effect of indinavir on the pharmacokinetics of voriconazole [21] and to demonstrate that ritonavir inhibited the metabolism of rifabutin [22].

9.2.3   Mechanistic Aspects

Drug interactions may be very complex. The mechanism of potential interaction is important to hypothesize from in vitro studies, previous clinical and preclinical studies, and experience with other related drugs. Such knowledge is essential to planning a good drug interaction study. Most studies are designed to evaluate the effect of a precipitant drug on an object drug. The precipitant drug may cause some physical or physiologic effect that alters the pharmacokinetics or pharmacodynamics of the object drug. Several questions need to be posed about the precipitant drug in relation to developing the study methods. What are the doses and administration schedules that are relevant to clinical practice? Is the interaction concentration dependent within the range of clinically achievable concentrations? Does the interaction take time to develop (e.g., P450 induction)? What is the primary goal of the study (e.g., to find the maximum potential interaction)? In some circumstances, one may be interested in whether the pharmacokinetics and/or pharmacodynamics of both drugs are affected by concomitant administration.

Multiple dosing of the precipitant drug is often desirable. The object drug may be administered as a single dose or in a multiple-dose regimen designed to achieve steady state. A single dose of the object drug may be appropriate when inhibition of elimination is suspected and safety concerns are substantial. In such cases, unpredictable accumulation would be avoided. One exception occurs when an object drug undergoes extensive first-pass metabolism and the precipitant drug inhibits this metabolism. Much greater systemic bioavailability may result even with single-dose administration.

Concerns about multiple-dose studies are exemplified by a study of voriconazole effects on cyclosporine pharmacokinetics. This study included renal transplant patients receiving treatment with cyclosporine that was continued throughout the study. Subjects received voriconazole or placebo for 7.5 days (period 1), underwent a washout period of at least 4 days, and then received the alternate treatment (voriconazole or placebo) for 7.5 days. Although 14 subjects were entered, only 7 completed the study and 7 were withdrawn during the voriconazole treatment. Voriconazole resulted in a mean 1.7-fold increase in cyclosporine exposure [23]. Although a multiple-dose regimen of the object drug may simulate clinical use and provide greater applicability, safety would favor a single-dose study in healthy subjects first. The addition of procedures to limit exposure to high concentrations during the interaction phase for a follow-up multiple-dose study needs to be considered. For example, the study could employ a dose reduction during the combination treatment. More extensive knowledge of the potential study outcomes, frequent and careful clinical monitoring, and perhaps real-time drug concentration monitoring may be necessary when the object drug is administered in a multiple-dose regimen.

9.2.4 Study Population

Drug interaction studies are most commonly performed in healthy volunteers. Healthy subjects are easier to recruit, the investigators can better control concomitant medications and activities, and study participation may be safer compared to patients with target illnesses. There is no compelling reason why performing a pharmacokinetic interaction study in healthy volunteers is less desirable than performing the study in a target population likely to receive both drugs, unless disease in the target population influences the magnitude of interaction or safety considerations prevent the use of healthy volunteers. The elderly are often cited as a group more susceptible to drug interactions. This is true because elderly patients receive more drugs and interactions only occur when two or more drugs are given concurrently [24]. In addition, geriatric patients and patients with organ failure may eliminate drugs more slowly and achieve higher concentrations than healthy/young counterparts. Administering reduced doses in these special populations designed to achieve exposure that is similar to that observed in healthy volunteers may reduce potential differences in exposure.

Interaction studies that involve pharmacodynamic assessments may or may not be best performed in the target population, depending on the nature of the pharmacodynamic effect. Suppose an object drug reduces wheezing and acute bronchospasm and increases forced expiry volume in 1 s (FEV-1) in patients with asthma. Administration of a precipitant drug in combination with the object drug leads to worsening of symptoms and lowering the FEV-1 in asthma patients. However, these effects are not seen in patients without asthma. Such an interaction would need to be studied in the target population.

One report of an interaction between a laxative polymer and digoxin found a pharmacokinetic interaction consistent with a 30% decrease in digoxin absorption. The concluding statement was "there was no consequence of this interaction on heart rate and atrial ventricular conduction." The study was conducted in healthy volunteers, and digoxin administration was not associated with changes in atrial ventricular conduction with or without the laxative administration. Although a small decrease in heart rate was noted following digoxin dosing, the laxative did not alter the observed change [25]. This study demonstrates the importance of using relevant pharmacodynamic parameters and a relevant study population. The pharmacodynamic parameter should be a validated surrogate marker and be sensitive to changes in response. Had the study been conducted in patients with atrial fibrillation and used a therapeutic dose of digoxin, changes may have been apparent. Discussions on specific issues relating to pharmacodynamic drug interactions are beyond the scope of this chapter because the endpoint parameters depend on the pharmacology of the specific drug class and the characteristics of the parameter itself.

9.3 Pharmacokinetic Interaction Studies

9.3.1 Interactions Affecting Drug Absorption

Drug interactions may involve absorption or other aspects of drug delivery. This chapter does not address pharmaceutical or physicochemical interactions that occur in vitro or ex vivo such as incompatibility involving intravenous admixtures or mixing within intravenous administration tubes. Drug interactions commonly occur with drugs that are administered orally. Most of these interactions involve the effect of a precipitant drug on gastric pH or physical interactions between the two drugs. If an acidic environment in the stomach is required for optimal dissolution, reduced absorption in the presence of drugs that increase gastric pH may occur. The interaction between acid suppressants (e.g., cimetidine or omeprazole) and ketoconazole or itraconazole is a classic example of this type of interaction [26, 27]. Interaction studies should be performed for drugs that have greatly reduced solubility at neutral pH compared to pH < 3. One must be careful to provide sufficient doses of the acid suppressant to increase gastric pH to >6 during the absorption period [28]. Continuous monitoring of gastric pH is recommended to ensure that the target pH is attained.

Many drugs bind or complex with other drugs, thereby preventing gastrointestinal absorption. Examples of this type of interaction include tetracycline and calcium carbonate, ciprofloxacin and aluminum antacids or iron products, and norfloxacin and sucralfate [29–31]. These interactions occur when both drugs are present in the stomach and upper gastrointestinal tract at the same time. Maximum interaction usually occurs when the precipitant drug is administered slightly before or at the same time as the object drug [30]. Although not well studied, differences in gastric pH, gastric emptying time, and transintestinal elimination of drug may influence the extent of these interactions.

In the infancy of pharmacokinetics, drug absorption after oral administration was regarded as a passive diffusion process affected by pH (portion unionized) and lipophilicity. We now know that the process is extremely complex and involves many transporters located in the basolateral and apical (lumen) sides of the gastrointestinal epithelium. There are numerous transporters on the apical membrane, some of which serve to facilitate absorption of drugs (e.g., peptide transporter 1 (pPEPT1) and organic cation transporters (OCT1/3)), and others serve to limit intestinal absorption (e.g., p-glycoprotein (pGP), multidrug resistance protein (MRP2)) [32]. Beta-lactam antibiotics are very hydrophilic drugs and would be expected to poorly diffuse across lipid membranes. However, some beta-lactams exhibit high oral bioavailability through vectorial apical to basal transport utilizing PEPT1 on the apical side and MRP3 on the basolateral side [32]. In contrast, fluoroquinolone antibiotics may inhibit PEPT1; however, this interaction is based on in vitro assessment of potential rather than clinical studies [33]. The best-known efflux transporter is probably pGP, which has a substrate specificity similar to CYP3A4. Substrate drugs are absorbed through the intestinal mucosa into the enterocytes and then transported back out into the intestinal lumen by pGP. There is an abundance of CYP3A4 present to metabolize the same substrate, resulting in a cycle of entry, efflux, and metabolism, which substantially limits bioavailability of some drugs. Strong inhibitors of pGP such as itraconazole or HIV protease inhibitors inhibit both pGP and CYP3A4 and result in very large increases in bioavailability of drugs like nadolol [34, 35]. In 2010 an International Transporter Consortium was formed and identified seven transporters of particular importance including pGP, BCRP, organic anion-transporting polypeptides (OATP1B1 and OATP1B3), organic cation transporter (OCT2), and organic anion transporters (OAT1 and OAT2). The list was updated to include multidrug and toxin extrusion proteins (MATE1 and MATE2K), multidrug resistance protein (MRP2, MRP3, and MRP4), and bile salt export pump (BSEP). Candidate probe substrates and inhibitors were proposed, although the substances often lack specificity for a given transporter [36].

The liver also operates with involvement of transporters. Drugs entering the portal circulation or in systemic circulation can be transported into hepatocytes by organic anion transport proteins (OATP1B1, OATP1B3, OATP2B1, OATP1A2, OAT1, OAT2) and sodium-taurocholate cotransporting polypeptide (NTCP). The drug can be extruded from the hepatocyte into blood by MRP3, MRP4, or MRP5. Finally, a drug can be transported into bile using several transporters including pGP, MDR3, MRP2, BCRP, and BSEP [34].

The most recent FDA guidance for drug interactions provides a decision tree for evaluating transporter drug interactions [10]. The guidance recommends that a cell-based assay be used to evaluate whether the drug is a substrate for pGP or BCRP, particularly if the drug is intended for oral administration. If the drug undergoes hepatic or biliary secretion to a significant extent (Cl $\geq$25% of total clearance), it is important to investigate whether the drug is a substrate for OATP1B1 and/or OATP1B3 using an in vitro system. If any of the screening results show that the drug is a substrate for these transporters, selected in vivo drug interaction studies are recommended. A list of known inhibitors and inducers for common transporters is provided in the FDA guidance. There should also be screening to determine if the drug induces or inhibits selected transporters. Assessment of effects on pGP is recommended, for example, if the drug inhibits or induces CYP3A4 in vitro [10].

9.3.2 Interactions Affecting Drug Distribution

Drug distribution may be affected by drug interactions. However, many studies conclude differences in volume of distribution that represent artifact rather than true differences. Changes in volume of distribution should be examined using intravenous dosing whenever possible. When oral administration is used, apparent changes in volume of distribution may represent changes in bioavailability. Comparisons should be made using steady-state volume of distribution (V_{ss}) only. Frequently V_{area} (also designated as V_z) is used for comparisons. However, this parameter is greatly affected by changes in the terminal elimination rate constant.

Steady-state volume of distribution may also be affected by experimental problems. Suppose a drug is well described using a three-compartment model when administered alone. The same drug is given after 10 d of rifampin treatment and the clearance is greatly enhanced. Drug concentrations are substantially lower following rifampin treatment, and the profile is best described using a two-compartment model. Presumably, the third exponential phase would remain present, but the concentrations may be undetectable with the assay used. V_{ss} is equal to mean residence time (AUMC/AUC) multiplied by systemic clearance (Cl) for an intravenous bolus dose, where AUMC is the area under the first moment of the plasma concentration-time curve. Although AUC would be decreased and Cl increased as a result of the interaction, these parameters would be affected minimally by missing the third exponential phase. However, the third exponential phase contributes a large portion of the total AUMC for the control treatment. Excluding this phase following rifampin treatment will cause an apparent decrease in the V_{ss}. Thus, problems fitting the control and interaction phases to the same model with equal reliability could result in apparent changes in V_{ss} when no true change occurred. Similar problems would occur with non-compartmental analysis, but the problem would not be as apparent.

Examples of drug interactions affecting distribution include the interaction between ceftriaxone and drugs that increase free fatty acid concentrations (e.g., heparin). Free fatty acids displace ceftriaxone from protein binding. In this example, there were profound physiologic changes due to cardiopulmonary bypass, administration of high-dose heparin and methylprednisolone, and intravenous fluids. Along with this there were profound changes in the free ceftriaxone concentrations and renal clearance [37]. On a positive note, the free (active) ceftriaxone concentrations would be highest during the operation and could boost efficacy as a prophylactic antibiotic; however, persistence with longer operations may be reduced. Such an interaction is generally not clinically significant because the increased free fraction (microbiologically active drug) results in no change in average steady-state unbound concentrations in plasma even though renal clearance is increased. In general, for drugs that are highly protein bound, protein displacement interactions may be clinically relevant when the object drug has a narrow therapeutic range and a small volume of distribution (<10 L/70 kg) [8, 9].

A potentially significant situation involves parenterally administered drugs that exhibit a high extraction ratio. Here nearly all of the drug that passes through the organ is removed or metabolized including both bound and unbound drugs. Displacement from protein binding will have no effect on the total clearance of the drug. However, the increased free fraction of drug may result in greater pharmacodynamic activity while the precipitant drug is present. For the interaction to be significant, the object drug must have a narrow therapeutic index so that the increase in free drug concentration will have toxicologic significance. Overall, protein binding displacement interactions are rarely clinically significant.

9.3.3 Interactions Affecting Renal Excretion

Changes in renal excretion of drugs can be subdivided into effects on filtration, secretion, and reabsorption. Glomerular filtration of drugs is limited by protein binding and only unbound drug is filtered. Drug interactions involving displacement of an object drug from serum protein will result in transiently higher unbound serum concentrations and lead to increased renal clearance for object drugs that have a low renal extraction ratio. The clinical significance of protein binding displacement is limited by the compensatory increase in renal clearance as lower total serum concentrations from increased clearance compensate for the increased free fraction.

Tubular secretion involves active transport of drugs from the serum to the tubular lumen mediated by a number of drug transporters. Separate transport systems are present for cationic and anionic compounds, but these transport systems have a very low degree of specificity. Various transport proteins are located on the basolateral side of the proximal tubular cells including OAT1/3, OAT2, OATP4C1, and OCT2, which are in the solute carrier family (SLC22A). These transporters mediate facilitated transport across an electrochemical gradient often exchanging for an ion (e.g., Na^+ and H^+) or another solute (e.g., dicarboxylate). The substrate is delivered from blood to the cytoplasm of proximal tubule cells [38]. OAT1/3 transport represents important pathways for secretion of many beta-lactam antibiotics, tetracycline, ciprofloxacin, acyclovir, adefovir, cidofovir, entecavir, stavudine, tenofovir disoproxil fumarate, and zidovudine. OCT2 is an important transporter for lamivudine and zalcitabine, although other anti-infective drugs are also substrates [39]. As with CYP450 enzymes, in vitro systems have been developed using probe drugs including furosemide for OAT1/3 and metformin for OCT2. These cell-based systems can be used to screen for potential inhibitor drugs [40]. In another system, adefovir and benzylpenicillin were used as probes for OAT1 and OAT3. Probenecid inhibited the uptake of both adefovir and benzylpenicillin, whereas para-aminohippurate (PAH) selectively inhibited adefovir uptake. These in vitro results were predictive of the interaction observed in humans, although a 47% increase in benzylpenicillin renal clearance induced by PAH was not expected. Penicillins, cephalosporins, and carbapenems are transported in the kidney by OAT1 and to a greater extent by OAT3 promoting the accumulation of the drugs in the cytoplasm of renal tubular cells.

Other transport proteins may be involved in extruding the drugs from the cell into the tubular lumen. A few members of these beta-lactam family have produced nephrotoxicity, and the possibility exists that this is related to transport and accumulation of too much drug in the proximal tubular cells. Antiviral drugs including adefovir, cidofovir, and tenofovir undergo transport by OAT3 and cause nephrotoxicity [41]. In the case of cidofovir, probenecid is used to reduce nephrotoxicity by inhibiting basolateral transport and intracellular accumulation in renal tubular cells [42].

Transporters are also integrated on the apical membrane of tubular cells and are involved in getting drug from the cytoplasm to the tubule lumen. Cation transport proteins include pGP, organic cation transporters (OCTN1/2), and MATE1/2. The organic anion transport proteins include MRP2/3, OAT4, and urate transporter (URAT1). OCTN, OAT4, and MATE1/2 are in the SLC family (SLC22A or SLC47A), whereas pGP and MRP are ATP-dependent active transporters [38]. Much less is known about the role of the apical efflux proteins in the context of drug interactions. Digoxin is a well-known substrate for pGP for which clarithromycin serves as an inhibitor. As pGP is found in many tissues, components of the interaction are difficult to dissect. However, clarithromycin coadministration with intravenously administered digoxin results in about 20% increased digoxin exposure (AUC) in part because of a 40% reduction in non-glomerular clearance [43]. From a toxicology perspective, accumulation of drug in the proximal tubule is a major determinate of kidney toxicity. Such accumulation is related to the balance of uptake across the basolateral membrane and trafficking across the apical membrane (extrusion and passive re-update). Minimizing update by inhibiting basolateral transport and being careful not to inhibit apical extrusion may be important to modulate nephrotoxicity risk.

Precipitant drugs may inhibit tubular secretion resulting in reduced renal clearance. Drugs that are extensively eliminated in the urine and have significant tubular secretion (renal clearance of free drug greater than 150% of glomerular filtration) are good candidates for studying this interaction mechanism. The normal glomerular filtration rate is about 120 mL/min, and the renal blood flow is approximately 1100 mL/min for a 70 kg adult. A drug can have a renal clearance approaching renal blood flow rate, as is observed with PAH, owing to its extensive tubular secretion. However, drugs that exhibit restricted to intermediate renal clearance are more susceptible to drug interactions involving inhibition of tubular secretion. The partitioning of a drug into red blood cells and the ability to diffuse out of red blood cells may also influence tubular secretion.

Probenecid may be administered with certain beta-lactam drugs to prolong their elimination rate. The beta-lactam agents most affected by this interaction have a high ratio of renal clearance to glomerular filtration rate and rely on the kidney as their major clearance organ. Before penicillin resistance was prevalent, a combination of probenecid and high-dose amoxicillin was used to provide single-dose treatment for uncomplicated gonorrhea [44].

To assess drug interactions involving renal excretion, collection of both urine and plasma (or serum) is required. A measure of the glomerular filtration rate before or during the study is helpful to explore the mechanism of interaction. Glomerular filtration rate (GFR) can be determined by radiolabeled ^{99m}Tc-diethylenetriamine pentaacetic acid clearance, ^{125}I–iothalamate clearance, inulin clearance, or creatinine clearance (with concurrent cimetidine treatment) [45–47]. Measurement of creatinine clearance also serves as a rough measure of GFR. However, overestimation of GFR is expected owing to a small component of tubular secretion. Although unusual, the tubular secretion of creatinine may be large. As cimetidine inhibits the tubular secretion of creatinine, concurrent treatment during urine collection can improve the estimate of GFR [47].

Competitive inhibition of tubular secretion is typically concentration dependent and is influenced by the concentration of the precipitant and object drugs. Concentration-dependent renal clearance of the object drug is established by collecting urine in intervals less than or equal to one half-life duration. Blood samples collected at the beginning and end of each urine collection interval are a minimum requirement, but more blood samples taken during the collection interval will provide a better estimate of plasma AUC. The renal clearance is calculated for each interval and would be expected to increase as drug concentrations (plasma AUC) decline. A precipitant drug may have only minor effect on the renal clearance when concentrations of the object drug are high, because saturation may already be present. However, a drug that potently inhibits tubular secretion should prevent the increase in renal clearance seen at low concentrations of the object drug. The precipitant drug must be present in sufficient concentrations throughout the observation period to observe inhibition. Thus, continuous infusion or frequent dosing of the precipitant drug may be required unless the half-life of the precipitant drug is long. An interaction study also may be planned using dosing regimens likely to be used in clinical practice. However, information about the mechanism of interaction may be lost. An assumption usually made in pharmacokinetics is that clearance of the object drug is stable during each assessment period. If there are large differences in peak and trough drug concentrations of the precipitant drug over the period in which the pharmacokinetics of the object drug is assessed, this assumption may be violated because the degree of inhibition depends on inhibitor concentration. Information about the mechanism of interaction may also be lost if urine is collected in only one interval to obtain the average renal clearance.

Tubular reabsorption is usually a passive process whereby drug present in the tubular lumen (high concentration) diffuses back into the capillary lumen and returns to circulation. The drug must be unionized to diffuse across the tubular membrane. Interactions occur from altered pH in the tubular lumen or from physical interaction between the precipitant and object drug within the tubular lumen. An independent measure of tubular secretion, filtration, and reabsorption is not possible in the clinical setting. Instead, only the overall renal clearance is measured, and the intrinsic clearance is compared to GFR to classify the elimination as net tubular reabsorption, filtration, or net tubular secretion.

9.3.4 Interactions Affecting Drug Metabolism

CYP enzymes metabolize many anti-infective drugs whose pharmacokinetics are affected by drugs that inhibit or induce these enzymes. Several anti-infective agents act as inhibitors (ritonavir, ciprofloxacin, etc.) or inducers (rifampin, rifabutin, etc.) of CYP enzymes. Goals for a metabolism interaction study are important to establish. The goal may be to determine if a clinically significant interaction is likely between two drugs or to determine more broadly if a drug serves as a precipitant drug involving a particular enzyme system. The precipitant drug should be administered in a clinically relevant, multiple-dose regimen with sufficient duration to achieve steady-state pharmacokinetic conditions. Longer durations of treatment may be required for time-dependent interactions. For example, the maximum induction with rifampin takes 10–13 days [48]. When no prior knowledge is available, multiple dosing for at least 1 week is usually sufficient. A longitudinal design in which the object drug is studied alone and then following treatment with the precipitant drug is preferred in the absence of prior knowledge about the interaction offset time. If the offset time is of interest, the object drug may be studied again one or more times after the precipitant drug is stopped.

More than 50% of drugs that undergo metabolism are metabolized primarily by CYP3A enzymes. These enzymes are induced by rifampin, rifabutin, phenytoin, carbamazepine, and barbiturates and are present in the gastrointestinal tract, liver, and other organs. CYP3A4 enzymes are responsible for first-pass metabolism of many drugs, and their inhibition may lead to pronounced increases in systemic bioavailability of orally administered object drugs that undergo first-pass metabolism. Precipitant drugs may induce or inhibit CYP3A4. Candidate object drugs are those that rely on metabolism by CYP3A4 enzymes for a substantial portion of their clearance. Midazolam is an excellent marker of CYP3A4 activity because its elimination depends almost entirely on hydroxylation by CYP3A subfamily of enzymes to form 1-hydroxy midazolam [49, 50]. Drugs that affect CYP3A activity in the gastrointestinal tract or liver may affect the apparent clearance of oral midazolam. N-demethylation of erythromycin is also catabolized by CYP3A and this metabolism occurs mostly in the liver. The intravenous administration of [^{14}C-N-methyl]-erythromycin and measurement of [14]CO_2 in breath provide a convenient marker of CYP3A4 activity in the liver (not gastrointestinal tract) [51–53] even though potential limitations of the test have been identified [54]. Cortisol is metabolized to 6β-hydroxycortisol by CYP3A4 isozymes. The measurement of urinary 6β-hydroxycortisol/cortisol ratio remains fairly stable without circadian differences. Agents that affect CYP3A4 enzyme activity usually cause changes in the 6β-hydroxycortisol/cortisol ratio [52, 53]. These markers are useful tools to identify induction or inhibition of CYP3A4, although changes in clearance may not correlate quantitatively among the different markers.

Other common metabolic enzyme pathways involve CYP1A2 and the polymorphic CYP2D6 and CYP2C19 isozymes. Probe drugs are caffeine and theophylline for CYP1A2 [55, 56], debrisoquin and dextromethorphan for CYP2D6 [57], and omeprazole and mephenytoin for CYP2C19 activity [58]. For caffeine and theophylline, changes in systemic clearance are usually evaluated. The measurement of paraxanthine/caffeine ratio in saliva at 6 h after caffeine intake also correlates with CYP1A2 activity [59]. CYP2D6 activity can be assessed by measuring changes in the dextromethorphan/dextrorphan ratio in urine [57]. CYP2C19 activity can be evaluated from the urinary S-mephenytoin/R-mephenytoin ratio after administration of racemic mephenytoin [60].

Markers of CYP isozyme activity are useful to evaluate whether a potential precipitant drug affects metabolism. There is also a need to evaluate whether a drug serves as an object drug resulting in toxicity, loss of therapeutic activity, or reduced effectiveness. Agents that are known to inhibit CYP1A2 (cimetidine, enoxacin), CYP3A4 (itraconazole, ketoconazole), CYP2D6 (quinidine, cimetidine), and CYP2C19 (omeprazole, fluconazole) are well known [61–65]. However, not all of these drugs have specific effects on only one isozyme. Rifampin, rifabutin, carbamazepine, and phenytoin are inducers of CYP3A4 and other enzymes [62, 63]. Lists of enzyme inhibitors and enzyme substrates can be found in recent publications [62, 64, 65].

If feasible, active or toxic metabolites in plasma and urine should be measured because the magnitude and direction of metabolite pharmacokinetic changes are often unpredictable. Multiple metabolic enzymes and pathways can confound predictions. The AUC of metabolite may be altered even if the metabolite is not a product of the affected pathway. Detectable changes in AUC of the parent drug may not be apparent if a minor metabolic pathway is affected or if compensatory changes in hepatic and renal clearance occur. Thus, there is a danger in concluding "no interaction" from data involving only the parent drug. Metabolic parameters such as the metabolic AUC ratio and the urinary recovery ratio of metabolite to parent drug can give useful information on mechanisms of interaction, particularly if the metabolite is eliminated exclusively by renal excretion.

9.3.4.1   Impact of Pharmacogenomics

Metabolic interactions are sometimes complicated by the existence of polymorphic enzyme expression. A recent trend in metabolic interaction studies is to characterize subjects by genotype and/or phenotype into extensive, intermediate, or poor metabolizers. In several of the studies reviewed, subjects were recruited without considering genotype or phenotype, leading to a very low number of subjects in less common metabolic groups [66–74]. Although more difficult and more expensive, the design would be improved by recruiting subjects based on genotype or phenotype with a target minimum number of subjects in each category. Larger clinical trial units should consider developing a subject database that includes genotype results for enzymes such as CYP3A5, CYP2C19, CYP2C9, and CYP2D6. Subject recruitment could be planned using a predictor panel concept similar to that used in microbiology to examine susceptibility against a panel of bacteria with categorized resistance mechanisms [66].

The impact of metabolic polymorphisms may vary substantially as demonstrated in the following examples. The effect of ritonavir on voriconazole exposure was studied in 20 subjects, which included 8 homozygous extensive metabolizers (EMs), 8 heterozygous EMs, and 4 poor metabolizers (PMs) based on CYP2C19 genotype. Total exposure ($AUC_{0-\infty}$) was increased 54% in homozygous EMs, 94% in heterozygous EMs, and 907% in PMs. Voriconazole Cl/F varied about eightfold during the placebo phase and part of this variation was from metabolizer status. Adding ritonavir resulted in about 70-fold variation from the highest Cl/F in a homozygous EM subject at baseline to the lowest Cl/F in a PM subject receiving ritonavir [67]. Findings have been mixed with CYP2C19. Moclobemide resulted in a significant increase in omeprazole AUC, an effect that was limited to EMs [70]. However, in the case of tacrolimus with administration of either lansoprazole or rabeprazole, an interaction was noted only in CYP2C19 PMs who also had the CYP3A5*3/*3 genotype. CYP3A activity becomes more important in these subjects as CYP3A4 and CYP3A5 exhibit a similar substrate profile [71]. Clarithromycin inhibited CYP2C19-mediated metabolism of omeprazole in EMs, IMs, and PMs to a similar extent. However, clinical implications become apparent after considering that coadministration of the two drugs in PMs resulted in 30-fold higher exposure (AUC) compared to the AUC in the EM group receiving omeprazole alone [72]. In contrast, oral contraceptives were shown to enhance carisoprodol AUC by 60% overall; however, there was no difference with respect to CYP2C19 genotype (EMs versus IMs) [73].

CYP3A4 is not polymorphic in expression; however, a small portion of the population expresses CYP3A5, which metabolizes essentially the same substrates as CYP3A4. Consequently, subjects expressing CYP3A5 tend to be EMs. Drugs that inhibit CYP3A4 may not have the same magnitude of effect on CYP3A5, which is typically less susceptible to inhibition [68]. Using grapefruit juice as an enzyme inhibitor of both CYP3A4 and CYP3A5, the urinary 6β-hydroxycortisol/cortisol ratio varied depending on CYP3A5 genotype. Likewise, genotype of MDR1 (pGP) was associated with urinary 6β-hydroxycortisol/cortisol ratio in a pattern that suggested that both polymorphisms affect this cortisol endpoint [69].

Some interactions are extremely complex as noted with the mixed inhibitor inducer HIV protease inhibitor combination, tipranavir/ritonavir. The combination produced weak induction of CYP1A1, moderate induction of CYP2C19, potent induction of pGP, and potent inhibition of CYP2D6 and CYP3A after multiple dosing [74].

Given the potential differences in the effects of metabolic interactions based on genotype, either measuring genotype or perhaps planning studies with genotype entry criteria should be considered.

9.3.5   Interactions Affecting Other Elimination Pathways

Some drugs are eliminated by fecal excretion and are excreted in bile or by transintestinal elimination. Enterohepatic recycling occurs when drugs are eliminated in bile as conjugates. Deconjugation may occur in the small intestine, thereby allowing for reabsorption of the parent drug. A precipitant drug that interferes with deconjugation will prevent enterohepatic recycling (reabsorption) and increase the apparent clearance. Potential examples of this interaction type involve antibacterial drugs and oral contraceptives [75]. Valproic acid (VPA) is a very lipophilic drug which undergoes conjugation to an acylglucuronide metabolite. The VPA-glucuronide can either be excreted by the kidney or hydrolyzed back to the parent drug in the liver. The enzyme responsible for VPA-glucuronide hydrolysis is an acylpeptide hydrolase found in liver cytosol. Thus VPA-VPA-glucuronide bidirectional cycling results in longer persistence of VPA in the body. This interesting pharmacokinetic phenomenon was explained after the interaction between VPA and meropenem was recognized, leading to markedly increased VPA clearance. Subsequently, meropenem and other carbapenem antibiotics were shown to inhibit this acylpeptide hydrolase [76].

Precipitant drugs that physically trap or bind another drug within the gastrointestinal lumen may also enhance the clearance of the object drug. Examples of this interaction include iron salts or aluminum hydroxide with doxycycline [77, 78].

9.4 Pharmacostatistical Techniques

Advances have been made in the past decade to facilitate detection and evaluation of drug interactions. The intent of this section is to focus on the recommended approaches for presenting and analyzing pharmacostatistical drug interaction data. In discussions below, the terms "test" and "reference" treatments refer to the administration of the object and precipitant drugs in combination (test) and administration of the object drug alone.

9.4.1 Statistical Analysis Approach

There are many approaches, both parametric and nonparametric, to analyze comparative pharmacokinetic data from drug interaction studies. The statistical strategy recommended by regulatory agencies in the United States [10] and Europe [11, 79], editors of clinical pharmacology journals [80], and others [81, 82] is to adapt the confidence interval approach used in average bioequivalence studies [14, 83]. A bioequivalence study is a type of comparative bioavailability study conducted to demonstrate that the shape and magnitude of blood or plasma concentration-time profiles produced by the drug formulations under study are sufficiently alike that therapeutic equivalence can be assumed. In drug interaction studies, the aim is to determine whether the interaction is clinically meaningful from differences in concentration-time profiles or other pharmacokinetic characteristics between test and reference treatments. Comparison between profiles is at a minimum based on maximal exposure (C_{max}) and overall exposure (AUC) and presented as a mean ratio (MR: test/reference). The two one-sided t-test is based on the null hypotheses: H_{01}, MR $\leq$ lower bound, and H_{02}, MR $\geq$ upper bound. The lower and upper bounds need to be specified in advance, and values of 0.8 and 1.25, respectively, are typically used for log-transformed pharmacokinetic parameters in bioequivalence and drug interaction studies. The alternate hypothesis is that the MR falls within the specified lower and upper bounds, H_{alt}: lower bound < MR < upper bound.

In traditional analysis, the null hypothesis stipulates that parameters for the object drug are equivalent for the test and reference treatment. When a significant difference is found, the null hypothesis would be rejected and a difference would be concluded. A small, clinically unimportant difference may be statistically significant at the 5% level of significance ($\alpha = 0.05$). The lack of significance does not necessarily imply "no interaction." In such cases, the statistical power, or probability of detecting a specified difference, must be considered. The specified difference should be a change that would be considered clinically important given the available pharmacodynamic and toxicologic information. A large, clinically important difference between treatments may not be statistically significant if sample size is small and within- and/or between-individual pharmacokinetic variability is large. Therefore, classical statistical approaches that attempt to confirm an interaction by rejecting the null hypothesis of "no difference" are inappropriate because the consumer risk is not controlled.

An equivalence approach is necessary to adequately address the risk to the consumer. Because a drug-drug interaction consists of different drug treatments, one should test the null hypothesis of "nonequivalence" by demonstrating "equivalence" or "lack of pharmacokinetic interaction," as first proposed by Steinijans et al. [84] In this manner the risk to the patient of a clinically relevant interaction can be defined within established limits. Generic drugs are approved on the basis of bioequivalence compared to a reference product. Risk to consumers is considered low for most drugs when substituting a generic drug that is considered bioequivalent. The same principle applies when a potential interacting drug is studied, and despite concomitant administration, the exposure to the object drug remains equivalent to the object drug given alone.

Two important assessment criteria must be defined before invoking the equivalence approach: [1] the range of clinically acceptable variation in pharmacokinetic response of the affected drug and [2] the risk to the consumer of incorrectly concluding a "lack of pharmacokinetic interaction." The range of clinically acceptable variation defines the equivalence range (clinical no-effect boundary). The range can be based on population average dose-related and/or individual concentration-response relationships derived from PK/PD models and other available information about the object drug that relates to the extent of difference in exposure caused by the interaction that is of no clinical consequence [10]. The consumer risk is the type I or α-error in statistics, which is usually set at 5%.

The equivalence method is based on two one-sided t-test procedure of rejecting the null hypothesis that the mean test/reference ratio is less than the lower equivalence limit or greater than the upper equivalence limit. At the 5% level of consumer risk, this procedure is operationally identical to the method of declaring equivalence (or lack of interaction) if the 90% confidence interval for the mean test/reference ratio is entirely within the specified equivalence range. More generally, the

$100 \times (1{-}2\alpha)\%$ confidence limits around the ratio (test/reference) of the means or medians of the test and reference treatments constrain the consumer risk to $100 \times (\alpha)\%$ as well as indicate the precision by the width of the confidence interval. In bioequivalence studies the accepted equivalence range is ±20%, which corresponds to a lower limit of 80% and an upper limit of 120% for original data or 125% for log-transformed data. A range of ±20% seems reasonable to assess product quality, but for drug interactions these limits may be wider or narrower depending on the patient population and the therapeutic index and pharmacokinetic variability of the object drug. For example, a range of clinically acceptable variation of 30% for changes in zidovudine AUC was suggested [85], whereas a range variation of 50% for changes in indinavir AUC was proposed [86]. Equivalence limits of the form (θ, $1/\theta$) have been proposed for data on both the original and logarithmic scales, where θ is the lower limit for the test/reference ratio [87]. The upper limit would be the reciprocal (e.g., limits of 0.8 and 1.25). No dose adjustment is required if the confidence interval falls within the no-effect boundary. Also, there is no requirement that the boundary must be symmetrical around a mean ratio of 1.0 or 100% [88]. For example, the lower bound could be 80% to ensure no loss of efficacy, but the upper bound could be set at 150% for a drug with low risk of toxicity.

Statistical inferences are made on either absolute (test-reference) or relative (test/reference) differences in the arithmetic means, geometric means (from logarithmic transformed data), harmonic means (from reciprocal transformed data), or medians of pharmacokinetic variables. Parametric analysis of variance (ANOVA) models appropriate for the study design are used to test differences in means (C_{max} and AUC), and nonparametric methods such as the Wilcoxon rank sum test or Wilcoxon signed-rank test are used to test differences in medians (t_{max}). If the study design is unbalanced from an unequal number of subjects in each sequence (crossover) or from missing data, assessments are based on least-squares means. Because clinicians prefer to think in terms of relative rather than absolute changes, pharmacokinetic differences are usually expressed as a ratio (test/reference). Confidence limits around these mean ratios for within-subject comparisons in crossover studies and between-group comparisons in parallel studies are constructed from the residual mean-square error (MSE) term in ANOVA. The ANOVA provides exact confidence limits for relative differences of geometric means if the distribution of variables is truly lognormal. Only approximate limits for relative differences of arithmetic means are possible, because ANOVA ignores variability in the reference mean and treats the reference mean as a constant instead of as a variable when dividing by the reference mean to convert a test/reference difference to a test/reference ratio, unless Fieller's theorem is applied [89]. Nonparametric approximate 90% confidence limits can be calculated for two-period, two-sequence crossover studies [90]. One should be cautious in concluding "no interaction" when approximate confidence limits generated from parametric or nonparametric techniques are within but near the equivalence limits. Also, inferences on mean data may not reflect how certain individuals in the study population respond to the interaction. A particular stratum of individuals may show an apparent interaction even though the overall mean data indicate no pharmacokinetic interaction.

9.4.2   Logarithmic Transformation of Pharmacokinetic Variables

All pharmacokinetic variables, except those such as t_{max} that depend on discreet sampling times, are logarithmically transformed before ANOVA [14, 84, 91]. Harmonic means have been proposed for inferences on half-life [92]. Transformation converts a multiplicative model to an additive model, which is the basis of ANOVA [ln(test/reference) = ln(test) – ln(reference)]. Decisions on t_{max} are best handled by nonparametric analysis. Most pharmacokinetic data have positively skewed distributions created by the truncation of these quantities at zero and have variances that depend on the mean. Transformation reduces the skewness and brings the distribution of data closer to normal. However, the main reason for transforming the data is to stabilize or make equal the within-subject (crossover study) or between-group (parallel study) variance and not to normalize the between-subject parameters [91]. Another advantage of transformation is that it is the best way to handle ratios for relative or proportional differences, and calculation of the associated confidence limits is straightforward.

For most studies the outcome will not change regardless of whether the original or log scale is used. There are two instances where conclusions can be opposite in a within-subject design [91]. If certain subjects with larger than average responses show larger than expected absolute differences, variability is increased on the original scale, whereas larger than expected absolute differences for smaller than average responses are expanded on the log scale. If this occurs, for example, when fast and slow metabolizers are studied together, then the within-subject variability and the relative mean changes can be different on the two scales.

9.4.3   Crossover Design and Analysis of Variance

The ANOVA for a crossover design includes the effects of sequence, subject within sequence, treatment, period, and, except for the 2, 2, 2 design, carryover. All effects except the sequence effect are tested by the MSE term. The sequence effect is tested against the subject-within-sequence effect. Any subgroup comparison of fixed effects (e.g., males and females) is tested with the subject mean-square term.

The sequence effect measures the difference between the groups of subjects defined by their sequence. In statistical parlance, a true sequence effect is known as the treatment-by-period interaction, which is a measure of the differential effect of the treatment (test-reference) in each of the periods. In the 2, 2, 2 design, the sequence effect is caused by three confounded sources: [1] a difference between subjects in the two sequences (i.e., group effects), [2] an unequal carryover of one treatment into the next period compared to the other treatment, or [3] a treatment-by-period interaction. In this case a significant sequence effect ($p < 0.1$) requires further explanation and evaluation of questions including: Was randomization appropriate? Was the washout period sufficient? Were trial conditions, analytical methodology, and clinical settings applied consistently? However, a true sequence effect (i.e., group effect) does not invalidate the determination of bioequivalence [93]. A sequence effect in the 2, 2, 2 design can be due to unequal carryover between treatments, in which case the analysis of period 1 data should be presented separately; carryover should be evaluated by checking the pre-drug plasma assay results.

The period effect measures the difference between study periods or alternatively the differential effect of the treatment in each of the sequences. In a 2, 2, 2 study, the period effect is completely confounded with treatment-by-sequence interaction. Any difference in treatment comparison (test-reference) between the two sequence groups cannot be distinguished from period effects. If there are carryover effects or if more than two periods are included, then the period effect and treatment-by-sequence interaction are not interchangeable. The period effect can be caused by equal carryover in each sequence from period to period, bias in analytical data if samples in each period were analyzed in different batches, differences in the study environment or procedures, and changes with time in stage of disease. As with sequence effect, carryover can be ruled out by checking the pre-dose drug concentrations before period 2.

In a 2, 2, 2 study, the presence of a treatment effect (i.e., period-by-sequence interaction) implies that differences between periods are in opposite directions for the two sequence groups (if P2 – P1 in S1 is negative, then P2 – P1 in S2 may be positive). The estimate of treatment differences will not be biased if a period effect is present.

If the test treatment is determined not to be bioequivalent, then a treatment effect may be expected; however, treatment effects may also be observed when a bioequivalent determination is made for products with low intra-subject variability. A significant treatment effect may be entirely ignored when equivalence criteria are met.

The MSE term is a measure of the intra-subject variability and is usually converted to a coefficient of variation (CV_W) to estimate the consistency of the magnitude of interaction among the subjects [94]. The CV_W is estimated as $100\% \times (e^{MSE} - 1)^{1/2}$ for logarithmic transformed data and as $100\% \times (MSE)^{1/2}/Y$ for original data, where Y is either the least-squares mean of the reference treatment or the combined mean of the two least-squares treatment means being compared. The goal of any within-subjects design is too minimize the CV_W. The interaction is considered highly variable for a particular pharmacokinetic parameter if the CV_W is >30%. The CV_W is a very informative parameter but is rarely reported in the literature. Values for a number of drugs orally administered in crossover bioequivalence studies have been tabulated by Steinijans et al. [95] The CV_W is important to know because the width of the confidence interval around the difference of treatment means, the calculation of post hoc power to detect these differences, and an estimation of sample sizes for planning future interaction studies require an estimate of CV_W.

There are a number of sources of variation in CV_W: the true intra-subject pharmacokinetic variation exhibited by a single person, analytical variability (measurement errors), within-batch variation in manufacture of the drug formulation, nonadherence to the medications, and the random subject-by-treatment interaction. This latter source is caused by random variability of treatments within subjects or within identifiable subgroups of the population studied. Each individual may behave differently to the test treatment, or subjects in subgroups may show similar variation within subgroups but different responses to the test treatment among subgroups. An example could be smokers responding differently from nonsmokers to one of the treatments. On the log scale, the random subject-by-treatment interaction is minimized if all subjects show the same relative change in the same direction.

9.4.4 Sample Size and Post Hoc Power Calculations

The sample size of the study needs to be planned with consideration of the purpose of the study. If the purpose of the study is to evaluate a potential drug interaction that is suspected based on preliminary data, the sample size can be somewhat conservative. However, if the goal is to demonstrate the lack of interaction for an individual drug when a member of the same drug class exhibits the interaction (class labeling), then the sample size should be larger. Estimations of sample size for a within-subject drug interaction study require a knowledge of CV_w for the interaction. CV_w values for drug interaction studies may be greater than those reported for drugs in bioequivalence studies [95] because not all subjects will respond to the precipitant drug to the same degree. Tables of sample sizes for 2, 2, 2 crossover designs to attain a power of 80% or 90% at the 5% nominal level for a given CV_w and expected relative difference in treatment medians or means are published for the multiplicative (logarithmic) model with equivalence ranges of 0.7–1.43 [96], 0.8–0.25 [87, 97], and 0.9–1.11 [96]. Similar tables are published for the additive (original) model [98] and for parallel designs [87, 99]. The minor influence of the between-subject coefficient of variability on sample size estimates for the 2, 2, 2 crossover design is demonstrated by Hauschke et al. [87]

Post hoc power calculations have limited utility but can be used for negative studies to estimate differences that could be detected with a certain power (usually 80% at the 5% significance level) or to estimate the power of the study to detect a specified difference (usually 20% difference from reference at the 5% significance level). These calculations require an estimation of the standard error of the difference in mean or medians. General equations for point hypothesis testing for original and logarithmic data using a central t-distribution are provided in references [89, 100]. General equations for interval hypothesis testing using a noncentral t-distribution for crossover and parallel designs are given in references [87, 89].

9.5 Pharmacokinetic Metrics and Characteristics

The major assumption in bioequivalence is that the Cl of the drug under investigation is constant over the course of the study and that AUC is a pure characteristic of extent of bioavailability (F). In drug interactions both clearance and bioavailability can change after oral administration. Therefore, changes in AUC can result from alterations in either parameter. Schall et al. [101] proposed the terminal elimination half-life ($t_{1/2,z}$) and the ratio of AUC/$t_{1/2,z}$ as characteristics for Cl and F, respectively, in drug-drug interaction studies. Interpretation involves looking at the ratio of $t_{1/2}$ (test/reference) and ratio of AUC/$t_{1/2,z}$(test/reference ratio). Note that AUC$_{0-\infty}$ should be used for a single-dose case and AUC$_{0-tau}$ should be used for steady state. Assuming a constant volume of distribution, if the $t_{1/2,z}$ ratio is >1, then the interaction results from reduced Cl, and if the $t_{1/2,z}$ ratio is <1, the interaction results from increased Cl. If AUC/$t_{1/2,z}$ (test/reference ratio) is >1, then the interaction is at least in part due to increased bioavailability. Finally, if AUC/$t_{1/2,z}$ (test/reference ratio) is <1, then the interaction is at least in part due to decreased bioavailability. However, interactions may result from a mixture of altered Cl and altered bioavailability.

Because AUC is a composite characteristic of Cl and F, and peak drug levels (C_{max}) reflect both rate and extent of absorption, these metrics can be used to indicate drug exposure [102]. AUC is the ideal metric for total systemic drug exposure and C_{max} is a measure of peak systemic exposure. The term drug exposure conveys more clinical relevance than the term "rate and extent of drug absorption" because drug safety and effectiveness are concerns in drug interaction studies.

9.6 Presentation and Interpretation of Drug Interaction Data

There are generally three ways to present comparative pharmacokinetic data for changes in the test treatment relative to the reference treatment: [1] a test/reference ratio expressed as a percentage; [2] an x-fold change, where x is the test/reference ratio; or [3] a percentage change [(test/reference ratio − 1) × 100%]. For example, an AUC ratio of 200% indicates a twofold increase and a 100% increase in AUC. Often x-fold changes are confused with percentage change, and the reader needs to be aware of which method of calculation was used.

Current thinking favors expressing the results in terms of a test/reference geometric mean ratio and the corresponding 90% confidence limits for AUC and C_{max} parameters. The use of 95% confidence limits should not be confused with 90% confidence limits. The former bounds will be wider and may lead to different conclusions in equivalence testing. Reporting the 95% confidence limits is another way of reporting a test of significance at the 5% level of significance. For example, AUC of bosentan increased 2.1-fold (95% confidence interval 1.5–2.7) after concomitant administration with ketoconazole [103]. The 95% confidence interval would be examined to determine if it includes the value 1.0, and if not, as in this case, a statistically significant interaction at the 5% level of significance ($p < 0.05$) would be concluded. Use of statistical testing for difference should not be used for the reasons cited in this chapter.

9.6.1   No-Effect Boundary

The "no-effect boundary" or acceptable range needs to be established a priori. If a drug interaction is concluded, the clinical significance of the interaction and recommendations on how to manage the interaction need to be formulated. The FDA guidance for drug interaction studies allows two approaches for developing a no-effect boundary [10]. The first approach is to describe the range of the selected exposure parameters over a range of doses that are normally used. The sponsor should include information on dose and/or concentration-response studies or PK/PD models to support the recommendation. If the exposure parameters remain within this range in the presence of a potential precipitant drug, the sponsor could conclude that "no clinically significant interaction is present." The second approach defaults to bioequivalence criteria where the 90% confidence interval for geometric mean exposure parameter ratio (test/reference) falls within 80–125% [10]. This latter approach is most commonly used.

The use of bioequivalence criteria should eliminate a substantial portion of studies that statistically conclude a drug interaction when only small clinically insignificant differences occur. As an example, digoxin steady-state AUC was 25.5 ng·h/ml after digoxin alone and 23.9 ng·h/ml after digoxin plus zaleplon (a hypnotic agent). From a test of significance (ANOVA, $p = 0.018$), a drug interaction would have been concluded. The geometric mean ratio (test/reference) was 93% with a 90% confidence interval of 89–98%, and this would more appropriately lead to a "no-effect" conclusion [104]. Potential problems with the equivalence approach include too small of a sample size and high variability. If the sample size is too small, confidence intervals tend to be wide, and this could result in a 90% confidence interval that falls outside of the "no-effect boundary" despite a mean ratio near 100%. Too large of a sample size with the bioequivalence approach does not cause adverse consequences other than excessive study costs and ethical issues of imparting risk to numbers of subjects greater than needed. For tests of difference, too small of a sample size will lead to low power and inability to detect an important drug interaction, and too large of a study population may cause detection of small, clinically insignificant changes.

Not only does the no-effect boundary need to be established a priori, use of unconventional ranges needs to be justified. In a study evaluating the effect of montelukast on digoxin, several problems are apparent. The authors used a no-effect boundary of 70–143% without appropriate justification. Digoxin exhibits a narrow therapeutic index and relatively low variability in exposure parameters in a healthy population.

The mean digoxin $AUC_{0-\infty}$ was 43.2 ng·h/ml for digoxin alone and 39.2 ng·h/ml for digoxin plus montelukast. Although the 90% confidence interval for $AUC_{0-\infty}$ was 70–118%, the authors concluded that montelukast has no effect on the pharmacokinetics of digoxin [105]. The use of this expanded no-effect boundary for a drug with a narrow therapeutic index is concerning. Moreover, the 90% confidence interval is too wide to fit within the range of 80–125%. The study involved a small sample size ($n = 10$) and did not address power.

In another study, which evaluated the effects of proton pump inhibitors on theophylline, the no-effect boundary was expanded to 70–143% for steady-state C_{max}, but not for steady-state AUC [106]. There is no pharmacokinetic basis to suspect a change in rate of absorption of theophylline from acid suppression, and the reason for the expanded boundary was not addressed. Because the observed 90% confidence limit for steady-state C_{max} fell within the range of 80–125%, conclusions remain appropriate. In some cases involving drugs (e.g., ethionamide) with moderate to high variability in exposure parameters, it may be difficult to obtain 90% confidence intervals that fall within the usual no-effect boundaries, requiring the use of large sample sizes or expanded boundaries [107].

An example of a study that used an expanded no-effect boundary and provided justification involved interactions between didanosine, indinavir, ketoconazole, and ciprofloxacin [108]. A no-effect boundary of 75–133% was used. The authors cited a study where the AUC of indinavir was increased 29% with clarithromycin administration, and the interaction was concluded to be not clinically significant. For ciprofloxacin, the authors cited the package insert and a publication and considered that a 48% increase in ciprofloxacin AUC in elderly subjects did not result in a recommendation for reducing the dose. For ketoconazole, the authors cited a study that reported a 59% increase in ketoconazole AUC when administered with food compared to fasting and considered that the labeling did not contain a recommendation for administering ketoconazole with food [108]. In another study, in which ketoconazole significantly increased the exposure of desloratadine, the interaction was concluded to be not clinically relevant as no changes in ECG parameters were observed [109]. Although such observation does not totally rule out clinical significance in special populations, the value of concomitant pharmacodynamic assessment is apparent.

9.6.2   Studies to Confirm Clinical Strategy

Another potential area of misinterpretation is when the doses and/or dosing intervals of the drug under investigation are different in the test and reference arms of the study. This may occur if the purpose is to obtain equivalent drug exposures over a specified time period in the absence and presence of an interacting drug. The magnitude of pharmacokinetic effect can appear smaller or larger if the control dose is larger or smaller. For example, 800 mg of indinavir every 8 h was estimated to give about the same AUC over 24 h as 400 mg indinavir every 12 h in the presence of 400 mg ritonavir every 12 h; from single-dose indinavir data, the magnitude of the interaction was actually about a fivefold increase in AUC if 400 mg of indinavir was used as the reference [110]. Depending on the purpose of the study, the analysis may compare the exposures between the two treatments; however, to avoid confusion, analysis should still be done with Cl or dose normalized AUC to characterize the extent of the interaction.

9.7 Summary

Since publication of the first edition of this chapter in 2001, issues still remain to be resolved concerning optimal design of drug interaction studies. Traditional issues, such as defining the research hypothesis (question of interest), determining the appropriate study population (healthy volunteers or patients), determining the study design (crossover, longitudinal or parallel, washout requirements, etc.), deciding between single-dose or steady-state, and deciding which pharmacokinetic and/or pharmacodynamic endpoints to evaluate, should depend on knowledge of the drugs involved, preliminary data on the potential interaction, and general knowledge of pharmacokinetics and drug interactions. Defining whether a drug interaction exists is now well accepted by regulatory agencies as an equivalence problem where endpoints are compared between the object drug given with and without the precipitant drug. The acceptable clinical no-effect boundary should be specified a priori, but allowing flexibility depending on the therapeutic index of the object drug and variability of the endpoints.

References

1. Ekins S (1996) Past, present, and future applications of precision-cut liver slices for in vitro xenobiotic metabolism. Drug Metab Rev 28(4):591–623
2. Decker CJ, Laitinen LM, Bridson GW et al (1998) Metabolism of amprenavir in liver microsomes: role of CYP3A4 inhibition for drug interactions. J Pharm Sci 87(7):803–807
3. Bonnabry P, Sievering J, Leemann T et al (2001) Quantitative drug interactions prediction system (Q-DIPS): a dynamic computer-based method to assist in the choice of clinically relevant in vivo studies. Clin Pharmacokinet 40(9):631–640
4. Rodrigues AD, Wong SL (1997) Application of human liver microsomes in metabolism-based drug-drug interactions: in vitro-in vivo correlations and the Abbott Laboratories experience. Adv Pharmacol 43:65–101
5. Koudriakova T, Iatsimirskaia E, Utkin I et al (1998) Metabolism of the human immunodeficiency virus protease inhibitors indinavir and ritonavir by human intestinal microsomes and expressed cytochrome P4503A4/3A5: mechanism-based inactivation of cytochrome P4503A by ritonavir. Drug Metab Dispos 26(6):552–561
6. Hochman JH, Yamazaki M, Ohe T et al (2002) Evaluation of drug interactions with p-glycoprotein in drug discovery: in vitro assessment of the potential for drug-drug interactions with p-glycoprotein. Curr Drug Metab 3(3):257–273

7. Benet LZ, Cummins CL, Wu CY (2003) Transporter-enzyme interactions: implications for predicting drug-drug interactions from in vitro data. Curr Drug Metab 4(5):393–398

8. Rolan PE (1994) Plasma protein binding displacement interactions--why are they still regarded as clinically important? Br J Clin Pharmacol 37(2):125–128

9. Sansom LN, Evans AM (1995) What is the true clinical significance of plasma protein binding displacement interactions? Drug Saf 12(4):227–233

10. FDA US (2012) Guidance for industry: drug interaction studies – study design, data analysis, implications for dosing and labeling recommendations; draft guidance. February 2012, Department of Health and Human Services, Food and Drug Administration, Center for Drug Evaluation and Research, 1–79

11. EMA (2012) Guideline on the investigation of drug interactions. Committee for Human Medicinal Products, European Medicines Agency, June 21, 1–59

12. Bjornsson TD, Callaghan JT, Einolf HJ et al (2003) The conduct of in vitro and in vivo drug-drug interaction studies: a Pharmaceutical Research and Manufacturers of America (PHARMA) perspective. Drug Metab Dispos 31(7):815–832

13. Gallicano KD, Sahai J, Shukla VK et al (1999) Induction of zidovudine glucuronidation and amination pathways by rifampicin in HIV-infected patients. Br J Clin Pharmacol 48(2):168–179

14. Ormsby E (1994) Statistical methods in bioequivalence. CRC Press, Boca Raton

15. Wang BS, Wang XJ, Gong LK (2009) The construction of a williams design and randomization in cross-over clinical trials using SAS. J Stat Softw 29(1):1–10

16. Fleiss JL (1989) A critique of recent research on the two-treatment crossover design. Control Clin Trials 10(3):237–243

17. Vuorinen J (1997) A practical approach for the assessment of bioequivalence under selected higher-order cross-over designs. Stat Med 16(19):2229–2243

18. Chow SC, Liu JP (1992) On assessment of bioequivalence under a higher-order crossover design. J Biopharm Stat 2(2):239–256

19. Nix DE, Di Cicco RA, Miller AK et al (1999) The effect of low-dose cimetidine (200 mg twice daily) on the pharmacokinetics of theophylline. J Clin Pharmacol 39(8):855–865

20. Grasela TH Jr, Antal EJ, Ereshefsky L et al (1987) An evaluation of population pharmacokinetics in therapeutic trials. Part ii. Detection of a drug-drug interaction. Clini Pharmacol Ther 42(4):433–441

21. Purkins L, Wood N, Kleinermans D et al (2003) No clinically significant pharmacokinetic interactions between voriconazole and indinavir in healthy volunteers. Br J Clin Pharmacol 56(Suppl 1):62–68

22. Cato A 3rd, Cavanaugh J, Shi H et al (1998) The effect of multiple doses of ritonavir on the pharmacokinetics of rifabutin. Clini Pharmacol Ther 63(4):414–421

23. Romero AJ, Le Pogamp P, Nilsson LG et al (2002) Effect of voriconazole on the pharmacokinetics of cyclosporine in renal transplant patients. Clini Pharmacol Ther 71(4):226–234

24. Cadieux RJ (1989) Drug interactions in the elderly. How multiple drug use increases risk exponentially. Postgrad Med 86(8):179–186

25. Ragueneau I, Poirier JM, Radembino N et al (1999) Pharmacokinetic and pharmacodynamic drug interactions between digoxin and macrogol 4000, a laxative polymer, in healthy volunteers. Br J Clin Pharmacol 48(3):453–456

26. Piscitelli SC, Goss TF, Wilton JH et al (1991) Effects of ranitidine and sucralfate on ketoconazole bioavailability. Antimicrob Agents Chemother 35(9):1765–1771

27. Blum RA, D'Andrea DT, Florentino BM et al (1991) Increased gastric pH and the bioavailability of fluconazole and ketoconazole. Ann Intern Med 114(9):755–757

28. Lebsack ME, Nix D, Ryerson B et al (1992) Effect of gastric acidity on enoxacin absorption. Clini Pharmacol Ther 52(3):252–256

29. Lehto P, Kivisto KT, Neuvonen PJ (1994) The effect of ferrous sulphate on the absorption of norfloxacin, ciprofloxacin and ofloxacin. Br J Clin Pharmacol 37(1):82–85

30. Nix DE, Watson WA, Lener ME et al (1989) Effects of aluminum and magnesium antacids and ranitidine on the absorption of ciprofloxacin. Clin Pharmacol Ther 46(6):700–705

31. Parpia SH, Nix DE, Hejmanowski LG et al (1989) Sucralfate reduces the gastrointestinal absorption of norfloxacin. Antimicrob Agents Chemother 33(1):99–102
32. Muller J, Keiser M, Drozdzik M et al (2017) Expression, regulation and function of intestinal drug transporters: an update. Biol Chem 398(2):175–192
33. Arakawa H, Kamioka H, Kanagawa M et al (2016) Possible interaction of quinolone antibiotics with peptide transporter 1 in oral absorption of peptide-mimetic drugs. Biopharm Drug Dispos 37(1):39–45
34. Shugarts S, Benet LZ (2009) The role of transporters in the pharmacokinetics of orally administered drugs. Pharm Res 26(9):2039–2054
35. Misaka S, Miyazaki N, Yatabe MS et al (2013) Pharmacokinetic and pharmacodynamic interaction of nadolol with itraconazole, rifampicin and grapefruit juice in healthy volunteers. J Clin Pharmacol 53(7):738–745
36. Momper JD, Tsunoda SM, Ma JD (2016) Evaluation of proposed in vivo probe substrates and inhibitors for phenotyping transporter activity in humans. J Clin Pharmacol 56(Suppl 7):S82–S98
37. Jungbluth GL, Pasko MT, Beam TR et al (1989) Ceftriaxone disposition in open-heart surgery patients. Antimicrob Agents Chemother 33(6):850–856
38. Yin J, Wang J (2016) Renal drug transporters and their significance in drug-drug interactions. Acta Pharm Sin B 6(5):363–373
39. Ivanyuk A, Livio F, Biollaz J et al (2017) Renal drug transporters and drug interactions. Clin Pharmacokinet 56(8):825–892
40. Ebner T, Ishiguro N, Taub ME (2015) The use of transporter probe drug cocktails for the assessment of transporter-based drug-drug interactions in a clinical setting-proposal of a four component transporter cocktail. J Pharm Sci 104(9):3220–3228
41. Hagos Y, Wolff NA (2010) Assessment of the role of renal organic anion transporters in drug-induced nephrotoxicity. Toxins 2(8):2055–2082
42. Hsu V, de LT Vieira M, Zhao P et al (2014) Towards quantitation of the effects of renal impairment and probenecid inhibition on kidney uptake and efflux transporters, using physiologically based pharmacokinetic modelling and simulations. Clin Pharmacokinet 53(3):283–293
43. Rengelshausen J, Goggelmann C, Burhenne J et al (2003) Contribution of increased oral bioavailability and reduced nonglomerular renal clearance of digoxin to the digoxin-clarithromycin interaction. Br J Clin Pharmacol 56(1):32–38
44. Megran DW, Lefebvre K, Willetts V et al (1990) Single-dose oral cefixime versus amoxicillin plus probenecid for the treatment of uncomplicated gonorrhea in men. Antimicrob Agents Chemother 34(2):355–357
45. Gaspari F, Perico N, Remuzzi G (1997) Measurement of glomerular filtration rate. Kidney Int Suppl 63:S151–S154
46. Brochner-Mortensen J (1985) Current status on assessment and measurement of glomerular filtration rate. Clin Physiol 5(1):1–17
47. Hellerstein S, Erwin P, Warady BA (2003) The cimetidine protocol: a convenient, accurate, and inexpensive way to measure glomerular filtration rate. Pediatr Nephrol 18(1):71–72
48. Baciewicz AM, Chrisman CR, Finch CK et al (2008) Update on rifampin and rifabutin drug interactions. Am J Med Sci 335(2):126–136
49. Wandel C, Bocker R, Bohrer H et al (1994) Midazolam is metabolized by at least three different cytochrome P450 enzymes. Br J Anaesth 73(5):658–661
50. Thummel KE, Shen DD, Podoll TD et al (1994) Use of midazolam as a human cytochrome P450 3A probe: ii. Characterization of inter- and intraindividual hepatic CYP3A variability after liver transplantation. J Pharmacol Exp Ther 271(1):557–566
51. Lown KS, Thummel KE, Benedict PE et al (1995) The erythromycin breath test predicts the clearance of midazolam. Clini Pharmacol Ther 57(1):16–24
52. Watkins PB, Turgeon DK, Saenger P et al (1992) Comparison of urinary 6-beta-cortisol and the erythromycin breath test as measures of hepatic P450iiiA (CYP3A) activity. Clini Pharmacol Ther 52(3):265–273

53. Hunt CM, Watkins PB, Saenger P et al (1992) Heterogeneity of CYP3A isoforms metabolizing erythromycin and cortisol. Clini Pharmacol Ther 51(1):18–23
54. Chiou WL, Jeong HY, Wu TC et al (2001) Use of the erythromycin breath test for in vivo assessments of cytochrome P4503A activity and dosage individualization. Clini Pharmacol Ther 70(4):305–310
55. Sarkar MA, Jackson BJ (1994) Theophylline N-demethylations as probes for P4501A1 and P4501A2. Drug Metab Dispos 22(6):827–834
56. Ziebell J, Shaw-Stiffel T (1995) Update on the use of metabolic probes to quantify liver function: caffeine versus lidocaine. Dig Dis 13(4):239–250
57. Anthony LB, Boeve TJ, Hande KR (1995) Cytochrome P-450iiD6 phenotyping in cancer patients: debrisoquin and dextromethorphan as probes. Cancer Chemother Pharmacol 36(2):125–128
58. Desta Z, Zhao X, Shin JG et al (2002) Clinical significance of the cytochrome P450 2C19 genetic polymorphism. Clin Pharmacokinet 41(12):913–958
59. Fuhr U, Rost KL, Engelhardt R et al (1996) Evaluation of caffeine as a test drug for CYP1A2, NAT2 and CYP2E1 phenotyping in man by in vivo versus in vitro correlations. Pharmacogenetics 6(2):159–176
60. Brockmoller J, Rost KL, Gross D et al (1995) Phenotyping of CYP2C19 with enantiospecific hplc-quantification of R- and S-mephenytoin and comparison with the intron4/exon5 g-->a-splice site mutation. Pharmacogenetics 5(2):80–88
61. Guengerich FP (1997) Role of cytochrome p450 enzymes in drug-drug interactions. Adv Pharmacol 43:7–35
62. Tanaka E (1998) Clinically important pharmacokinetic drug-drug interactions: role of cytochrome P450 enzymes. J Clinical Pharmacy Ther 23(6):403–416
63. Lomaestro BM, Piatek MA (1998) Update on drug interactions with azole antifungal agents. Ann Pharmacother 32(9):915–928
64. Caraco Y (1998) Genetic determinants of drug responsiveness and drug interactions. Ther Drug Monit 20(5):517–524
65. Shannon M (1997) Drug-drug interactions and the cytochrome P450 system: an update. Pediatr Emerg Care 13(5):350–353
66. Bradford PA, Sanders CC (1993) Use of a predictor panel to evaluate susceptibility test methods proposed for piperacillin-tazobactam. Antimicrob Agents Chemother 37(12):2578–2583
67. Mikus G, Schowel V, Drzewinska M et al (2006) Potent cytochrome P450 2C19 genotype-related interaction between voriconazole and the cytochrome P450 3A4 inhibitor ritonavir. Clini Pharmacol Ther 80(2):126–135
68. Isoherranen N, Ludington SR, Givens RC et al (2008) The influence of CYP3A5 expression on the extent of hepatic CYP3A inhibition is substrate-dependent: an in vitro-in vivo evaluation. Drug Metab Dispos 36(1):146–154
69. Li D, Abudula A, Abulahake M et al (2010) Influence of CYP3A5 and MDR1 genetic polymorphisms on urinary 6 beta-hydroxycortisol/cortisol ratio after grapefruit juice intake in healthy Chinese. J Clin Pharmacol 50(7):775–784
70. Cho JY, Yu KS, Jang IJ et al (2002) Omeprazole hydroxylation is inhibited by a single dose of moclobemide in homozygotic em genotype for CYP2C19. Br J Clin Pharmacol 53(4):393–397
71. Miura M, Inoue K, Kagaya H et al (2007) Influence of rabeprazole and lansoprazole on the pharmacokinetics of tacrolimus in relation to CYP2C19, CYP3A5 and MDR1 polymorphisms in renal transplant recipients. Biopharm Drug Dispos 28(4):167–175
72. Furuta T, Ohashi K, Kobayashi K et al (1999) Effects of clarithromycin on the metabolism of omeprazole in relation to CYP2C19 genotype status in humans. Clini Pharmacol Ther 66(3):265–274
73. Bramness JG, Skurtveit S, Gulliksen M et al (2005) The CYP2C19 genotype and the use of oral contraceptives influence the pharmacokinetics of carisoprodol in healthy human subjects. Eur J Clin Pharmacol 61(7):499–506

74. Dumond JB, Vourvahis M, Rezk NL et al (2010) A phenotype-genotype approach to predicting CYP450 and p-glycoprotein drug interactions with the mixed inhibitor/inducer tipranavir/ritonavir. Clinical Pharmacol Ther 87(6):735–742

75. Zachariasen RD (1994) Loss of oral contraceptive efficacy by concurrent antibiotic administration. Women Health 22(1):17–26

76. Suzuki E, Nakai D, Ikenaga H et al (2016) In vivo inhibition of acylpeptide hydrolase by carbapenem antibiotics causes the decrease of plasma concentration of valproic acid in dogs. Xenobiotica 46(2):126–131

77. Nguyen VX, Nix DE, Gillikin S et al (1989) Effect of oral antacid administration on the pharmacokinetics of intravenous doxycycline. Antimicrob Agents Chemother 33(4):434–436

78. Neuvonen PJ, Penttila O (1974) Effect of oral ferrous sulphate on the half-life of doxycycline in man. Eur J Clin Pharmacol 7(5):361–363

79. Muller HJ, Gundert-Remy U (1994) The regulatory view on drug-drug interactions. Int J Clin Pharmacol Ther 32(6):269–273

80. Waller PC, Jackson PR, Tucker GT et al (1994) Clinical pharmacology with confidence. Br J Clin Pharmacol 37(4):309–310

81. Fuhr U, Weiss M, Kroemer HK et al (1996) Systematic screening for pharmacokinetic interactions during drug development. Int J Clin Pharmacol Ther 34(4):139–151

82. Kuhlmann J (1994) Drug interaction studies during drug development: which, when, how? Int J Clin Pharmacol Ther 32(6):305–311

83. Pidgen AW (1992) Statistical aspects of bioequivalence--a review. Xenobiotica 22(7): 881–893

84. Steinijans VW, Hartmann M, Huber R et al (1996) Lack of pharmacokinetic interaction as an equivalence problem. Int J Clin Pharmacol Ther 34(1 Suppl):S25–S30

85. Gallicano K, Sahai J, Swick L et al (1995) Effect of rifabutin on the pharmacokinetics of zidovudine in patients infected with human immunodeficiency virus. Clin Infect Dis 21(4):1008–1011

86. De Wit S, Debier M, De Smet M et al (1998) Effect of fluconazole on indinavir pharmacokinetics in human immunodeficiency virus-infected patients. Antimicrob Agents Chemother 42(2):223–227

87. Hauschke D, Kieser M, Diletti E et al (1999) Sample size determination for proving equivalence based on the ratio of two means for normally distributed data. Stat Med 18(1):93–105

88. Huang SM, Lesko LJ, Williams RL (1999) Assessment of the quality and quantity of drug-drug interaction studies in recent NDA submissions: study design and data analysis issues. J Clin Pharmacol 39(10):1006–1014

89. Chow S, Liu JP (2000) Design and analysis of bioavailabilty and bioequivalence studies, second edition, revised and expanded. Marcel Dekker, New York

90. Wijnand H (1996) Some nonparametric confidence intervals are non-informative, notably in bioequivalence studies. Clin Res Reg Affairs 13:65–75

91. Midha KK, Ormsby ED, Hubbard JW et al (1993) Logarithmic transformation in bioequivalence: application with two formulations of perphenazine[erratum appears in j pharm sci 1993 dec;82(12):1300]. J Pharm Sci 82(2):138–144

92. Roe DJ, Karol MD (1997) Averaging pharmacokinetic parameter estimates from experimental studies: statistical theory and application. J Pharm Sci 86(5):621–624

93. Zintzaras E (2000) The existence of sequence effect in cross-over bioequivalence trials. Eur J Drug Metab Pharmacokinet 25(3–4):241–244

94. Hauschke D, Steinijans WV, Diletti E et al (1994) Presentation of the intrasubject coefficient of variation for sample size planning in bioequivalence studies. Int J Clin Pharmacol Ther 32(7):376–378

95. Steinijans VW, Sauter R, Hauschke D et al (1995) Reference tables for the intrasubject coefficient of variation in bioequivalence studies. Int J Clin Pharmacol Ther 33(8):427–430

96. Diletti E, Hauschke D, Steinijans VW (1992) Sample size determination: extended tables for the multiplicative model and bioequivalence ranges of 0.9 to 1.11 and 0.7 to 1.43. Int J Clini Pharmacol, Ther Toxicol 30(Suppl 1):S59–S62

97. Hauschke D, Steinijans VW, Diletti E et al (1992) Sample size determination for bioequivalence assessment using a multiplicative model. J Pharmacokinet Biopharm 20(5):557–561
98. Liu JP, Chow SC (1992) Sample size determination for the two one-sided tests procedure in bioequivalence. J Pharmacokinet Biopharm 20(1):101–104
99. Chow SC, Wang H (2001) On sample size calculation in bioequivalence trials.[erratum appears in j pharmacokinet pharmacodyn. 2002 feb;29(1):101]. J Pharmacokinet Pharmacodyn 28(2):155–169
100. Gallicano K, Sahai J, Zaror-Behrens G et al (1994) Effect of antacids in didanosine tablet on bioavailability of isoniazid. Antimicrob Agents Chemother 38(4):894–897
101. Schall R, Hundt HK, Luus HG (1994) Pharmacokinetic characteristics for extent of absorption and clearance in drug/drug interaction studies. Int J Clin Pharmacol Ther 32(12):633–637
102. Tozer TN, Bois FY, Hauck WW et al (1996) Absorption rate vs. exposure: which is more useful for bioequivalence testing? Pharm Res 13(3):453–456
103. van Giersbergen PL, Halabi A, Dingemanse J (2002) Single- and multiple-dose pharmacokinetics of bosentan and its interaction with ketoconazole. Br J Clin Pharmacol 53(6):589–595
104. Sanchez Garcia P, Paty I, Leister CA et al (2000) Effect of zaleplon on digoxin pharmacokinetics and pharmacodynamics. Am J Health Syst Pharm 57(24):2267–2270
105. Depre M, Van Hecken A, Verbesselt R et al (1999) Effect of multiple doses of montelukast, a cyslt1 receptor antagonist, on digoxin pharmacokinetics in healthy volunteers. J Clin Pharmacol 39(9):941–944
106. Dilger K, Zheng Z, Klotz U (1999) Lack of drug interaction between omeprazole, lansoprazole, pantoprazole and theophylline. Br J Clin Pharmacol 48(3):438–444
107. Auclair B, Nix DE, Adam RD et al (2001) Pharmacokinetics of ethionamide administered under fasting conditions or with orange juice, food, or antacids. Antimicrob Agents Chemother 45(3):810–814
108. Damle BD, Mummaneni V, Kaul S et al (2002) Lack of effect of simultaneously administered didanosine encapsulated enteric bead formulation (videx EC) on oral absorption of indinavir, ketoconazole, or ciprofloxacin. Antimicrob Agents Chemother 46(2):385–391
109. Banfield C, Herron J, Keung A et al (2002) Desloratadine has no clinically relevant electrocardiographic or pharmacodynamic interactions with ketoconazole. Clin Pharmacokinet 41(Suppl 1):37–44
110. Hsu A, Granneman GR, Cao G et al (1998) Pharmacokinetic interaction between ritonavir and indinavir in healthy volunteers. Antimicrob Agents Chemother 42(11):2784–2791

Index

© Springer International Publishing AG 2018
M.P. Pai et al. (eds.), *Drug Interactions in Infectious Diseases: Mechanisms and
Models of Drug Interactions*, Infectious Disease,
https://doi.org/10.1007/978-3-319-72422-5

Printed in the United States
By Bookmasters